AF182018

BREAD-MAKING QUALITY OF WHEAT:
A CENTURY OF BREEDING IN EUROPE

European Cereal Atlas Foundation

The European Cereal Atlas Foundation, ECAF, is a non-profit organisation with the aim of promoting scientific research on cereals and their grains by compiling an Atlas of Cereal Growing in Europe as well as other publications about cereals and by stimulating the scientific study of cereals which anticipates the creation of such publications.

ECAF Board

Ir L.A.J. Slootmaker, Chairman, Kesteren, The Netherlands
Dr Ir A. Darwinkel, Secretary, Research Station for Arable Farming and Field Production of Vegetables, Lelystad, The Netherlands
Ir N.M. de Vos, Treasurer, Wageningen, The Netherlands
Prof Dr A. Brönnimann, Member, Swiss Federal Research Station for Agro nomy, Zürich-Reckenholz, Switzerland
Dr F.G.H. Lupton, Member, Cambridge, United Kingdom
Ir D.A. Donner, Member, Plant Research International, Wageningen, The Netherlands

Bread-making quality of wheat

A century of breeding in Europe

Part One:

Developments in bread-making processes

B. Belderok

Part Two:

Breeding for bread-making quality in Europe

J. Mesdag & D. A. Donner

Editor on behalf of ECAF: D. A. Donner

Springer-Science+Business Media, B.V.

Library of Congress Cataloging-in-Publication Data

Bread-making quality of wheat: a century of breeding in Europe / by B. Belderok and J. Mesdag & D.A. Donner.

p.cm.

ISBN 978-90-481-5493-7 ISBN 978-94-017-0950-7 (eBook)
DOI 10.1007/978-94-017-0950-7

1. Bread. 2. Wheat. I. Belderok, B. (Bob) II. Mesdag, J. (Hans) III. Donner, D. A. (Dingena A.)

TX769 B7743 2000
641.8′15–dc21

00-033080

ISBN 978-90-481-5493-7

Printed on acid-free paper

Editor (on behalf of ECAF): D. A. Donner, Wageningen.
Language editing: Ian Cressie (Cressie Communications, Oosterbeek)
Country maps by Lucas Janssen, Wageningen.

All Rights Reserved
© 2000 Springer Science+Business Media Dordrecht
Originally published by Kluwer Academic Publishers in 2000
No part of the material protected by this copyright notice may be reproduced or utilized in any form or by any means, electronic or mechanical, including photocopying, recording or by any information storage and retrieval system, without written permission from the copyright owner.

Contents

Part Two:
Breeding for bread-making quality in Europe
by J. Mesdag & D. A. Donner

General introduction to the Nordic countries

Norway

Sweden

Finland

Denmark

Germany

Foreword

Wheat has a long history of serving as an important food crop to mankind. Especially in the Northern Hemisphere, it has been appreciated as a major source of energy through its carbohydrates, and in more recent times for its supply of valuable proteins. This combination of carbohydrates and proteins gives wheat its unique properties for making breads of different kinds of tastes.

During the course of history, the quality of wheat has improved steadily, undoubtedly for a long time by accident, and for reasons little understood. Over the last 150 years our knowledge has increased on farming and crop husbandry, on bringing about improvements through goal-oriented plant breeding, and on milling and baking technology, leading to the standards that we enjoy today. This process will certainly continue as our knowledge of the genetic reservoir of wheat species increases.

The European Cereal Atlas Foundation (ECAF) maintains the aim of increasing and disseminating knowledge about cereal crops. Within that scope ECAF has decided to publish a book on the history of bread wheat in Europe, the development of associated bread-making technology, and the breeding of bread wheats during the twentieth century.

As ECAF is a Dutch foundation, its Board is particularly pleased to have found three Dutch scientists willing to contribute to this volume. Two of them have served wheat science in the Netherlands for their entire scientific careers, spanning a period starting around 1955 and lasting for several decades of very productive wheat science development.

Dr Bob Belderok, who contributed the first part of this book, has long experience in the Institute for Cereals, Flour and Bread (IGMB) of TNO, Wageningen (now part of TNO Nutrition and Food Research International, Zeist), and is an internationally recognised expert on wheat milling and baking technology. He has worked at the interface of technological development and plant and grain quality characteristics for many years. ECAF acknowledges TNO's contribution through making it possible for Dr Belderok to utilise data and information collected during and after his service with them.

Ir Hans Mesdag spent his entire scientific life in the cereal department of the Foundation for Agricultural Plant Breeding (SVP), now merged into Plant Research International, part of Wageningen University and Research Centre. His careful experimens with wheat germplasm gathered from all corners of the globe and with genotype-environment interactions that influence the bread-making quality of breeding lines – the most promising of which were made available to Dutch plant breeding companies – made him renowned among his practical wheat-breeding colleagues.

Ir Dingena Donner, after a substantial period of international wheat breeding on several continents, has served ECAF as a member of the Board and has maintained the secretariat of the editorial committee preparing this book. Her editing work and her contribution to the second part of the book, as co-author with Hans Mesdag, has been of critical value to this publication.

The Board of ECAF is much indebted to Dr Cees Mastenbroek, himself a successful wheat breeder and one who has been acknowledged for his international activities for the benefit of plant breeding. As ECAF's former chairman, he played a great part in the early initiatives to write this book. As chairman of the editorial committee, his continued interest had an important influence on the final content of the book.

Many other specialists have made contributions: they are acknowledged for their specific efforts at the end of the appropriate chapter for each country. In addition, ECAF is much indebted to Mr Ian Cressie of Cressie Communication Services for his correction of the English manuscript and to Kluwer Academic Publishers for the publication of this book.

I sincerely hope that this book, explaining the state of the art of breeding for bread wheat at the beginning of the twenty-first century, and with its description of what has been the history of purposive wheat breeding for bread-making quality during the twentieth century, will help the next generation of wheat quality engineers – farmers, technologists, biochemists, breeders, geneticists, plant physiologists and many others – to further improve wheat, one of the world's tastiest basic foods.

Kesteren, June 2000

On behalf of the Board of ECAF
Leo A.J. Slootmaker, chairman

Glossary

Statistics

Anonymus (1927) La distribution du froment dans le monde. Rome: Institute of International Agriculture.
Anonymus (1947–1998) FAO Yearbook of Food and Agricultural statistics; production, Volumes 1–52. Rome: FAO.

Pedigrees

Notation according to Purdy et al. (1968) [Purdy LH, Loegering WQ, Konzak CF, Peterson CJ, Allen RE (1968) A proposed standard method for illustrating pedigrees of small grain varieties. *Crop Science* 8: 405–406.]
Varieties noted with $*^2$ have been used as a parent and as a backcrossing parent (e.g. Table H.2)

Parameters used in tables

- protein content: percentage protein in the grain, on a dry matter basis
- loaf volume: always expressed in ml/100 gram flour
- *Glu-1* score: unless mentioned otherwise, always according to Payne (1987) see chapter 7, part I
- year: refers to the year in which the variety has officially been registered

Abbreviations used in tables

ww	winter wheat	abs	absolute
sw	spring wheat	rel	relative
prot cont	protein content	sel	selection
br	breeder	qual	quality
br st	breeding station	tkw	thousand kernel weight
cv(s)	cultivar(s)	pelsh	Pelshenke
prod	production	farin	farinograph

PART ONE

Developments in bread-making processes

B. BELDEROK
Former head of cereal research of
TNO Nutrition and Food Research Institute,
Zeist, The Netherlands

1. Historical introduction

1.1. Introduction

Since time immemorial, cereals have played an important role in human nutrition. They have been grown by man for thousands of years, beginning long before the Christian era. Botanically, cereals belong to the grass family (Gramineae), which is subdivided into several genera.

In temperate zones, one can readily find representatives of the genera like wheat, barley, rye and oats. Wheat and barley are grown on relatively fertile soils, whereas rye and oats are more adapted to poor (acid and dry) soils. Maize is the most important cereal crop in subtropical zones, although it is grown in temperate zones too.

In tropical zones, rice, sorghum and various kinds of millet are grown. Rice is the main crop in areas where water is sufficiently available (South-East and East Asia). In areas where water is limited (Africa), mainly sorghum and maize are cultivated, whereas millet is a 'famine crop' that is able to thrive under extremely dry conditions. Rice, sorgum and millet are as important for the population of tropical areas as wheat is for the population of temperate areas.

Wheat, rice and maize are spread widely over the world. The importance of wheat is mainly due to the fact that its seeds can be ground into flour, semolina, etc., which form the basic ingredients of bread and other bakery products, as well as pastas (e.g. macaroni, spaghetti).

1.2. Evolutionary history of the hexaploid bread wheats

Wheat in its present-day form has gone through a long and interesting evolution (Maynard Smith, 1958; Riley, 1975). The origin of the genus *Triticum* (wheat) is found in Asia and parts of Africa, in the area that stretches from Syria to Kashmir, and southwards to Ethiopia. This is where cultivated wheats gradually evolved from wild plants, in the very distant past long before the birth of Christ.

The genus *Triticum* consists of several species, which can be divided into three basic natural groups, each distinguished by the number of chromosomes in the generative and vegetative cells. Pollen and unfertilized egg cells contain 7, 14 or 21 chromosomes. In vegetative cells this number is doubled. Consequently, diploid, tetraploid and hexaploid wheat species carry $2 \times 7 = 14$; $4 \times 7 = 28$ and $6 \times 7 = 42$ chromosomes, respectively.

Today's commercial wheats (durum wheat and common or bread wheat) are products of natural hybridization of ancestral types, none of which nowadays is still of any commercial importance. The best known ancestor of our modern cultivated wheats is *Triticum monococcum,* einkorn wheat, a diploid wheat containing two sets of seven chromosomes. The haploid set of the chromosomes of a distinct diploid species is called a genome. The genome of *monococcum* is called the A-genome. Genetically, plants of this species can thus be described as AA plants.

Archeological findings indicate that a spontaneous cross between a wild einkorn wheat and an unknown wild grass species with seven generative chromosomes may have occurred sometime before 8000 B.C. This second set of chromosomes, from the wild grass, clearly differs from the A-genome and is indicated by the letter B. As the chromosomes of the A- and B-genomes are not able to pair during meiosis, the hybrid with $2 \times 7 = 14$ chromosomes is not fertile. Through spontaneous doubling of chromosomes, however, a fertile allopolyploid originated with 28 chromosomes, 14 derived from *T. monococcum* and 14 from the unknown grass parent. The genome of this tetraploid species is called the AB-genome. Genetically, plants of this species can be described as AABB plants.

The wild tetraploid *Triticum dicoccoides* and the cultivated tetraploids *Triticum dicoccum* and *Triticum durum* developed in a similar way. *T. dicoccoides* and *T. dicoccum* are often referred to as emmer wheats. *T. durum* is grown widely nowadays as durum or pasta wheat, with numerous cultivars in Mediterranean countries and North America.

The origin of the 42 chromosome common or bread wheats, *Triticum aestivum*, is more remarkable. It was presumed for a long time that these wheats were allohexaploid hybrids of the tetraploid *T. dicoccoides* and some diploid species with 14 somatic chromosomes. The discovery of a technique to experimentally produce polyploids by treatment with colchicin made it possible to confirm this assumption and to identify the other parent. It proved to be the species *Aegilops squarrosa,* also called *Triticum tauschii*, a species without any economic value that grows as a weed on the borders of wheat fields in the Near East. Aegilops has a somatic chromosome number of 14. These chromosomes belong to the D-genome (the term C-genome had already been used to characterize other Triticum species). Therefore, the genome of *T. aestivum* is called the ABD-genome, and plants of this species can be characterized by the formula AABBDD.

The artificially produced allopolyploid hybrids were fully fertile and resembled bread wheats in appearance. When crossed with bread wheats, a

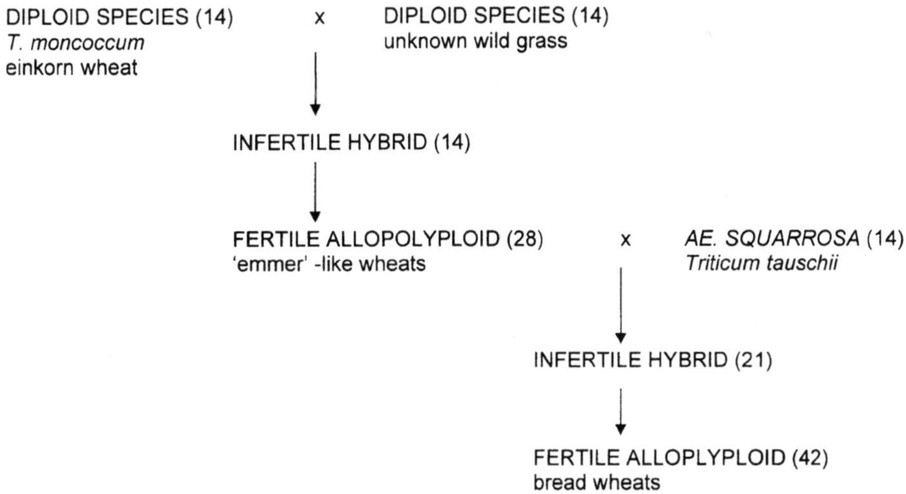

Figure 1.1. The probable evolutionary history of the hexaploid bread wheats according to Maynard Smith (1958) (somatic chromosome numbers are shown between parentheses).

fertile offspring arose. Thus, it was possible to reproduce experimentally the process summarized in Figure 1.1, which illustrates the evolution of bread wheat.

T. aestivum probably generated spontaneously somewhere on the Iranian highlands or nearby areas. Archeological finds indicate that this took place some 6000 years B.C.

The unique milling and baking properties of common bread wheat are not found among the diploid and tetraploid wheats. Since only the hexaploid group possesses the D set of chromosomes, derived from *Aegilops squarrosa*, the desirable quality characteristics of bread wheats have been attributed preponderantly to the presence of this third genomic component.

1.3. Dispersion of wheat in ancient times[1]

The wild and primitive wheat species from which cultivated wheat has evolved are characterized by a brittle rachis of the ear, which causes the spikelets to separate; also, the grains are strongly enclosed by the chaff. For man, these are undesirable features that cause problems during harvesting and threshing. Moreover, the number of grains per ear in a primitive species is low. Therefore, the older cultivated wheats with these unwanted characteristics

[1] For more information see Peterson (1965a).

have gradually been replaced by types with naked kernels, which are easier to thresh and have higher yields.

Archeological discoveries not only proved that einkorn and emmer were grown as crops some 4000 years B.C., but also that these crops were widely spread over northern Africa and western Europe.

The most primitive cultivated wheat, the diploid einkorn, has been a minor crop, grown on a smaller scale than emmer; it often has been merely a contaminant of emmer. Grains of emmer have been detected in the ruins of Troy, situated at the entrance of the Dardanelles, a city which was destroyed and rebuilt several times between 2000 B.C. and the first centuries after the birth of Christ. The cultivated tetraploid emmer became the most important wheat and as a result of human migrations around 4000 B.C. its cultivation spread from the Middle East to Egypt and Ethiopia, and to large areas around the Mediterranean Sea. There it remained the most widely grown wheat until somewhat after 300 B.C., when wheats of the type of *T. durum*, with naked seeds and a much wider adaptability, began to appear.

Thousands of years of cultivation, continuous gene mutation and cross-fertilization have resulted in a tremendous variability in tetraploid wheats derived from the *T. dicoccoides* wild type. These wheats have been used to make both unleavened and leavened bread, and for several other purposes, some of which are still common today (e.g. bulgur, couscous). During the time of the Roman Empire, most of the wheats carried from the colonies to Rome were of the *dicoccum/durum* species. In the course of a few centuries, around the beginning of the Christian era, *T. durum* took the place of *T. dicoccum* as the most important species of wheat.

Of all cultivated tetraploid wheats, today *T. durum* types are by far the most important, even though they are grown on only 10% of the whole area under wheat, the remaining 90% being the hexaploid bread wheats. Durum wheat is particularly suited to the production of pasta products (macaroni, spaghetti, etc.).

Literature on the dispersion of hexaploid bread wheats over Europe is scarce. One may reasonably assume that *T. aestivum* came to western Europe from its area of origin during the second and first millenium B.C. through the Mediterranean region. During the same period in Europe north of the Alps, *T. dicoccum* was replaced in many regions by a hexaploid wheat, *T. spelta*. The changeover followed the same course that *T. dicoccum* had taken earlier: through the Rhine valley as far as Britain and Scandinavia. Spelt wheat in turn was replaced by the common or bread wheat. In the late Middle Ages, the hexaploid wheats *T. spelta* and *T. aestivum* occurred widely all over Europe.

The high chromosome number of hexaploid wheats contributed to their wider adaptability than species such as einkorn and emmer. This explains the great dispersion from southwestern Asia in various directions.

By crossing and selecting, man has produced numerous cultivars of hexaploid wheat and it is still the most important wheat species grown in Europe, North America and Australia today. In particular the flour of *T. aestivum* is not only suited for baking bread; it is also used for making biscuits, various confectioneries and all sorts of pastas.

1.4. Bread making

Man began to cook cereals long before starting to make bread. Some sort of porridge was probably the first form in which cereals were prepared for food. This technique of preparation is still practiced by many cultural groups, especially in Africa. Porridge was, in many cases, succeeded by flat, unleavened cakes. These primitive forms of 'bread' still persist here and there, sometimes having ritual significance: various porridges, the Scottish oat cake, Indian chappaties and Jewish unleavened bread.

People from Sumeria, in the southern part of Mesopotamia, were the first to bake leavened bread. About 6000 years B.C., they started to mix sour dough with unleavened dough. Sour dough is generated during the natural yeasting process of flour and water, during which carbon dioxide is formed, which in turn causes the dough to rise.

The Sumerians transmitted their way of preparing bread to the Egyptians some 3000 years B.C. The Egyptians perfected the system and started to use yeast generated from brewing beer. Moreover, they developed a baking oven which made it possible to bake several loaves at once.

The production of wheat loaf bread was achieved successively by the Egyptians, the Greeks and the Romans and was considered by them to be a sign of a high degree of civilization. Bread must have been eaten in large quantities by well-to-do Jews and Egyptians. In most families, women and girls ground the grain with hand mills and prepared both leavened and unleavened bread.

Bread also represented a substantial part of the daily food in ancient Athens, as well as later in Rome and other Italian towns. The wheat needed for its preparation was mainly imported from elsewhere in the Mediterranean region. In Imperial Rome, much of the baker's trade consisted of contract work for the city or town. This meant supplying bread to people in public institutions, the army, prisons, other official bodies and to citizens – when a free issue was being contemplated.

1.5. Cereals in Europe before and during the Middle Ages

In the Mediterranean region, long before recorded history, various cereals formed a substantial part of the human diet. For some time, barley was grown more extensively than wheat and apparently it was considered to be more valuable for human consumption, albeit closely followed by wheat.

A considerable amount of information is available about the crops grown in Britain during the period between 1200–1700 A.D. (McCance & Widdowson, 1956). Wheat, rye, barley and oats were the principal grain crops, beans and other pulses were grown as well. People depended on locally grown products to a much greater extent than in later centuries.

Throughout the Middle Ages, in central and eastern Europe, rye was more important than the various species of wheat available at the time. Bread made from rye, however, was mainly considered as food for the poor. When rye was in short supply, it was supplemented with flour from oats, peas, beans and sometimes even acorns. Rich people ate bread prepared from a mixture of wheat and rye flour. In the Mediterranean basin, however, rye was hardly known. Around the middle of the 18th century, wheat became more and more popular and gradually most people began to eat wheat bread, although the poorest still had to content themselves with rye bread.

Cereals were not only needed for making bread and other foodstuffs but also to brew beer. Coffee and tea were not known until the 17th century, available water was of poor quality and wine was expensive. It is therefore not surprising that beer, albeit a weak brew, was a common drink for a large part of the population.

1.6. Amsterdam – the granary of Europe[2]

Up to the 13th century, cereal production in northwestern Europe was sufficiently high as to fulfill the needs of the population. As the numbers of inhabitants increased, it gradually became necessary to import cereals from countries with a surplus, for example the Baltic states.

An intensive commercial network was set up between the larger cities of northwestern Europe (London, Boston (England), Bruges, Antwerp and, especially, Amsterdam) and the more remote areas of the Mediterranean and the Baltic. As far as the exchange of cereals was concerned, contacts between Amsterdam and the Hanseatic cities of north Germany were the most valuable. Connections between these cities had been established as far back as the

[2] For more information see Heyder (1979); Schama (1987).

13th century, when Dutch cargo ships sailed along the coast to Hamburg and, by the end of that century, they even dared to extend their journeys to Jutland, the Sont and the Baltic. Hamburg was considered a transit harbour between west and east. Several merchants from Hamburg settled in Amsterdam and at the beginning of the 15th century even formed their own guild.

It is not surprising that in Amsterdam, through contacts with German merchants, a growing interest developed in commercial activities with the German home cities, especially after Dutch sailors became familiar with the trading routes. That is how the first commercial traffic between Amsterdam and Hamburg, Lübeck, and several other German cities was established. Later on, these cities were no longer visited as business was done directly with the countries bordering the Baltic.

The main products imported into Holland from Russia and the Baltic states were wheat, rye, wood, furs, bacon, beer, linen, cloth and wool; while simultaneously butter, cheese, wine, (Flemish) cloth, tin, lead, salt and later on herring (common food for our ancestors) were exported from Holland into other countries in northwestern and northeastern Europe.

Cereals, especially, were traded in such large quantities that Amsterdam was said to be the 'Granary of Europe'. At the end of the 16th and the beginning of the 17th century some 1000–1200 ships sailed through the Sont every year, of which more than half was loaded with cereals for Amsterdam.

From Amsterdam, Flanders could be reached by interior waterways. In the 14th, 15th and most of the 16th century, Bruges and Antwerp had been the staple markets for goods (including cereals) imported from countries on the Mediterranean and from France, southern Germany and England. These products were reloaded in Bruges and Antwerp and shipped to Amsterdam to be sold on markets there.

The central position of Amsterdam in the European grain trade lasted until the end of the 18th century. Developments in other countries, not the least of which were in northern America and Russia, led to a situation in which Amsterdam lost its leading position in the trade.

1.7. Landraces

Ever since wild *aestivum* wheats were first domesticated, new forms of wheat have arisen from time to time in farmers' fields through hybridization, gene mutation and natural selection. Those types that were more adapted to cultivation than the stock from which they arose tended to survive. Farmers probably applied some kind of conscious or unconscious selection by discarding apparently poor plants or by choosing healthy ones for sowing. As a result, landraces or groups of landraces developed.

Until late in the 19th century, wheat varieties of the countries of north-western Europe consisted mainly of such landraces. They were well adapted to the regions where they were grown, for example by being sufficiently frost resistant. However, they were often mediocre in features such as yield or strength of straw. As landraces are usually quite heterogeneous, it has often been possible to select distinct types with improved yield, strength of straw or other agronomic properties.

Landraces are generally named after the region from where the seed was obtained or after the region where they originally were grown. Little is known about types of landraces cultivated before the 17th century. Most information dates from the 19th and 20th centuries (Zeven, 1986 and 1990).

Landraces can be categorized into landrace groups. As an example, the most important groups cultivated in the Netherlands during the 19th and 20th centuries were:

- *Zeeuwse landrace group.* Races belonging to this group covered French and Belgian Flanders, the Dutch province of Zeeland, parts of Essex, Great Britain and probably adjacent regions. They were awnless, had white glabrous glumes and white grains. Their frost resistance was low.
- *Ruwkaf Essex landrace group.* Landrace material of this group was introduced from Great Britain into the Netherlands. It was marked by awnless ears, pubescent white glumes and white grains. The original material was probably grown in Essex.
- *Gelderse landrace group.* This group was introduced into the Netherlands in the 19th century from regions around Cleve, Goch and Geldern in western Germany. The same material was introduced into other parts of Germany and has, therefore, covered a large part of western Europe. Originally the wheat of this group came from eastern Europe. The group is characterized by awnless ears, red or white glabrous glumes and red grains.
- *Squarehead landrace group.* Squarehead is a wheat type that was discovered in 1865 in Great Britain and later introduced into the Netherlands. The period of cultivation was too short to allow complete adaptation.

1.8. The second half of the nineteenth century

The 19th century was marked by considerable social and technical changes (Jacob, 1956). In 1800 the world population was less than one thousand million people. Fifty years later this number had increased to 1.3 thousand million. Another fifty years brought the total world population to about 1.8 thousand million. This growth has not stopped, and in the year 2000, at the end of the 20th century, the world population reached already 6 thousand million. This increase in population, especially in Europe, created an important incentive to strive for higher production levels in agriculture, including animal production. The application of mineral fertilizers to reduce specific nutritional deficiencies largely contributed to growth in production.

At the same time, enormous economic and industrial changes were taking place. The possession of overseas colonies meant an important source of income for the state budget of many European countries and it offered the opportunity to spend more money on the development of an infrastructure to the benefit of trade and industry. A considerable increase in the use of the steam engine during the 1880s led to a network of railroads and an improvement in the navigability of the main waterways.

The changes that took place in North America during this period markedly affected the wheat trade in Europe. Wheat was brought to the northern part of the present-day United States by Russian Mennonites in the 1830s and early 1840s. Here, production became important after immigrants started to cultivate the Great Plains in subsequent decades: North and South Dakota, Nebraska, Kansas, Oklahoma, Texas, New Mexico, Colorado, Wyoming and Montana. A similar development took place in Canada. There, wheat had been imported from France at the beginning of the 17th century. In the following period, provinces such as Manitoba, Saskatchewan and Alberta developed into very important wheat producing areas.

This rapid increase in wheat production was partly due to immigrants having at their disposal vast areas of virgin arable land, especially in the 19th century. These settlers came from various European (relatively poor) countries expecting to build a better life in the New World. Farms were very large by European standards. Many immigrants brought seeds of crops, including wheat, with them from their native lands. This provided a considerable range of wheat varieties for trial on farms in the new territories.

At more or less the same time, mechanization in agriculture made a great leap forward (Jacob, 1956; Peterson, 1965b). The use of huge horse-drawn reapers (McCormick), self-binders and steam-powered threshing machines made it possible to harvest vast areas of wheat in a short time. In the 20th century, these machines were replaced by combine harvesters, integrating the functions of reaping and threshing in one single operation.

The construction of a railway network across the North American continent was an efficient way to open up the western part of the United States. Wheat could then be transported cheaply to the Atlantic and the Pacific coasts, from where it was shipped to Europe. Most Canadian wheat was transported to the Atlantic coast via Winnipeg and the Great Lakes. Some was transported by train to Vancouver and from there shipped to Europe.

The enormous scale on which wheat was produced in North America made it possible to offer the product on the European market for relatively low prices. In addition, North American wheats had higher protein levels and excellent baking qualities compared to most European wheats. Therefore flour mills in Europe, especially those in England, where industrialization had recently flourished, showed a clear preference for American and Canadian wheat. Practically all wheat used for bread making, in most of the north and western European countries, was imported from the New World, hence the term 'American Wheat Imperium'. This situation lasted for most of the 20th century. Even in 1950, about 75% of the bread wheat processed in Western Europe was obtained from the United States and Canada.

In the second half of the 19th century, both in the Old and the New World, a start was made with the deliberate breeding of wheat varieties with improved agronomic and technological qualities such as earliness, disease resistance, high yield and good baking properties. This work was based on the theory of evolution as developed and propagated by Lamarck and Darwin.

As an example, the history of the famous Canadian variety Red Fife is described briefly. At first breeding of common red wheat in Canada consisted mainly of bringing seed from Europe and subsequent selection by individual farmers. About 1842, David Fife, a farmer in Ontario, discovered a plant of spring growth habit in a plot of winter wheat that had been grown from seed received from Europe. The progeny of this plant gave rise to a high yielding variety possessing excellent milling and baking qualities that came to be known as 'Red Fife'. It was introduced into Western Canada about 1882, and from then until 1908 it was the leading variety in that region. The variety was introduced into the United States in about 1860 and proved to be a great success there too. Red Fife, via the variety Yeoman, is an ancestor of many improved British and French wheat varieties.

After about 1880, the trade in high-quality seeds developed in Europe and several breeders and their families, such as De Vilmorin, Rimpau and Von Lochow grew to prominence. The breeder Hyalmar Nillson, who in 1886 became the first director of the breeding station of the Sveriges Utsädeförening in Svalöf (Sweden), was famous for his consistent application of line selections in wheat. In the Netherlands, L. Broekema started to cross English Squarehead types, with stiff straw and high yields, and the Zeeuwse Witte, a local variety with relatively good baking qualities. The result was the famous variety Wilhelmina, with white grains, which was introduced on the market in 1901.

1.9. The twentieth century

The rediscovery of Mendelian laws in 1900 by De Vries, Correns and Von Tschermak resulted in a new understanding in genetics. As a result wheat breeding was modernized, offering new possibilities and prospects.

In Europe, a growing tendency to become independent of wheat imports from the USA and Canada developed. To achieve this goal, the activities of breeding stations all over Europe were increased, resulting in a great number of new varieties. Much attention was paid to yield capacity, disease resistance and tolerance to climatic stress factors; milling and baking qualities were also taken into account. European breeders were well aware that the task of equalling the quality of wheats from across the ocean was a hard one.

During the first years after the Second World War, Europe spent all its energy on rebuilding its economy. Scientific research during the second half of the 20th century indicated that increasing application rates of nitrogen fertilizer, especially its split application, resulted in grain yields considerably higher than had been achieved up till then. In addition, the use of herbicides and fungicides contributed significantly to this production increase. In France, for instance, average wheat yield in the years 1950–1955 was 2100 kg per ha; in the period 1980–1985 yields increased to an average of 5180 kg per ha. Ten years later (1994–1995) the average yield per hectare was 6450 kg. Similar developments took place in other countries.

In 1957, six countries (Belgium, Germany, France, Italy, Luxemburg and the Netherlands) signed the Rome Convention, in which they agreed to cooperate economically in the European Community. The Rome Convention included a separate chapter on the subject of agriculture. After a transitional period, the Common Market for agriculture became a fact in 1967. For most products a so called market plan was introduced, such as for cereals, dairy products, sugars, meat and oil-producing seeds.

One of the basic principles of the new agricultural policy was that of communal preference. This meant that customers in the European Community (EC) should give preference to products from one of the other member countries. In order to give wheat from EC-countries a better chance on the market, the price of wheat from 'third countries' was kept artificially at a higher level than that of EC wheat. Moreover, when possible, better prices were paid for wheat with good bread making quality. This stimulated breeders to pay more attention to the milling and baking quality of their breeding material.

In 1973, Denmark, Ireland and the United Kingdom entered the Community, followed in 1981 by Greece, in 1986 by Portugal and Spain and in 1995 by Austria, Sweden and Finland. The Europe of Fifteen, now referred to as the European Union (EU) covers a market of some 375 million consumers.

Liberalization of the world trade in the 1990s led to a new Common Agricultural Policy (known as CAP reforms) in the European Union, with substantial changes in the subvention system for agricultural products. In order to reach a more market-oriented price system, intervention in the pricing of most agricultural products, including cereals, was reduced to a minimum.

References

McCance RA, Widdowson EM (1956) Mediaeval and early modern England (1200–1700). In: McCance RA, Widdowson EM (eds), Breads, white and brown; their place in thought and social history. Pitnam Med Publ Co, London: 10–27.

Heyder M (1979) Amsterdam – korenschuur van Europa. Amsterdam: Stadsdrukkerij van Amsterdam, 104 pp.

Jacob HE (1956) Sechstausend Jahre Brot. 2nd ed. Rohwolt Verlag, Hamburg: 335–378.

Maynard Smith J (1958) The theory of evolution. Penguin Books, Middlesex: 226–227.

Peterson RF (1965a) Species. In: Peterson RF (ed), Wheat: botany, cultivation and utilization. Leonard Hill Books, London and Intersc Publ, New York: 1–16.

Peterson RF (1965b) Harvesting and threshing. In: Peterson RF (ed), Wheat: Botany, Cultivation and Utilization. Leonard Hill Books, London and Intersc Publ, New York: 266–278.

Riley R (1975) Origins of wheat. In: Spicer A (ed), Bread: social, nutritional and agricultural aspects of wheaten bread. Appl Sc Publ, London: 27–76.

Schama S (1987) In: Knopf A (ed), The embarrassment of riches; an interpretation of Dutch culture in the Golden Age. New York: 168.

Zeven AC (1986) Landrace groups of bread wheat (*Triticum aestivum* L em Thell). Acta Horticulturae 182: 365–376.

Zeven AC (1990) Landraces and improved cultivars of bread wheat and other wheat types grown in the Netherlands up to 1944. Agric Univ Papers 90.2, Wageningen, 103 pp.

2. The wheat grain

2.1. Anatomy[1]

The fruits of most plants contain one or more seeds which, at ripeness, can be easily separated from the rest of the fruit tissue. For *Gramineae* this is different: a fertilized egg cell in the ovary develops into a single seed, comprising the whole fruit. Fruit wall (pericarp) and seed coat are united (Figure 2.1), as a result the seed and fruit cannot be separated. This type of fruit, which is characteristic for all grasses, including cereals, is given the botanical term of *caryopsis.*

Figure 2.2a shows a whole wheat grain of a length of about 5 mm. The kernel has a somewhat vaulted shape with the *germ* or *embryo* (the future plantlet) at one end, and a bundle of hairs, which is referred to as the *beard or brush* at the other end. Figure 2.2b shows a cross-section of the grain, in which the *endosperm* is clearly visible. The endosperm is rich in starch and contains the proteins that will form the gluten at dough making. The ventral, flat side of the grain has a rather deep *crease.* The endosperm is surrounded by the fused *pericarp* and *seed coat.* The endosperm mainly contains food reserves, which are needed for the growth of the seedling.

Figure 2.2c shows an enlarged view of part of the endosperm. The section consists for the greater part of prismatic cells, filled with starch granules embedded in protein. In the centre of the grain (not shown in this picture) these cells are round. Only the outer endosperm, the *aleurone layer*, has a different structure: it consists of a single layer of cells of cubic shape. The 'inner endosperm', i.e. the endosperm without the aleurone layer, is referred to as *mealy* or *starchy endosperm.* The aleurone layer is rich in proteins and enzymes, which play a vital role in the germination process.

During the first stage of milling, the outer layers of the wheat grain, i.e. the *bran*, are separated from the mealy endosperm. The fracture is located right under the aleurone layer. This means that bran is made up of the fused pericarp, plus the seed coat, plus the aleurone layer.

Wheat grains have either a dark, orange-brown appearance (red wheats or red-seeded wheats) or a light, yellowish colour (white wheats or white-seeded wheats). There was a time when many countries preferred white-seeded wheats. In Europe this preference has gradually disappeared and mainly red-coloured varieties are grown nowadays. The main reason for this

[1] For more information see Evers (1988); Hoseney (1986); Kingswood (1975).

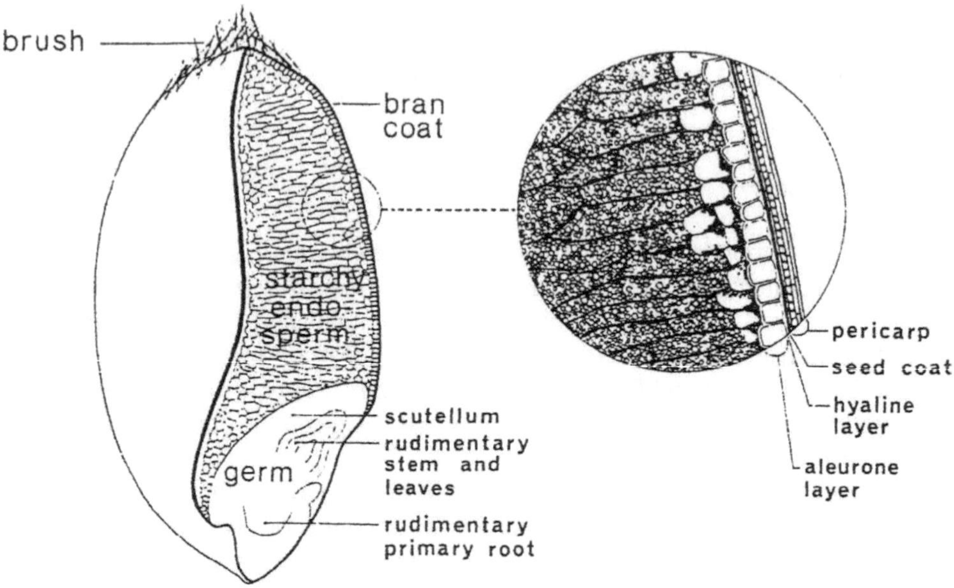

Figure 2.1. Wheat grain cut lengthwise through crease.

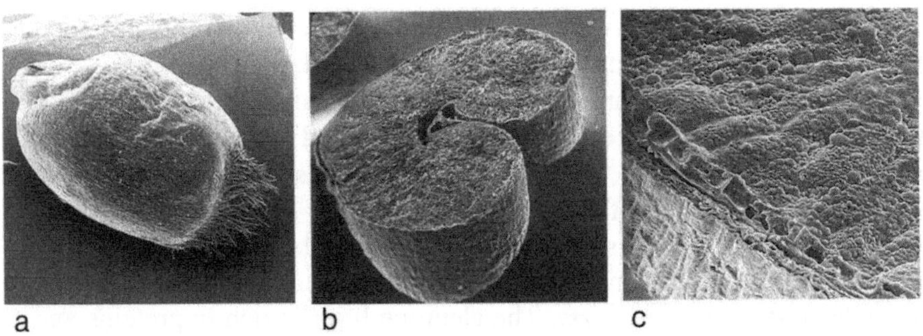

Figure 2.2. Scanning electron microscope images of wheat grains: (a) intact wheat grain, at the right the beard, at the left the germ; (b) cross section of a grain, at the right the crease; (c) detail of the cross section picture.

change is that white wheats are more susceptible to pre-harvest sprouting than the majority of the red wheats. White wheat varieties are grown particularly in Australia, and to a lesser extent in Canada and the USA.

In brief, the wheat grain is constituted by three distinct parts: the germ (embryo), the bran and the mealy endosperm. Wheat grains contain 2–3% germ, 13–17% bran and 80–85% mealy endosperm (all constituents converted to a dry matter basis).

2.2. Chemical composition[2]

The moisture content of commercial lots of wheat may vary between 12 and 18%, depending on the weather during harvest.

Until recently, the composition in percentages of wheat fibres were based on analysis of crude fibre material according to a method developed by Weende. The values resulting from this analysis tend to be rather low. A more accurate method was inroduced later by Van Soest: the Neutral Detergent Fibre (NDF) method. It measures the 'dietary fibres' and results in higher values that correspond more closely to reality. The fibre contents mentioned in Table 2.1 have been measured using the NDF method. As a consequence, the values for carbohydrates are lower than those for the method of Weende.

Table 2.1. Chemical composition of the whole wheat grain and its various parts (converted to percentages on a dry matter basis)

	Whole grain	Mealy endosperm	Bran	Germ
Proteins	16	13	16	22
Fats	2	1.5	5	7
Carbohydrates	68	82	16	40
Dietary fibers	11	1.5	53	25
Minerals (ash)	1.8	0.5	7.2	4.5
Other components	1.2	1.5	2.8	1.5
Total	100	100	100	100

Most of the mealy endosperm consists of food reserves: 82% carbohydrates (mainly starch), 13% proteins and 1.5% fats. The contents of minerals (ash) and of dietary fibres are low, 0.5% and 1.5%, respectively.

More than half the bran consists of fibre components (53%). Proteins and carbohydrates each represent 16% of total dry matter. The mineral content is rather high (7.2%). The two external layers of the grain (pericarp and seed coat) are made up of empty cells and are dead. The cells of the inner bran layer, i.e. the aleurone layer, are filled with living protoplasts. This explains the rather high levels of protein and carbohydrate in the bran.

Finally, the germ is rich in proteins, fats, carbohydrates and dietary fibres. The mineral level is also rather high (4.5%).

[2] For more information see Pomeranz (1988).

2.3. Grain hardness

> Note that the terms hard wheat and soft wheat, used in official EU regulations, do not have the same meaning as described in this section. EU officials use the term 'hard' wheat exclusively for durum or macaroni wheats, and the term 'soft' wheat for *aestivum* or bread wheats, irrespective of the hardness of their grains.

Wheat varieties vary widely in grain hardness. Two categories are distinguished: varieties with 'hard' and varieties with 'soft' grains. The kernels of hard wheats often, but not always, have a dark, shiny aspect and they look vitreous. Soft grains have a more opaque, floury appearance.

Grain hardness is a character of particular interest to the milling and baking industries. Hard grains exhibit more resistance to grinding than soft wheat grains; the several methods for determining grain hardness are based on this property (see Section 8). In general, hard wheat grains have a higher protein content than soft ones. Growing conditions also play a role.

Hard wheats result in more damaged starch than soft wheats in the milling process. In bread, this damage largely contributes to the soft texture and the gentle, pleasant mouthfeel of the crumb; moreover starch damage has a retarding effect on the bread going stale. Soft wheat flours, on the other hand, have less damaged starch. They are more suited for the production of biscuits, cakes, crackers, wafers, etc. The lower protein content and the finer granulation of a soft wheat flour undoubtedly contributes to its suitability for these products.

The inheritance of grain hardness has been documented by Symes (1965, 1969), who in crosses between soft and hard wheat cultivars demonstrated that in most cases one single gene is responsible for grain hardness. However, in some cases grain hardness is caused by more than one gene. For a wheat breeder, these are important conclusions, since the conversion of a soft wheat variety into a hard one, and vice versa, should be relatively easy by means of crossing and back-crossing.

Photographs taken with a SEM (Figure 2.3) clearly show that differences in hardness can be related to differences in the interior structure of the endosperm cells (Belderok, 1973).

Figure 2.3a shows a European wheat variety with soft grains that are low in protein content (10.9% on dry matter basis). Elliptical starch granules are clearly visible. The surface of these granules is generally smooth but dents can be seen on a few spots. Probably these dents have been caused by neighbouring starch granules that were in the way during the growing process. Between the starch granules, one can see a small number of round, angular particles of smaller size and lighter shade. The light-coloured particles are

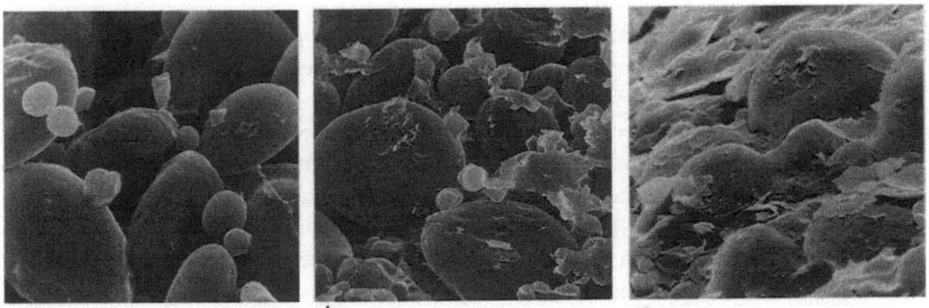

Figure 2.3. Scanning electron microscope images of endosperm cell contents: (a) image of a soft European wheat variety; (b) image of a middle-hard Northern-Spring wheat; (c) image of a hard and vitreous Australian Prime-Hard wheat.

what are known as *protein bodies*. The image suggests that, inside the cell, starch granules and protein bodies exist as discrete particles without much coherence. Furthermore, one can discern small voids, which are actually filled with air. Such a loose structure scatters light falling in many directions and causes the mealy, floury appearance of soft wheat grains.

Figure 2.3b shows the inside of an endosperm cell of a US spring wheat variety, with a medium-hard grain structure and a protein content of 16.5% of total dry matter. In spite of a relatively high protein level, cell elements are more or less the same as in Figure 2.3a. However, the distribution of the protein is both in the form of protein bodies and a paste spread between the starch granules. This causes a higher degree of consistency of cell content.

Figure 2.3c shows an extremely hard, vitreous Australian Prime-Hard wheat with a protein content of 17.8% (dry matter basis). The protein bodies do not appear as separate particles, but as a paste that is spread between the starch granules, covering them completely. Obviously the content of this cell is much more coherent than that of the cells shown in Figures 2.3a and 2.3b. The compact structure and the absence of voids between the starch granules explain the vitreous appearance of Prime-Hard wheat, since most light passes right through the grains and is not reflected.

References

Belderok B (1973) Bedeutung der Kornhärte als Selektionskriterium bei der Weizenzüchtung auf Backqualität. Getr Mehl u Brot 27: 361–366.

Evers AD (1988) Microscopic structure of the wheat grain. In: Pomeranz Y (ed), Wheat: chemistry and technology (vol I), 3rd edn. St Paul, MN: Am Assoc Cereal Chem: 47–95.

Hoseney RC (1986) Structure of cereals. In: Hoseney RC (ed), Principles of cereal science and technology. St Paul, MN: Am Assoc Cereal Chem: 1–31.

Kingswood K (1975) The structure and biochemistry of the wheat grain. In: Spicer A (ed), Bread: social, nutritional and agricultural aspects of wheaten bread. London: Appl Sci Publ: 47–65.

Pomeranz Y (1988) Chemical composition of kernel structure. In: Pomeranz Y (ed), Wheat: chemistry and technology (vol I), 3rd edn, St Paul, MN: Am Assoc Cereal Chem: 97–158.

Symes KJ (1965) The inheritance of grain hardness in wheat as measured by the particle size index. Aust J Agric Res 16: 113–123.

Symes KJ (1969) Influence of a gene causing hardness on the milling and baking quality of two wheats. Aust J Agric Res 20: 971–979.

3. Milling of wheat

3.1. Introduction[1]

Most cereal grains are used either for the production of animal feed or for the milling of flour for human consumption.

If used as an *animal feed*, the grains have only to be crushed to obtain a granulated product. Any type of mill may be used for this purpose. In most cases it will be a hammer mill, which in one processing step fragments the grains and pulverizes them by a beating action of hammers.

The grinding of wheat for *human consumption*, however, has to meet higher standards and is called *milling*. The main purpose of milling is to completely separate bran and germ from the mealy endosperm and to thoroughly pulverize the mealy endosperm into middlings, semolina and flour (definitions of these substances are given in Section 3.4). This is achieved through the shearing and scraping action of millstones (in windmills and watermills) or mill rollers (in a modern flour-milling factory). In this process it is essential to extend slightly the floury particles to such a degree that their internal structure is made loose. In German this is called 'Auflösung'. The water absorption capacity at doughmaking will be higher in a flour with a loose structure and it will form a more elastic and stronger dough than in a flour with a compact structure. Milling of flour for human consumption is achieved in three steps:

1. cutting open the grains,
2. scraping bran from semolina and other floury substances, and
3. reducing the size of the semolina and middlings to form flour.

3.2. Conditioning or tempering

In European mills, grains of soft-kerneled wheats are usually brought to moisture contents of 16% and hard-kerneled wheats to 17.5%. Higher moisture levels are required for milling extremely hard wheats. Duration of tempering varies from about 8 hours (for soft wheats) to 24 hours (for hard wheats).

A *tough bran* and a *friable mealy endosperm* are prerequisites for an efficient separation of the two. As the bran becomes progressively tougher and less brittle with increasing moisture content, milling will produce flour that

[1] For more information see Hoseney (1986); Bass (1988).

is less contaminated by bran dust, thus being whiter and having a lower ash content. The mealy endosperm is rendered mellower and more friable with the addition of the correct amount of moisture. However, excessive moisture causes flaking, whereas insufficient moisture leads to undue pulverization. Since an ideal moisture distribution is achieved when the bran is slightly damper than the endosperm, a second light tempering (about 0.5% moisture) is often applied shortly (less than 1 hour) before milling.

To improve the physical state of the grains for milling, the actual milling process is preceded by *conditioning* or *tempering* of the wheat. This process involves the addition or removal of moisture (wetting or drying, respectively) for definite periods of time, in order to obtain the desired moisture content in the mass of grain, and the desired distribution of moisture throughout each kernel.

3.3. Milling

3.3.1. Milling in a traditional mill

In windmills and watermills, cereals are ground between a pair of horizontally placed millstones. The lower or bed stone is fixed to the floor of the milling room. The upper stone rotates on a central axis. The two stones are approximately 1.5 m in diameter, but the thickness of the stones may vary from mill to mill. Wheat is poured from a bin in the upper storey of the mill into a hole (the eye) in the centre of the upper stone (Figure 3.1). In this way the wheat is positioned centrally between the two millstones. The upper-side of the lower stone and the under-side of the upper stone are fluted (Figure 3.2).

The distance between the two stones is in the centre about 1 cm, in the periphery less. During the first few rotations the grains are cut and broken by the edges of the flutings, which work like scissors. Thereafter, the milling good goes in a spiral movement towards the periphery of the stones, where the next step of the process takes place: the inner part of the grains, i.e. the mealy endosperm, is separated from the outer part, i.e. the bran. The large and medium-large endosperm particles that are released during this process are called semolina. The diameter of these particles can vary from about 300 μm to 750 μm.

Finally, the ground particles end up in the outer 15 cm of the millstones, where the distance between the stones is the smallest. Here, the semolina is rubbed into dust-like particles: flour. Floury particles ground on a millstone have a diameter of about 200 μm or less.

If the stones are properly set, the tough bran will remain intact. The flour that is ejected from between the millstones will be a mixture of bran, pure

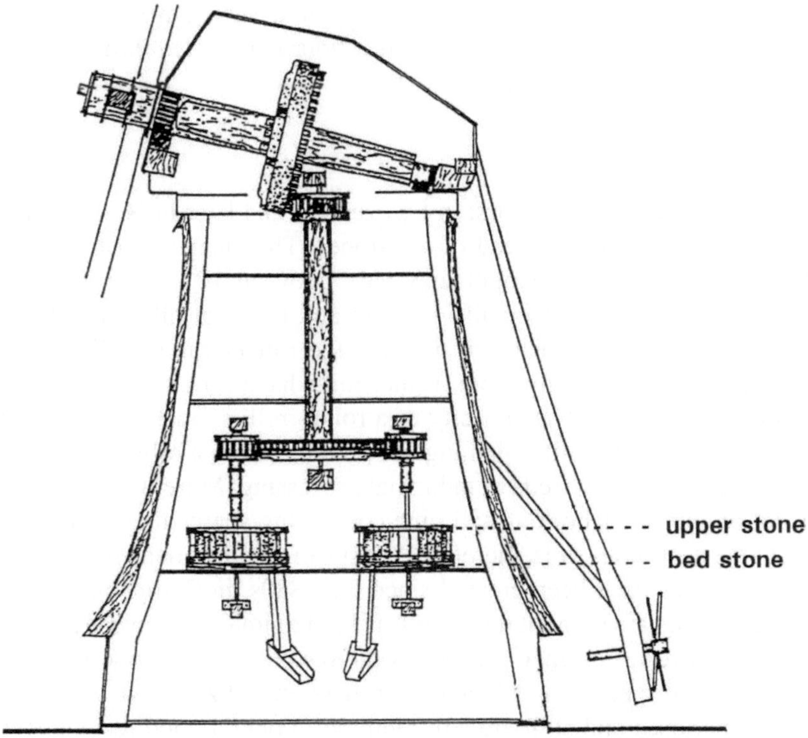

Figure 3. 1. Driving gear of a stone-mill.

Figure 3. 2. Carves on the milling surfaces of the two millstones (H. Titulaer, Mühlsteinbau, Geldern 1-Veert D).

flour and semolina that has not been fully ground. Such a mixture is referred to as *whole-wheat flour*. This whole-wheat flour may be led through a round sieve (a bolter) in order to obtain pure flour.

3.3.2. Milling in a roller-mill

Around 1850 the idea arose, first in Switzerland and later in Hungary, to use porcelain or steel rollers instead of millstones. The diameter of these rollers is 25 or 30 cm and their length may vary between 100 and 150 cm. The rollers are positioned horizontally in pairs and rotate at different speeds in opposite directions. The distance between the rollers can be adjusted. Porcelain rolls are not used any more. Roller-mills have certain advantages over traditional stone-mills. The capacity of a roller-mill is much larger than that of a windmill or watermill. The range of products from a roller-mill is larger and better geared to the needs of industrial processing. Moreover, a roller-mill does not require the laborious and time-consuming sharpening of the furrow edges of the millstones. It is, therefore, not surprising that many modern flour-milling factories have been created since the 1880s, in Europe as well as in North America. Life without these mills is hard to imagine nowadays. Roller-milling is a cleverly designed process. Although most of the work in a modern mill is done by special machines (e.g. rollers and plansifters) considerable milling experience and skill are still required to obtain good results. Roller-milling comprises several grinding steps, each being followed by a sifting operation. Broadly speaking, the grinding stages may be grouped into two successive sytems: the break system and the reduction system.

The break system. Primarily, the break system aims to bring about a virtual separation between the bran and the mealy endosperm. It is carried out with the aid of corrugated or fluted iron rollers, called *break rollers* (Figure 3.3). They operate in pairs, revolving in opposite directions and at different speeds (the speed differential is approximately 1:2). The process includes successive milling steps, in which the corrugations of the rollers become progressively finer and the milling gap (distance between the rollers) smaller. Under each pair of rollers a set of horizontal sieves (*plansifters*) is placed so that the milled product is graded according to size. End products of the break system are mainly coarse offals (bran), germs, coarse endosperm particles (semolina, 300–750 μm and middlings, 125–300 μm) and a certain amount of flour (diameter less than 125 μm).

The reduction system. Primarily, the reduction system aims to reduce coarse endosperm particles (semolina and middlings) to flour fineness. It involves a series of reduction steps in which, as in the case of breaking, the rollers are set

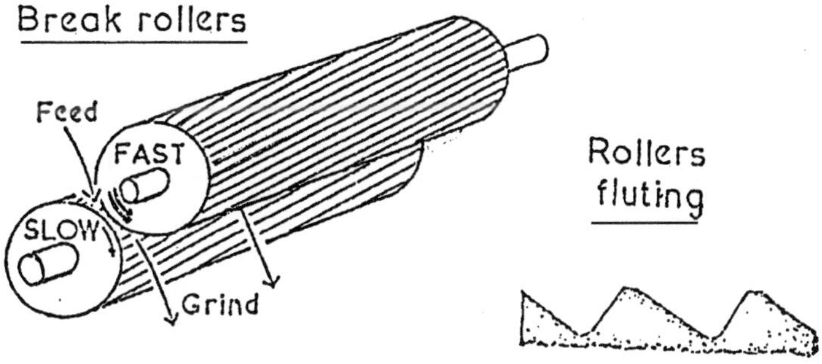

Figure 3. 3. The milling of wheat with two break rollers.

progressively closer at each successive processing step. The reduction rollers, also, operate in pairs. They have smooth, slightly roughened (mat) surfaces; and speed differentials are 4:5 (in Europe) or 2:3 (in USA and Canada). Each reduction is again followed by plansifters, through which the ground material is sifted according to particle size. The reduction process is repeated several times until ultimately most of the mealy endosperm has been converted to flour.

End products of the reduction system are mainly: fine offals (shorts) and fine endosperm particles (flour).

To illustrate the milling steps in some more detail, the simple *flow diagram* of the experimental laboratory mill in Figure 3.4 will be discussed here.

Conditioned wheat is fed to a pair of corrugated chilled-iron rollers known as the *first break-rollers*, one of which revolves at two-and-a-half times the speed of the other. The space between the first break-rollers is such that the grains are broken mainly into relatively coarse pieces with a minimum of crushing, so as to avoid powdering the bran (powdered bran cannot be separated from flour).

The material released from the first break-rollers passes to a sifter machine (*plansifter*), which separates the particles according to size by means of a stack of horizontal sieves of increasing degrees of fineness. The sieve sizes are (from top to bottom) 750, 300 and 125 μm, thus dividing the grinding stock into four fractions: coarse materials (larger than 750 μm), semolina (between 300 and 750 μm), middlings (between 125 and 300 μm) and flour (smaller than 125 μm).

The finest material (flour) leaves the mill and passes directly to the flour bag. Intermediate-sized particles, those larger than flour particles, are sent to the first reduction rollers (semolina) or (the smaller particles, middlings) to the third reduction rollers.

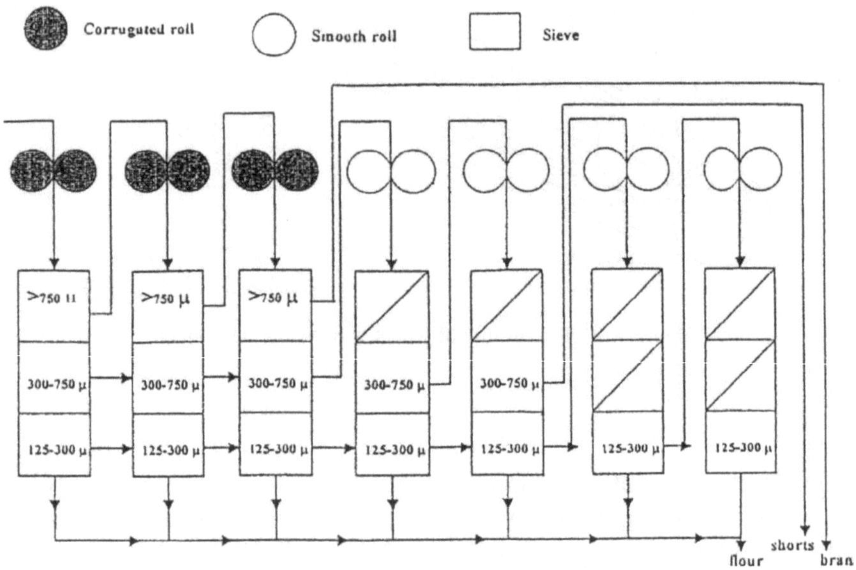

Figure 3. 4. Flow diagram of an experimental mill.

The coarsest particles are fed to the *second break-rollers*, which have finer flutings and are set closer than the first break-rollers. Again, the resulting grinding stock is divided into four sieve fractions: coarse material, semolina, middlings and flour.

After passing through the *third break-rollers*, the coarse material (overtails of the top sieve) consists of bran flakes from which almost all endosperm has been removed. The second and third break-rollers are meant to scrape all endosperm off the bran skin, releasing the mealy endosperm in large particles of semolina and middlings and leaving the bran in large flat flakes so that it can be easily removed. Note that the production of flour is not the main object of the break system.

The coarse endosperm particles released from the break-rollers are fairly pure, i.e. free from bran fragments. They are fed to *reduction rollers*, where smooth rollers crush these particles into flour. One roller revolves at 1.25 times the speed of the other. The smooth rollers desintegrate the endosperm, whereas most of the tougher bran and germ particles are merely somewhat flattened and can thus be sifted out. The reduction rollers (designated as first, second, third, and fourth reduction rollers) are set progressively closer together.

Thus, the main object of the reduction system is the production of flour. In addition to flour, fine offals are obtained, called *shorts*. The simple mill diagram in Figure 3.4 shows only one offal. This byproduct consists of small bran fragments to which some endosperm is still attached. It also contains germ. In contrast to bran, germ can be easily separated from the endosperm, due to its putty-like consistency. It will flatten rather than pulverize under the force of the rollers.

Laboratory roller-mills exist in many sizes and capacities. The smallest can mill 100 g of wheat into flour within a few minutes. The largest have capacities of about 100 kg/h or more. All these mills, even the smallest, produce flour which is comparable to flour of commercial mills. They may be used in ordinary laboratory analyses such as those carried out by milling and breeding companies (including baking trials on a micro-scale).

3.4. End-products of milling and their application

Semolina is a milling product consisting of coarse particles of mealy endosperm with a diameter of 300–750 μm. *Middlings* are a product consisting of intermediate-sized endosperm particles with a diameter of 125–300 μm. Both are used as ingredients in porridge and for the production of pastas. *Flour* consists of fine endosperm particles of less than 125 μm diameter. It is used for making white bread, confectionary products, and certain pasta products. Besides fine endosperm particles or flour, *meal* contains a certain amount of flaky bran and, sometimes, germ. It is used for making light and dark types of brown bread. In so-called *whole-meal flour* and *whole-meal bread*, the endosperm, bran and germ are present in the same proportion as they are in wheat grains.

3.5. Milling quality

In a commercial mill, the aim is to obtain a maximum yield of flour with a 'healthy' white colour. As described in Section 2, mealy endosperm generally has a very low ash content (about 0.5%), whereas bran and germs are characterized by much higher ash levels (about 7.2% and 4.5%, respectively). In other words, the lower the ash content, the whiter the flour.

A laboratory-scale mill may give a rough indication of the milling quality of wheat. A middle-sized mill with an hourly capacity of several kilogram is well equipped for giving reliable milling figures. Larger mills with higher capacities will provide more detailed information; they are used exclusively in the laboratories of flour-milling factories.

The *milling quality*[2] of wheat is determined by successively measuring the following characteristics in a standard milling experiment:

a) the amount of grain that can be milled per time unit,
b) the flour yield of each milling passage, and
c) the moisture and ash content of each milling passage.

The values of (b) and (c) need to be converted to a standard moisture content (approximately 14%). The successive flour fractions are combined till an extraction rate of 70% has been reached. If an extraction rate of 70% cannot be reached, a small quantity of shorts is added. Finally, the ash content of this mixture is determined.

> The ash content of flour at a standard milling or extraction rate (in our case 70%) is a reliable indication of the milling quality of the wheat. The lower the ash content, the higher the flour yield in a commercial mill will be. The ash content of flour may vary between 0.45% and 0.60%.
>
> Theoretically, a reverse technique may be used whereby the flour yield at a standard ash content (e.g. 0.45%) is taken as a measure for milling quality. The thus determined extraction rates vary from 67% to 75%. This method is generally used in commercial milling practice.

3.6. Soft wheat versus hard wheat

In the milling of *soft wheats*, the mealy endosperm is fragmented indiscriminately. Therefore, the flour particles consist mainly of fragments of single cells or fragments of clusters of two or three neighbouring cells. The interior of the cells more or less bulges. When rubbed between thumb and forefinger, the flour has a soft, 'woolly' feel. The flour of soft wheats contains many loose starch granules and protein particles due to the low degree of cohesion between starches and proteins. The flour easily sticks together and sifts with difficulty, tending to close the apertures of sieves.

In the milling of *hard wheats*, fracturing of the grain follows the middle lamellas of the mealy endosperm cells. As a consequence, flour particles consist of one or more intact cells, some with and some without adhering wall material. As the cohesion between starch and protein is much stronger in hard than in soft endosperm wheat, the flour contains hardly any loose starch granules and protein particles. Hard wheats yield a coarse, gritty flour which is free-flowing and easily sifted. When rubbed between thumb and forefinger,

[2] For more information see Russell Eggit & Hartley (1975).

a more or less 'sandy' structure can be felt. The flour yield of a hard wheat is generally a few percent higher than that of a soft wheat.

3.7. Durum wheat milling

Semolina from durum wheat is the preferred raw material for the manufacture of pastas (macaroni, spaghetti, noodles). In durum milling, the objective is to produce a maximum yield of highly purified semolina. Although the same sequence of operations is involved in the production of semolina as in flour, the milling systems differ in their design.

Durum wheat milling involves thorough cleaning and proper conditioning of the grains, light and careful grinding, and thorough purification. In a *purifier*, a shallow stream of impure semolina passes over a large sieve that is shaken rapidly backward and foreward. An upward air current through the sieve draws off very light material to dust collectors and tends to hold bran particles on the surface of the moving semolina so that they drift over the tail of the sieve.

Cleaning, breaking, sifting and purifying systems are, therefore, much more elaborate and extensive than in flour mills. On the other hand, the reduction system is much shorter in durum mills because the primary product is removed and finished in granular condition; reduction rollers are only required for reduction of the tailings from the various parts of the break system.

References

Bass EJ (1988) Wheat flour milling. In Pomeranz Y (ed), Wheat: chemistry and technology, Vol II, 3rd edn. St Paul, MN: Am Assoc Cereal Chem: 1–68.

Hoseney RC (1986) Dry milling of cereals. In Hoseney RC (ed), Principles of cereal science and technology. St Paul, MN: Am Assoc Cereal Chem: 133–152.

Russell Eggit PW, Hartley AW (1975) Extraction rates – Milling processing implication. In Spicer A (ed), Bread: Social, nutritional and agricultural aspects of wheaten bread. London: Appl Sci Publ: 219–233.

4. Survey of gluten proteins and wheat starches

4.1. Gluten proteins

4.1.1. The composition of wheat proteins

The protein content of wheat grains may vary between 10% and 18% of the total dry matter. Wheat proteins are classified according to their extractability and solubility in various solvents. Classification is based on the classic work of T.D. Osborne at the turn of the last century. In his procedure, sequential extraction of ground wheat grains results in the following protein fractions:

- *albumins*, which are soluble in water;
- *globulins*, which are insoluble in pure water but soluble in dilute NaCl solutions, and insoluble at high NaCl concentrations;
- *gliadins*, which are soluble in 70% ethyl alcohol, and;
- *glutenins*, which are soluble in dilute acid or sodium hydroxide solutions.

Albumins are the smallest wheat proteins, followed in size by globulins. The separation of albumins and globulins turned out to be not as clear as initially suggested by Osborne. Gliadins and glutenins are complicated high-molecular weight proteins.

Most of the *physiologically active proteins* (enzymes) in wheat grains are found in the albumin and globulin groups. In cereals, the albumins and globulins are concentrated in the seed coats, the aleurone cells and the germ, with a somewhat lower concentration in the mealy endosperm. The albumin and globulin fractions cover about 25% of the total grain proteins.

Gliadins and glutenins are *storage proteins* and cover about 75% of the total protein content. The wheat plant stores proteins in this form for future use by the seedling. Gliadins and glutenins are mainly located in the mealy endosperm and are not found in the seed coat layers nor in the germ.

Storage proteins in wheat are unique because they are *technologically active*. They have no enzyme activity, but they have a function in the formation of dough as they retain gas, producing spongy baked products.

4.1.2. The function of gluten proteins

During mixing of a wheaten dough, gliadins and glutenins absorb a certain amount of water; the thus hydrated constituents are then transformed into a coherent protein mass, called *gluten*. Starch, added yeast and other dough components are embedded in this gluten skeleton. The bread-making quality of a dough depends chiefly on two factors: the quantity and quality of its gluten.

Gluten quantity. The protein content of wheat grains mainly depends on the amount of nitrogen available to individual plants during their growth. Another factor is the number of grains that can profit from this nitrogen: a high grain number tends to be coupled with a low protein level of the individual grain and vice versa. The protein level of the grains can be raised considerably by increasing the amount of N-containing fertilizers. Split nitrogen applications are common practice nowadays: a first application is given in spring (winter wheat in early spring, spring wheat at sowing), followed by a second and sometimes a third during the period from stem elongation until just before flowering. The aim of the first application is to support initial growth and to stimulate tillering of the crop. The nitrogen of the second, and perhaps third, application mainly favours the protein uptake of the grains.

Gluten quality. Yeast produces carbon dioxide out of sugars in the dough. Cultivars with a good bread-making quality have a dough that is able to produce and retain a large amount of carbon dioxide in small alveoles. This is an essential condition for the production of a well-shaped loaf of bread with a medium-large to large volume. Some wheat varieties do not respond, with regard to baking quality, in the same way to a higher protein level in the grains. The quality of the gluten from the plant is mainly determined by its genetic background; it cannot be changed by changes in environmental conditions or by the use of fertilizers. Higher protein levels will generally result in higher baking quality, provided that the gluten quality is good. However, higher levels of N in a variety with a poor protein quality will give only a very restricted improvement in the baking quality (Figure 4.3).

4.2. Wheat starch[1]

4.2.1. Composition of the wheat starch

Cereal grains store energy in the form of starch. The amount of starch contained in a wheat grain may vary between 60% and 75% of the total dry weight of the grain. Starch occurs in seeds in the form of granules. Wheat has two types of starch granules: large (25–40 μm) lenticular and small (5–10 μm) spherical ones. The lenticular granules are formed during the first

[1] For more information see Hoseney (1986a).

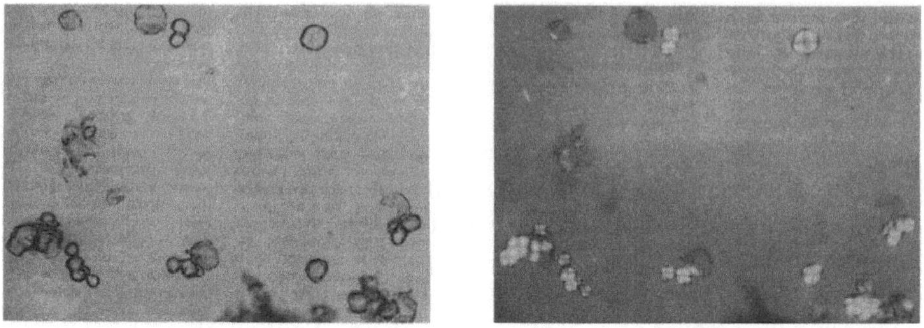

Figure 4.1. Microscopic view of native and damaged starch granules. Left view under normal light, right view under polarized light.

15 days after pollination. The small granules, representing about 88% of the total number of granules, appear 10–30 days after pollination.

Starch is basically a polymer of glucose. Chemically, at least two types of polymers are distinguishable: *amylose*, a predominantly linear polymer, and *amylopectin*, a highly branched polymer. The molecular weight of amylose is around 250 000 (1500 glucose molecules) but varies widely. The structure of this polymer was assumed to be mainly linear, but this appears to be true for only part of the amylose; the remainder is slightly branched.

Amylopectin is branched to a much greater extent than amylose. So much that, on the average, the unit chain in amylopectin is only 20–25 glucose molecules long. Amylopectin has a molecular weight of about 10^8. It is truly a huge molecule – one of the largest found in nature, with 595 000 glucose units, or 30 000 chains with an average degree of polymerization of 20. Amylopectin is thought to be randomly branched. The ratio of amylose to amylopectin is relatively constant, at about 23.

4.2.2. Damaged starch

In the process of milling, part of the starch present in the endosperm is mechanically crushed; the miller calls this *damaged starch*. Under a microscope, damaged starch granules can be easily recognized from native starches, the former swell in water so that their outline becomes unsharp (Figure 4.1). These granules are called 'ghosts'. Undamaged starch granules hardly swell in water and keep their clear outline completely. Damaged granules can easily be coloured with a solution of Congo red or a very dilute (0.02–0.03%) iodine solution. Moreover, damaged granules have lost their crystalline structure and, as a consequence, no longer show a double fraction in polarized light.

The term 'damaged starch' may suggest that this is an undesirable property. On the contrary, flour with good bread-making characteristics needs a certain amount of damaged starch because it has two properties important in bread making:

- it absorbs water quickly and in much greater quantities than undamaged starch, and;
- it can be digested by amylolytic enzymes at dough temperatures to give various breakdown products, including maltose and dextrins (Figure 4.2).

In certain cases it may even be advantageous to increase the proportion of starch damage and thus the water absorbing capacity of a flour, i.e. to make the highest number of loaves possible from a certain amount of flour. This is only useful in countries where bread is sold by its total weight and not by its dry matter weight.

Two factors determine the proportion of starch damage. The first is the wheat-kernel hardness. Milled under the same conditions, a hard, vitreous grain will sustain more starch damage than a soft, mealy kernel. The second factor is the milling procedure. At the start of the milling process, bran and mealy endosperm are separated. The endosperm particles (semolina) are then reduced to flour. Starch damage can be achieved either under low milling pressure combined with a large number of milling passes, or under high milling pressure and less passes. In most flour types, the percentage of damaged starch varies between 4% and 10% of the total dry matter.

As far as loaf volume and other bread-making characteristics are concerned, a proportion of starch damage of some 7% of the dry matter content of the flour is considered to be optimal. On the other hand, the damaged starch content should not exceed the 9%, because the dough and the bread crumb will then become sticky, causing problems in the slicing of the loaf.

The percentages in this section have been determined according to the *67–30 method* (Approved Methods of the American Association of Cereal Chemists, AACC, St Paul, Minnesota, 1990). In most UK laboratories the *Farrand method* (Cereal Chemistry 41, 1964, 98–111) is used, which gives values that are considerably higher than those of the AACC method.

4.2.3. Enzymes[2]

Because cereals store their excess energy as starch and contain relatively high levels of this substance, it is not surprising that the starch-degrading enzymes have been studied extensively.

Cereals contain two types of amylases: *α-amylase and β-amylase.* Alpha-amylase is an enzyme that breaks glucosidic bonds on a more or less random basis. The result of the enzyme action is a rapid decrease in the size of large starch molecules and thereby a rapid reduction of the viscosity of a

[2] For more information see Hoseney (1986b).

starch suspension. Enzymes work much faster on damaged starch than on native starch, although given sufficient time, they will also degrade undamaged starch granules.

β-amylase is an enzyme that breaks down starch at the ends of the polymers and releases maltose molecules (maltose is a disaccharide sugar containing two glucose residues). It, thus, reduces a linear chain to maltose. β-amylase cannot pass a branch point on the starch chain.

If the amylose were completely unbranched, one would expect that only maltose would be produced from amylose. Actually only about 70% of the amylose is converted to maltose, indicating that some branching does occur. With amylopectin, only about 50% is converted to maltose, the remainder being a large molecular-weight (10^4) β-limit dextrin.

The combination of α- and β-amylase degrades starch quite rapidly and much faster and more completely than either alone. Each break that α-amylase makes, produces new ends for β-amylase to attack. A mixture of the two enzymes does not, however, completely degrade starch, as neither enzyme can break the branch points in amylopectin. In general, a combination of the two enzymes results in about 85% conversion of starch to sugar.

There are two types of α-amylase, *green amylases* and *malt amylases*. Green amylases occur in immature wheat grains. During ripening, the activity of this enzyme rapidly decreases, dropping to practically zero at harvest-ripeness. Malt amylases have a clearly different chemical composition, but the same working mechanism, as green amylases. They hardly occur in sound, intact wheat grains. However, when ripe wheat grains start germinating the activity of malt-amylase enzymes increases rapidly. If this happens under natural conditions in the field it is called *sprouting in the ear*.

In some varieties, the activity of the green α-amylases does not diminish during ripening. Sometimes, it even increases. The result is that ripe grains of these varieties have a relatively high level of α-amylase, even when they have not sprouted. This makes the wheat unsuitable for bread making. Some well-known examples include the English varieties Maris Huntsman, Fenman, Kinsman, Mardler and Norman. Each of these is derived from the old Belgian variety Professeur Marchal, a variety that was frequently used by English wheat breeders during the 1960s, until this deleterious characteristic was identified.

Unlike α-amylase, β-amylase is found in sound intact cereal grains. In general, the level does not increase much as a result of germination.

4.2.4. Sprouting damage

One last aspect of grain quality relevant to bread making is resistance to sprouting. Under rainy conditions, the grains of some cereal varieties germinate in the ear either before or at harvest-ripeness, known as sprouting in the ear. Some wheat varieties never exhibit any sign of sprouting, even in very poor weather, whereas other varieties are so liable to sprouting that their grains even germinate after very light rains. Most varieties are intermediate in this respect.

Research has shown that just-matured wheat grains are usually not able to germinate. They first have to go through a rest period, called *dormancy*, which ends more or less gradually, depending on the variety. Research has also shown that the length of the dormant period depends on the weather before harvest: high temperatures and low relative air humidities in the dough-ripe stage result in a shortening of dormant period; low temperatures and high relative air humidities result in lengthening of the period. Thus, it is possible that grains of a variety that is resistant to sprouting in normal years undergo such a shortening of the dormant period due to hot and/or dry weather in the dough-ripe stage that they will start to germinate because of rainfall just before or at harvest time.

Bread made from sprouted wheat has an inferior quality because of the high α-amylase activity (Section 4.3.1). Therefore, particularly in countries with a rainy climate, varieties with high sprouting resistance should be chosen for bread making.

At germination, the endosperm reserves are transformed into soluble components that serve as food for the young plant. This process is initiated by an increase in α-amylase activity in the grains, although no external signs of germination are yet visible. Only one or two days after the start of the sprouting process, small rootlets appear at the base of the grains.

The α-amylase activity in grains and flour has proved to be a good indicator of the degree of sprouting. Alpha-amylase activity can be measured rapidly and accurately with the Hagberg-Perten test (Section 8).

4.3. Components of baking quality

Broadly speaking, baking quality is determined by factors relating to *gas production* and *gas retention*.

4.3.1. Gas production

In wheat milling, a certain percentage of the starch (4-10%) will be damaged by the pressure and shearing action of the rollers (Section 4.2.2). Damaged starch can be easily digested by amylase enzymes at dough temperatures to give the breakdown products, dextrins and maltose sugars.

In the later stages of dough fermentation, yeast enzymes transform the maltose molecules into carbon dioxide (CO_2) and alcohol. The CO_2 gas aerates the dough; the alcohol evaporates in the oven during baking. Insufficient gas production is not a serious defect as it may be easily corrected by adding extra sugars and/or malt enzymes in the formula.

The dextrins in the bread crumb contribute to the gentle mouthfeel and the velvety texture of the crumb; moreover, they have a retarding effect on the staling of the bread crumb.

During bread baking, a high α-amylase activity is undesirable as an excess of this enzyme will break down too much starch into dextrins, causing the crumb to become sticky and unpleasant to chew; the bread is also difficult to slice. On the other hand, the amylase activity should not be too low, either. For a proper fermentation of the dough, a certain amount of available sugar is needed as substrate for the yeast.

4.3.2. Gas retention

Gas retention depends on the visco-elastic properties and the quantity of the gluten. During mixing, gluten forms a skeleton around the starch granules and part of the endosperm cells. This skeleton holds the CO_2 formed during fermentation. The better the extensibility and the elasticity of the gluten, the better it will hold the CO_2 enclosed in the dough and so the more the dough will rise. When, subsequently, the dough is heated in the oven, the protein skeleton will be fixed, and with it crumb structure and loaf volume. Good gluten characteristics are found in most of hard, high-protein wheats. In North American literature, these wheats are often called *strong wheats*.

The gluten of soft, low-protein wheats is not strong enough to prevent CO_2 from escaping readily; this would lead to small loaves of coarse crumb structure. Soft, low-protein wheats with poor gluten characteristics are often referred to as *weak wheats*. Obviously, strong wheats are not required for the baking of unleavened bread – a type that is eaten in the Middle East and India.

The best test to demonstrate bread-making quality is the baking process itself (Figure 4.3). The loaf obtained in such a test will be judged according to several criteria: volume, crust colour, crumb structure and crumb colour. A very high correlation exists between loaf volume and the total value of the other bread characteristics. Therefore, loaf volume is mostly used as an indicator of baking quality.

The lines in Figure 4.4 represent the relationships between loaf volume and protein content of the flour from three wheat varieties, of good, intermediate, and low baking quality. Samples of the Swedish spring-wheat variety Ring give rise to much larger loaves at a given protein content than those of the German spring-wheat variety Peko. Thus, the gluten quality of the

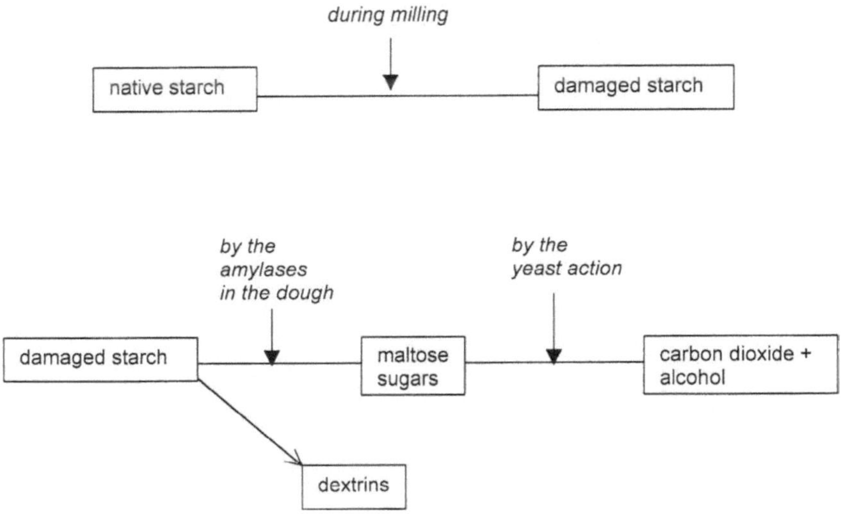

Figure 4.2. Role of starch during milling of wheat and baking of bread.

Figure 4.3. A loaf baked from a flour sample of a breeding line with a good baking quality (right) and a loaf baked from a flour sample of the old commercial variety Orca (left) with a poor baking quality; both grown in a trial field in the Netherlands.

former variety is better than that of the latter. The Dutch variety Orca holds an intermediate position.

N.B. The relationship between protein content of the flour and loaf volume obtained in a baking test provides the best indication of the bread-baking quality of a flour sample.

A baking test is too elaborate and too expensive to be used by a breeder during the early stages of selection. Moreover, usually the quantity of grains per selected line is not large enough for a standard baking test. Therefore,

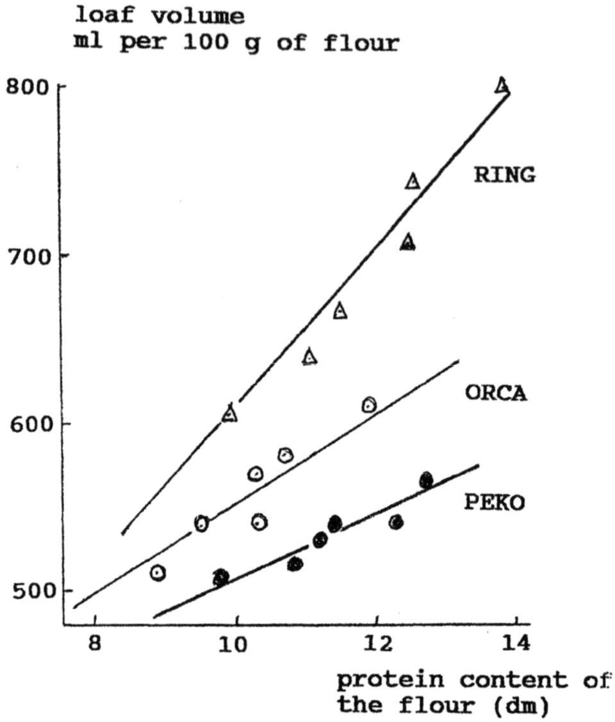

Figure 4.4. The relation between loaf volume and protein content of the flour for three wheat varieties of good (△), intermediate (○) and poor (●) baking quality.

quite a number of indirect tests for gluten quality have been developed, some of which are of special interest to breeders: the wheatmeal fermentation test (Pelshenke test); the Zeleny sedimentation test; the SDS sedimentation test; the Chopin alveograph test; and the mixograph test. A concise description of these tests is to be found in Section 8.

4.4. Soft wheat versus hard wheat

Soft-wheat flours have a relatively low protein content and form a soft gluten with poor gas-retaining properties. They contain a low percentage of damaged starch (4–6% on a dry matter basis) and, therefore, have low water-absorbing capacity. They yield doughs of inferior handling quality (which give trouble in machine baking), and are critical in their mixing and fermentation requirements, so that they are more likely to fail in breadbaking. Soft-wheat flours require less mixing and fermentation than hard-wheat flours to give good results in the making of biscuits and/or biscuit-like products.

Hard-wheat flours contain a relatively high content of proteins and form a tenaceous, elastic gluten with good gas-retaining properties. They are capable of being baked into well-risen, shapely loaves of bread, possessing a good crumb structure. They have a relatively high percentage of damaged starch granules (6–9% on a dry matter basis) and, therefore, require a considerable amount of water to make a dough of proper consistency. Their doughs have excellent handling qualities and are not critical in their mixing and fermentation requirements; for this reason they yield good bread over a wide range of baking conditions.

References

Hoseney RC (1986a) Cereal starch. In: Hoseney RC (ed), Principles of cereal science and technology. St Paul, MN: Am Assoc Cereal Chem: 33–68.

Hoseney RC (1986b) Minor constituents of cereals. In: Hoseney RC (ed), Principles of cereal science and technology. St Paul, MN: Am Assoc Cereal Chem: 89–110.

5. Bread making

5.1. Introduction

In addition to its milling properties, wheat needs certain other characteristics to make it suitable for the production of a particular food item. As wheat is predominantly milled for bread making, the general term *baking quality* usually refers to the specific properties required for the production of leavened bread. Bread is a traditional food, and the baking profession has a long tradition. Most European bakeries are small, even sole-trader enterprises with their production unit annexed to the main selling point, the bakery shop. This is common in France, Spain, Italy, Germany and eastern European countries. Nevertheless, large-scale bakeries with several points of sale do also exist in Europe, predominantly in the UK and Scandinavia. Differences between recipes, processes, types of bread, size of the business and equipment are too large to speak of an average European baker. If one, nevertheless, wishes to characterize the European baker, one could describe him or her as a small entrepreneur usually with one point of sale, reasonably well equipped, and product- rather than process-oriented. In Section 5 attention will be paid to (1) traditional bread-making processes as they were practised in most European countries until the middle of the 20th century; and (2) the technological developments that have come into use thereafter.

5.2. Traditional processes[1]

By way of example, the bread-making procedure, as it has been practiced in most Dutch bakeries during the 20th century, will be described.

5.2.1. Formula

The basic ingredients (by weight) are: 100 parts of flour, 50–60 parts of water, 1–2 parts of yeast, and 1–2 parts of salt. Flour is the major structural element. It is responsible for forming a visco-elastic dough that retains gas. As described in Section 4, bread flour is preferably from hard, high-protein wheats.

[1] For more information see Lupton & Derera (1981); Hoseney (1986).

The yeast may be compressed fresh yeast or active dry yeast. Its major role is to convert fermentable carbohydrates into CO_2 and ethanol. In addition to its gas-producing ability, yeast has a marked effect on the rheological properties of the dough.

Salt has two major functions. Firstly it improves the taste: bread made without salt is quite tasteless. Its second function is to influence the dough's rheological properties: salt makes the dough stronger.

One or more of the following *optional ingredients* may be added in dough making: sugar (0.5–2% flour weight), malt flour or malt extract (0.1–1% flour weight), and fat (1–3% flour weight). Sugar is a source of fermentable carbohydrate for the yeast and it provides a sweet taste to the bread.

With regard to amylase activity, as described in Section 4, a certain level of these enzymes is necessary to obtain bread of an optimal quality. A lack of enzymes in the wheat may be compensated for by adding enough malted flour or barley flour to reach a Falling Number level between 180 and 250 (Section 8.2.4).

Fat or emulsifiers are often added to the dough to make the crumb softer and to keep the bread fresh longer. Moreover, they cause an increase in volume of some 10% compared to bread made without such shortenings. Traditionally, lard, a byproduct of pig- and cattle-butchering, was used for this purpose. Nowadays, specific additives that may be derived from any kind of oil or fat are developed for bread making. It is also possible to use an emulgator, such as lecithin.

Very often, the miller adds *flour improvers* to flour at levels of a few parts per million (w/w). These improve dough strength considerably, resulting in bread with a higher loaf volume and a better crumb texture. Oxidative flour improvers are commonly used by modern bakeries to obtain satisfactory baking products. Two well-known flour improvers are potassium bromate and ascorbic acid. Potassium bromate has been frequently used in the past, but as it may adversely affect human health it has been replaced almost completely by the harmless ascorbic acid, better known as vitamin C. Flour for commercial bread making in Europe is improved with ascorbic acid. It may seem odd to refer to ascorbic acid as an oxidizing agent, whereas chemically it is a reducing agent. In flour, however, it plays a reducing role due to its rapid enzymic conversion to dehydroascorbic acid, the true oxidizing agent.

5.2.2. Mixing and dough formation

The bread-making process comprises three stages: mixing ingredients into a dough; fermentation periods alternated with punches (punches are a variety of mechanical treatments of the dough intended to drive out occluded gas); and, finally, baking. Within this framework there is large variation in the duration and rate of mixing, as well as in the number and duration of fermentation periods. Dough mixing produces a homogeneous mixture of all the formula ingredients. The most important aim is to form a coherent visco-elastic mass that is typical of a wheat flour dough. The events that take place during dough mixing are together indicated as *dough development*. Mixing takes 10–20 minutes; the dough leaves the mixer at a temperature of 25–30 °C.

5.2.3. First fermentation

At the end of the mixing process, the dough still does not have the gas-retaining properties required to produce a well-risen loaf of bread. These properties are developed during one or more resting periods, called the fermentation time. During fermentation, the dough becomes more flexible and elastic and the yeast gradually becomes fully active. In conventional bread making, the first fermentation period (bulk fermentation) takes about 30 minutes.

5.2.4. Dividing into pieces, rounding and intermediate proof

After the first fermentation, the dough will be divided into pieces of either 50 g or 400 g. Each piece will be rounded, either by hand or mechanically, to give it a ball shape. During this scaling in pieces and rounding, most of the CO_2 that has been formed in the first fermentation period will be squeezed out of the dough. This loss of gas will be compensated by submitting the dough pieces to an extra fermentation period of 30 minutes, referred to as 'intermediate proof' in British literature.

5.2.5. Moulding and final proof

The next step is to mould the dough pieces in order to give them the desired loaf shape. The moulding action consists of sheeting the dough and rolling the sheet into a cylinder, which is put into pans in the case of tin bread. During the following rest period (the final proof) of about one hour, the dough will undergo a substantial increase in volume. However, in many European countries bread may also be baked on the floor of the oven. In that case, the dough cylinders will be placed side by side on a metal plate, which, after a final proof of again one hour, will be fully risen.

5.2.6. Baking

At the end of the final proof, the risen dough is put into an oven and baked for 15–45 minutes, depending on the size of the dough pieces. Oven temperature is usually 230–250 °C. During baking, the dough volume will continue to increase. The better the quality of the flour, the more the dough will increase in volume. The crust of the loaf and the texture of the crumb will be fixed during the baking stage at 55–65 °C by gelatinization of the starch, which then binds part of the water. After baking, the loaves have to cool down to room temperature. The shape of floor-baked bread is less regular than that of tin bread. On the other hand, floor bread has a stronger crust and more aroma.

5.3. Technological developments[2]

5.3.1. Increased utilization of European wheat

In 1969 the European Union (EU) consisted of six countries with a total human wheat consumption of 38 MT of *aestivum* wheat per year, of which 8.3 MT were imported from non-EU countries (e.g. USA, Canada). The imported wheat was primarily used in the flour-milling industry.

In 1995 the EU comprised 15 countries with a total human consumption of 64.5 MT *aestivum* wheat per year; only 2.5 MT being imported from other countries. This means that the contribution of EU wheat to total human consumption increased considerably over those 25 years. This change was prompted by the EU agricultural policy and had no technological basis.

The higher utilization of EU wheats for bread making has also had its influence on the breeding, growing and marketing of wheat varieties with better bread-making qualities. In various countries, farmers grow an increasing proportion of high quality wheats, and millers have been able and willing to buy these wheats for making bread flour. However, the baking quality of European wheat varieties is not as good as that of US and Canadian wheats, so the milling and baking industry has had to adapt to the new situation.

5.3.2. Addition of gluten to flours

Up to 1970, almost all cereal starch produced in the EU was maize starch. In the following 15 years, most of the maize starch factories were closed and large production units for wheat starch were erected. For the European milling and baking industry, the pleasant consequence of this was that large quantities of gluten appeared on the market as a byproduct of starch. This gluten is mainly used to improve the baking quality of flour made from European wheat.

The addition of gluten to flours is especially desirable for the production of light-textured brown bread and wholemeal bread. There is a large market

[2] For more information see Chamberlain (1975); Williams (1975); Sluimer (1994).

in Europe for these high-fibre products. In wholemeal bread, quantities of gluten up to 5% of flour weight are added.

During the last few decades of the 20th century, the consumption pattern of bread has changed drastically. In 1984, 72% of the bread consumed in the Netherlands was a standard white bread of a low fibre content. In 1990, the consumption of brown bread, including whole-meal and multi-grain breads amounted to 72% of total bread consumption. A similar picture can be seen in Germany. In other countries the change in pattern is less pronounced.

5.3.3. Mechanical dough development

During the first half of the 20th century, lengthy and complex bread-making processes were customary in the UK, for example fermentation and/or proofing of several hours was quite usual. As a consequence, bakers had to get up in the middle of the night to provide their clients with fresh bread for breakfast the next morning.

Especially since World War II, many efforts have been made to reduce night work in bakeries. It soon became clear that fermentation and proofing times could be reduced considerably by using intensive mixing machines. The development of a protein skeleton requires less time with intensive mixing than with conventional mixing. This is referred to as *mechanical dough development.*

The term 'intensive mixing' primarily refers to the high speed mixers: rather robust machines that have mixing arms with relatively simple movements which make more strokes per minute. The capacity of the motor is approximately four times higher than that of a traditional mixer. As the development of the dough is better, the subsequent fermentation process and proofing are shorter. An optimal dough may be obtained within 4–12 minutes. This type of mixing machine easily found its way into European bakeries.

The most advanced technology in this field is widely found in the UK (Chamberlain, 1975; Williams, 1975). Prior to the introduction of mechanical dough development, the great majority of bread produced in commercial bakeries was made by a process with a first fermentation period of three or more hours. One should bear in mind that very strong Canadian wheat that needed a long fermentation time, was incorporated in the miller's grists. Also, British types of yeast were less active than Continental yeasts. By using very powerful high-speed mixers, it became possible to skip the first fermentation completely. The whole bread-making process from mixing to the end of baking was reduced to $1\text{-}1^1/_2$ hours. The loaf appearance, the crumb structure and the taste of the bread were more than satisfactory to the British consumers.

It was the British Baking Industries Research Association (BBIRA) – now part of the Flour Milling and Baking Research Association – which, between

1958 and 1962, developed the *Chorleywood Bread Process* (CBP) and made it ready for use. Chorleywood is the name of the place where the BBIRA-institute is located. The procedure is also referred to as the *Tweedy Process*, after the name of the mixing machine that was initially used. This high-speed mixer has a curious shape, with a hexagonal plate at the base of a cylindrical mixing chamber. The plate can make 250–500 revolutions per minute. Obviously the capacity of the mixer has to be very high, i.e. 25–50 kWh per 100 kg flour. By using this type, a dough with the required characteristics may be obtained within a very short time (2–3 minutes).

5.3.4. Dough retarding – frozen dough

Another way to reduce night work and still produce fresh bread at any time and at any place is to use cooled or completely frozen doughs. These techniques are frequently used in Europe nowadays. A clear distinction should be made between *dough retarding* practiced in craft bakeries using doughs with a shelf life of one or two days, and the *frozen dough* production in a centralized bakery for a number of outlets with a shelf life of 1–2 weeks.

Dough retarding was developed for the small traditional bakery to reduce night work. By preparing a dough up to the moulding one or two days in advance, it is possible to start working less early in the morning. After the dough pieces have been put in tins, they are kept in a retarder (refrigerator) at temperatures between 0 °C and –10 °C. As yeast is hardly active at temperatures below 0 °C, the rising process is practically brought to a halt. The retarder may be programmed to start warming up at a fixed time, so that the dough is ready for baking when the baker normally starts work. However, when the dough is kept in the retarder for more than two days, the quality of the baked bread will deteriorate considerably.

The process of dough retarding requires certain adjustments in formula and methods of preparation. The flour should be of high baking quality and the dough must not have undergone a full rise before it is cooled down. In other words, the baker has to start with 'green doughs' instead of working with well-risen doughs. In addition, a short and less intensive mixing process should be applied.

The production of frozen doughs has resulted from the requests of supermarkets and croissant shops that bake their own bread products to stimulate sales. In this concept, the dough is prepared in a centralized bakery and frozen after it has been shaped into the required pieces. The pieces are then transferred to the selling points, where they will be thawed in a special piece of equipment. Brought back to the right temperature, the final proof will start.

The advantage of this concept is that no dough-making machinery is needed at the point of sale, just the equipment for thawing the frozen doughs and giving them a final proof, and an oven.

Enterprises specialized in the production of frozen doughs are able to deliver high-quality dough pieces with excellent baking characteristics owing to rigidly controlled processing. Recently, fully-proofed frozen dough products have entered the market. These products can be baked directly from the frozen state. A product of good quality made in this way is the croissant. Fully-proofed frozen dough for loaves and baguettes is also commercially available.

References

Chamberlain S (1975) Advances in bread-making technology. In: Spicer A (ed), Bread: Social, nutritional and agricultural aspects of wheaten bread. London: Appl Sci Publ: 259–272.

Hoseney RC (1986) Yeast-leavened products. In: Hoseney RC (ed), Principles of cereal science and technology. St Paul, MN: Am Assoc Cereal Chem: 203–244.

Lupton FGH, Derera NF (1981) Soft wheat breeding in Europe and Australia. In: Yamazaki WT, Greenwood CT (eds), Soft wheat: production, breeding, milling and uses. St Paul, MN: Am Assoc Cereal Chem: 99–127.

Sluimer P (1994) Personal communication, Wageningen.

Williams A (ed) (1975) Breadmaking – The modern revolution. London: Hutchinson Benham, 252 pp.

6. Manufacturing of other wheat products

6.1. Biscuit (cookie) making[1]

6.1.1. Introduction

One of the largest outlets for soft-wheat flours is the biscuit industry. Biscuits are made from soft-wheat flour and are characterized by a formula high in sugar and shortening, and relatively low in water. Apart from these major ingredients; baking powder, emulsifiers, flavourings, and dough conditioning agents are also used. The diversity of biscuit products is rather great. Biscuits are dry products with a moisture content of less than 10%. As a consequence, they have a very long shelf-life. In the United States, similar products are called 'cookies'. In addition to their requiring soft-wheat flour, most of these products have in common that they are leavened chemically; yeast is rarely used in these soft-wheat products.

6.1.2. Chemical leavening

By far the most popular leavening agent is sodium bicarbonate ($NaHCO_3$), or *baking soda.* Its popularity is based on a number of advantages: it is cheap, non-toxic, easy to handle and it gives no off-flavour in the end products. In dough, sodium bicarbonate can exist as free CO_2 or in one- or two- ion forms, HCO_3^- or $CO_3^=$. At pH < 5, nearly all bicarbonate in the dough is dissociated into the leavening gas CO_2. At a pH between 5 and 8, only part of the CO_2 is in the gaseous state. No leavening gas is available at a pH > 8. Many soft-wheat dough products have a final pH of around 7, so if significant quantities of gas are to be obtained, the dough must contain acid components that lower pH. Many ingredients used in baking are sources of acid (e.g. acidic fruits). When the other ingredients are acidic, one can use sodium bicarbonate by itself to obtain leavening. If the formula does not contain acid, a combination of baking soda and an acid (i.e. *baking powder)* should be used. Monocalcium phosphate is often used as an acid component in many applications: it reacts readily at room temperature and is widely used in baking powders. However, a number of other acid-producing salts exists, some reactive at room temperature (e.g. cream of tartarate, the mono-potassium salt of tartaric

[1] For more information see Hoseney (1986a,b,c); Yamazaki & Greenwood (1981).

acid) and others requiring oven temperatures for reaction (e.g. sodium acid pyrophosphate and sodium aluminium phosphate).

6.1.3. Production

The first stage in the production of biscuits is the mixing of the ingredients into a dough of even composition and uniform physical properties. The physical properties of the dough vary according to its composition, and to some extent to the way the dough has been prepared. The next step is the preparation of separate dough pieces of a desired shape ready for baking. This may be achieved in three ways: cutting, moulding or depositing. In commercial practice, biscuits are baked in long tunnel ovens. Typically, the baking zone of such an oven is about 1 m wide and 45–90 m long. The biscuits are generally baked on a solid-steel belt that conveys the product through the oven at a rate that will produce the desired baking time.

6.1.4. Dividing up the dough

The best way to classify biscuits is by the way the dough is divided into dough pieces.

Cutting-machine biscuits. The main difference between this method and the other methods of dividing the dough is that the desired pieces are punched from a continuous sheet of dough. This sheet passes underneath a roller on which stamps with protruding edges are mounted. Dough pieces are cut from the sheet. The dough still surrounding the cut-out pieces, called 'scrap', is lifted from the conveyor belt and removed to be returned to the hopper of the dough sheeter. This dough contains much more water than the other two types of biscuit doughs. The dough used for this method must have sufficient tensile strength and extensibility to enable it to be smoothly stretched into a sheet and to transport the scrap without problems. For this reason the gluten has to be 'developed'.

Rotary-moulded biscuits. For this type of biscuit, the dough is forced into moulds on a rotating roll. As the roll completes a half-turn, the dough is extracted from the cavities and placed on a belt for baking. The dough for rotary-moulding has to be short, so that all the moulds can been filled uniformly. On the other hand, it should have sufficient coherence and not be sticky, so that the moulded dough pieces can easily be pulled out of the moulds. To achieve this, more shortening is used than in dough-cut biscuits. Doughs for rotary-moulded biscuits are characterized by fairly high sugar and shortening levels and very low amounts of water (less than 20% of the total weight of flour, including its moisture content). The dough is crumbly, lumpy and stiff, with virtually no elasticity. The cohesion of this type of dough mainly comes from the shortening used. The gluten in the dough should not develop during mixing.

Deposited biscuits. Unlike cutting and moulding, the dough for deposited biscuits is portioned directly on a steel belt from the oven or on baking plates. A depositor works as follows: a hopper is filled with dough, which is pressed from the hopper to a nozzle by a set of counter-rotating rollers. The dough is then extruded through the nozzle and cut to size, usually by a wire. Hence, the American name 'wire-cut biscuits'. Nozzles of different shape enable the final product to be varied. Deposited biscuits rise and spread as they are baked; the final size of the biscuit is determined by the dough formula and the flour used. Depositing requires an easily deformable dough without elasticity. The relatively soft dough must be cohesive enough to hold together, yet short enough to separate cleanly when cut by the wire. A typical formulation may contain 50–70% sugar, 50–60% shortening and up to 15% eggs (all proportions based on flour weight).

6.1.5. Biscuit quality

Biscuit quality can be summarized in two general terms. Firstly is the size of the biscuit, both the width and the height. If the biscuit spreads too much, it cannot be packaged without breaking. If the spread is too little, then the package box will not be completely filled. Secondly is how the biscuit bites. Good quality biscuits must have a tender bite. This is obtained by the use of extra fat or shortening. Soft-wheat flours are superior for these products. The lower level of protein in combination with weak glutens imparts tenderness to the baked biscuits. Processing and machine qualities are excellent and consumers find the eating qualities attractive.

6.2. Pasta making[2]

6.2.1. Introduction

Pasta or paste is a wheat-based product which is made from a dough that has not undergone fermentation or baking. The storage life of dried pasta products is very long, up to several years at relative air humidities below 70%. It is

[2] For more information see Hoseney (1986d).

possible to prepare a meal quickly by boiling the pasta in water or baking it in fat.

The legendary Marco Polo (1254–1323) has been given credit for introducing noodles from China into Italy. Macaroni and lasagna were well-known in Italy in the 14th century. In the 15th century, consumption of pasta products greatly leaped forward in Italy, possibly because:

- durum-type wheat with hard and vitreous grains, growing in the southern parts of Italy, proved to be optimal for the preparation of alimentary pastes; and
- the climate, especially around Naples, was very suitable for drying pasta in the open air, due to the presence of humid air from the sea during daytime and dry air from inland at night.

Up to the early 19th century, pasta was produced on a small scale, almost exclusively by housewives and cooks. Nevertheless, the production of pasta on an industrial scale started in the late 18th century around Naples and Genoa. There, pasta-making machinery appeared on the market halfway through the 19th century. At the turn of the 19th to the 20th century, Italian-type pasta-making factories were established in other countries as well.

In these factories, wheat semolina and water were mixed by a mixing machine into a loose, crumbly dough, then transferred batch-wise to a second machine and kneaded into a compact, smooth product. The dough was then transferred to an extruder for extrusion into a continuous dough. Eventually it was cut into pieces of the desired shape. Pasta production in these first factories was carried out in batches and was not continuous.

Since 1935, pasta manufacturing has changed drastically. Production in batches has been superseded by a continuous process. In the late thirties, the first continuously working screw press was introduced. Whereas the capacity of batch presses was no more then 150 kg/h, the average capacity of a modern continuous press is about 1000 kg/h. Screw presses combine dosing of ingredients, mixing and kneading dough, and dividing dough into pieces in one instrument: four unit operations that used to be carried out by four different machines.

6.2.2. Raw material

Pastas are generally made from semolina derived from *Triticum durum*. However, in some countries (e.g. Germany and the Netherlands) a considerable amount of flour from mixtures of durum and bread wheats or even only bread wheat is used for pasta making. *Aestivum* wheat for bread making is *not* appropriate for pasta production. A hard *Triticum aestivum* with a fairly high protein content is needed for pasta making. Not only protein content, but especially protein quality is of major importance.

In many countries, egg macaroni products are manufactured as well; these contain 3 eggs per kg pasta. Vegetable macaroni products are also made (e.g. in Italy and France), containing tomato or spinach.

6.2.3. Dough making

Pasta formula is very simple: it usually consists of only semolina (and/or flour) and water. Enough water is added to obtain a dough with a moisture content of 30%, which gives a very dry dough. The water content is about half that of bread dough. When mixed, pasta dough is formed to balls of about 1–2 cm diameter. Then, the dough is kneaded again for 15–20 minutes into a smooth homogeneous mass.

6.2.4. Extruding

From the mixer, the dough enters an auger that kneads and exerts pressure as the dough moves down to the barrel of the extruder. The dough is pressed to the specific holes of a die at temperatures of 40–50 °C. At higher temperatures there is a risk of denaturation of the protein, which should be avoided. In the die, pressures of 80–120 atm occur. The combined effect of kneading and pressure produces a smooth homogeneous dough that can be extruded. A considerable amount of heat is produced in that process, therefore the extruder barrel is jacketed and cooled with water. A cutting machine is positioned behind the dies of the short goods, to cut the extruded dough into adequate pieces. Long goods are only cut after drying. Pasta products may be categorized into four groups:

- *Long goods*, the most common type, include macaroni (tube-shaped, hollow, and more than 2.5 mm but less than 6.5 mm in diameter) and spaghetti (cord-shaped, not tubular and between 1.5 mm and 2.5 mm in diameter);
- *Short goods*, which include elbow macaroni, shells and fusilli. The number of sizes and shapes that can be produced is virtually unlimited;
- *Curled pasta*, which includes products such as nests and skeins; and
- *Specialty items*, which include lasagne, manicotti and stuffed pasta.

6.2.5. Drying

The extruded product still contains about 30% moisture and must be dried to about 12% before it is fit for transportation and storage. If it dries too fast, the product may check. *Checking* is the formation of numerous hairline cracks in the product that makes it appear opaque and decreases its strength. Drying that is too slow may also cause problems. The transport system of pasta in dryers is different for short and long goods. Long goods are transported on sticks mounted on a chain that passes through the dryers; short goods are dried in drums or belt dryers; whereas curled pasta products are dried on screens. Just before packing, long goods are cut into 27 cm lengths and bent parts are then removed.

6.2.6. Pasta quality

Uncooked pasta should be mechanically strong so that it will retain its size and shape during packaging and transport. When cooked in boiling water, the product should maintain its shape without splitting or falling apart. After it is cooked, it should give a firm bite (known as 'al dente') and its surface should not be sticky. The cooking water should be free of starch. Finally, the pasta should be resistant to over-cooking. Although it is said that 'real' pasta should be made from durum-wheat semolina, actual conditions in a large part of the world impose the use of bread-wheat flours, some of them with rather poor properties for pasta production. However, nowadays there are installations that, by the use of high temperatures, make it possible to use flour from mixtures of durum and bread wheats or from bread wheats only for the production of long or short pasta. The resulting product is of a higher quality, i.e. more solid and thus less sticky and more resistant to over-cooking, than was formerly possible.

6.3. Other uses of wheat

Other wheat products exist than bread, biscuits and pastas. A detailed description of their production is beyond the scope of this text. However, for completeness, the most important of these products are listed here.

Yeast-leavened baked foods. Strong (hard-wheat) flour is the basic ingredient for most baked products in which the gluten network holds the CO_2 gas produced by yeast fermentation and the steam formed between the laminations in Danish and puff pastries. Crusty wheat rolls are eaten in many European countries every day, as well as traditional French bread (e.g. 'baguettes') and croissants in France. Because the shelf-life of these products is only a few hours, many bakers produce these types of bread twice a day. Also, there is a wide variety of soft rolls (buns, brioches, etc.) with a shelf-life that is about one day longer than the crusty rolls and croissants. Danish and puff pastries are made up of layers of dough and layers of shortening or butter. To obtain the layers, the dough is worked to a sheet, then half of it is covered with fat and the dough folded to sandwich the fat. The dough is then repeatedly sheeted and lapped, creating many alternating layers of dough and fat. The shelf-life of Danish and puff pastries is at most 8 hours. The well-known (cake) doughnut, which is very popular in English-speaking countries, also belongs to the group yeast-leavened baked products.

Chemically-leavened baked foods. Soft-wheat flour is used for many chemically-leavened baked products such as snack crackers; tea breads (muffins, crumpets, scones); sponges and cakes; Swiss rolls; fruit tarts; and a wide variety of confectionery products, often filled and/or decorated with products like whipped cream, custard cream, fondant or fruit. Soft-wheat flour appears to be uniquely suited for these goods: it gives products that are more tender, and larger in size and that have a superior internal structure to those made from flours of other classes of wheat. The low protein content and water absorbing capacity, and the fine granulation undoubtedly contribute to its acceptance.

Non-baked foods. Soft-wheat flour is also the basic ingredient for the production of many breakfast cereals: wheat flakes, shredded biscuits, puffed cereals, extrusion cooked cereals, wheat germs, etc. Some of these cereals require cooking in milk, some are ready to eat (Hoseney, 1986e). Beer is usually made from barley malt. However, wheat malt is sometimes used for the manufacturing of a good beer with a characteristic odour and taste, for example 'Weizen Bier'.

Non-food uses. The use of wheat to manufacture starch and gluten in Europe is restricted to a few countries. The most important are the Netherlands, Germany, France, England, Ireland and Spain. The production capacity rapidly increased during the 1980s. In 1990, some 152 000 t of gluten (more than 50% of world production) and approximately 1 Mt of wheat starch were produced in Western Europe, for which 2 Mt of wheat were needed. About 80% of the gluten produced, is used as a dough strengthener in bread-flour mixes. Thus, expensive wheats of good baking quality and high protein content can be replaced by cheaper wheats of medium protein content fortified with gluten. Gluten can also be used in the production of pet foods and breakfast cereals. Wheat starch is the basic material for a large number of industrial products, among them are glucose, modified starches, ethanol and adhesives. One should not forget that a substantial part of the European wheat production is used for animal husbandry. Some figures for the EU may illustrate this: of the 64.5 Mt of *aestivum* wheat produced in 1990, 24.3 Mt were used as animal feed. It should be mentioned that part of this wheat was produced and consumed on the same farm. The rest of it was processed by the mixed-feed industry. The only requirements that wheat must meet to be used as animal feed are that its moisture content must not exceed 16% and that the grains must be of good external quality, i.e. have no visible fungal or heat damage.

References

Hoseney RC (1986a) Soft wheat products. In: Hoseney RC (ed), Principles of cereal science and technology. St Paul, MN: Am Assoc Cereal Chem: 245–275.

Hoseney RC (1986b) Chemical leavening. In: Hoseney RC (ed), Principles of cereal science and technology. St Paul, MN: Am Assoc Cereal Chem: 246–251.

Hoseney RC (1986c) Cookies. In: Hoseney RC (ed), Principles of cereal science and technology. St Paul, MN: Am Assoc Cereal Chem: 251–255.

Hoseney RC (1986d) Pasta and Noodles. In: Hoseney RC (ed), Principles of cereal science and technology. St Paul, MN: Am Assoc Cereal Chem: 277–291.

Hoseney RC (1986e) Breakfast cereals. In: Hoseney RC (ed), Principles of cereal science and technology. St Paul, MN: Am Assoc Cereal Chem: 293–304.

Yamazaki WT, Greenwood CT (1981) Soft wheat uses in Europe. In: Yamazaki WT, Greenwood CT (eds), Soft wheat: production, breeding, milling and uses. St Paul, MN: Am Assoc Cereal Chem: 209–266.

7. Genetic basis of quality in bread wheat

7.1. Introduction

Bread wheat is a hexaploid species (see Section 1) containing three different but related genomes of seven chromosomes, arbitralily named A, B and D. In all likelihood, these genomes originate from the three ancestors of our modern bread wheat. The unique milling and baking properties of common bread wheats are not found among diploid and tetraploid wheats. Since only the hexaploid group possesses the D set of chromosomes, the desirable quality characteristics of bread wheat have been attributed mainly to the presence of this third genomic component. More recent research has revealed, however, that the other genomes are also involved. The cytogenetics of wheat chromosomes has been studied extensively. As a result of these studies, each of the 21 different chromosomes has been given a chromosome number (1 to 7), followed by the genome assignment (A, B or D). The Arabic numerals 1 to 7 represent the homoeologic chromosomes, i.e. chromosomes with more or less the same genetic structure.

7.2. Grain hardness

Wheat varieties vary widely in their grain hardness. Two main categories can be distinguished: varieties with 'hard' and varieties with 'soft' grains. The kernels of hard wheats often, but not always, have a dark and shiny aspect, and a vitreous appearance. The kernels of soft wheats have a more opaque and floury appearance. Grain hardness is a property of particular interest for the milling and baking industry (for details see Sections 2.3, 3.5 and 3.6).

The inheritance of grain hardness has been dealt with by Symes (1965, 1969), who in crosses between soft and hard Australian wheat cultivars demonstrated that, in most cases, one single gene was responsible for the difference in hardness between the two parents, although modifying genes could also play a role. Doekes & Belderok (1976) found that major factors for grain hardness were located on chromosome 5D of each of the hard wheat varieties Cheyenne, Hope and Timstein, and, moreover, on chromosome 7B of Timstein. The presence of only one of these chromosomes with a gene for grain hardness was sufficient to make the wheat hard. For a wheat breeder this is an important conclusion, since it means that the conversion of a soft-wheat variety into a hard one, and vice versa, is relatively easy.

7.3. Milling quality

Berg, a wheat breeder at Weibulsholm (Sweden) from 1911 until his death in 1946, initiated research on milling quality. He showed that grittiness of flour is mainly a varietal characteristic. He started systematic breeding work on quality and transferred improved milling quality from Hungarian wheats with hard grains to Swedish winter-wheat material with soft grains. His variety Eroica, released in 1943, was the first north European wheat variety with hard grains.

Fajersson (Berg's successor in 1946) carried on the research on milling quality. A number of wheat varieties with excellent milling properties, e.g. Starke, Starke II, Holme, Walden, Ring, and Pompe, were bred by Fajersson and marketed in Sweden and neighbouring countries.

Svensson (1981) presented a thesis at Weibulsholm about varietal and environmental effects on milling quality of wheat. He showed that, for many properties of milling quality, the variation between varieties was greater than that attributable to environment.

The work carried out by these three authors stimulated many European breeders to pay more attention to the improvement of the milling quality of their new varieties.

The main factors determining the milling quality of wheat grains are *ease of milling* and *yield of flour*. High flour yield is promoted if the grain is short, plump, almost spherical in shape and smooth, without a deep crease. Cookson (1975) subdivided the complex concept 'milling quality' into the following three requirements, which are still valid:

- the wheat grain must give a good yield of flour (min 72%);
- the flour must have a good colour, and;
- the wheat must have the capacity to give a reasonable level of damaged starch upon roller-milling.

The identity of the chromosomes involved in the inheritance of milling characteristics were investigated by Doekes & Belderok (1976). They used aneuploid substitution lines of Cappelle Desprez, Cheyenne, Hope and Timstein into the recipient variety Chinese Spring. The latter is a soft red spring wheat with poor milling and baking characteristics. The four donor varieties have satisfactory milling and baking properties. Cappelle Desprez is a soft red winter wheat, Cheyenne is a hard red winter wheat, and Hope and Timstein are hard red spring wheats.

By comparing the milling properties of the parental varieties with those of each of the substitution lines, it was possible to determine on which of the chromosomes the genes affecting flour yield are localized. The substitution lines yielded milling values similar to the poor-milling recipient Chinese

Spring, with the exception of the 7B and 5D substitution lines of Cheyenne, in which the flour yields were at the same level as those of the high-yielding Cheyenne parent.

7.4. Bread-making quality

7.4.1. Endosperm proteins

As described in Section 4.1, three-quarters of grain proteins consist of gluten. These proteins can be divided into *gliadins* and *glutenins*, according to their extractability and solubility in various solvents. They are located in the mealy endosperm and are not found in the seed coat layers, nor in the germ.

Gliadins and glutenins are storage proteins. They make wheat unique because of their function in the formation of dough, as they retain the gas necessary for the production of spongy baked goods. The two fractions impart different properties to the dough: the glutenins can provide both elasticity and extensibility, whereas gliadins are viscous and confer extensibility. Large glutenin molecules are tough and elastic with little extensibility. By contrast small glutenins are weakly elastic but much more extensible. About 50% of the gluten are gliadins; low-molecular-weight (LMW) glutenins amount to 40% of the total and high-molecular-weight (HMW) glutenins to 10%.

The gliadins occur in a complex mixture of small and simple proteins, with molecular weights ranging from 11 000 to 63 000. Any variety may contain up to 50 different gliadins. Gliadins do not have a intermolecular disulphide-bonded subunit structure. They do of course have intramolecular disulphide-bonds. They have a globular structure, aggregated into micro-fibrils, that are folded into the glutenin network. Clusters of gliadin genes are called gliadin blocks (Metakovski, Novoselskaja, Kopus, Sobko & Sosinov, 1984; Sozinov & Poperelya, 1982). Additional gliadin loci outside these blocks have also been described (Metakovski, 1991).

The glutenins are much larger aggregates of high molecular weight (up to several million), based on more than 19 different subunits connected by disulphide bonds. Glutenins are considered to be the most important components of wheat protein with respect to baking quality.

Genetic analyses show that:

- Genes controlling the HMW subunits of glutenin occur on the long arms of chromosomes 1A, 1B and 1D. The loci that contain the HMW glutenin subunit genes are collectivily called *Glu-1* and individually *Glu-A1, Glu-B1* and *Glu-D1* for chromosomes 1A, 1B and 1D, respectively.

- The short arms of chromosomes 1A, 1B and 1D contain the genetic information for the ω- and γ-gliadins, as well as for the LMW subunits of glutenin.
- Genes for the α- and β-gliadins occur on the short arms of chromosomes 6A, 6B and 6D (see also Section 7.4.4).

The albumin and globulin fractions are believed to be of no or only minor importance with respect to baking quality, although some molecules have disulphide-bonded subunit structures and are known to be deleterious. By far the most important are the glutenin fractions.

7.4.2. HMW glutenin subunits

The glutenin aggregates can be broken down into their component subunits by treatment with an agent that breaks disulphide bonds (such as 2-mercapto ethanol), and an agent that disrupts hydrophobic interactions and hydrogen bonds, such as the anionic detergent sodium dodecyl sulphate, SDS. Following these treatments, the subunits can be separated by electrophoresis in polyacrylamide gels containing SDS (SDS-PAGE).

The technique of determining the glutenin subunit composition of wheat cultivars is described in Section 8.2.5. In this determination, proteins are fixed and stained with a Coomassie Brilliant Blue solution. After destaining with an ethanol-acetic acid solution, the protein subunits become visible as dark-coloured bands. These are numbered according to their mobility from 1 (lowest mobility) up to 12 (greatest). Some new bands have since been discovered with mobilities between 1 and 12 (Figure 7.1).

The genes that control the synthesis of HMW subunits of glutenin are located at three loci, one each on the long arms of chromosomes 1A, 1B and 1D. There is considerable variation in the pattern of the HMW subunits due to different alleles at each of the gene loci. Three alleles were identified for the A1 locus (subunit 1, 2*, and the null genotype); at least seven alleles or pairs of alleles for the B1 locus (6 + 8, 7, 7 + 8, 7 + 9, 13 + 16, 17 + 18, 20) and at least four for the D1 locus (2 + 12, 3 + 13, 4 + 12, and 5 + 10). A subunit pair is a combination of subunits with closely linked genetic locations. The numbering of the subunits is according to Payne and Lawrence (1983). Additional alleles, such as 21* *(Glu-A1)*, 14+15 *(Glu-B1)* and 2.2+12 and 2+10 *(Glu-D1)* have since been described by other authors.

Initially, different numbering and lettering systems were used by several research groups to describe the subunit patterns obtained by SDS-PAGE. This made it difficult to compare one group's work with that of others. Payne & Lawrence (1983) proposed a simple, universally usable nomenclature. They recommended that their allele numbers be referred to by all research groups, thus enabling the work of different laboratories to be compared without dif-

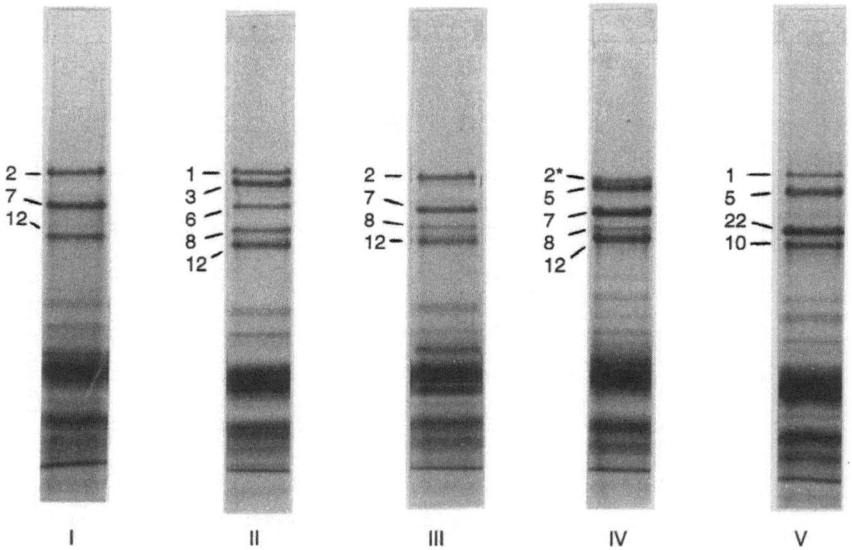

Figure 7.1. SDS-PAGE patterns of some European/Dutch wheat cultivars: (I) Arminda, (II) Ritmo, (III) Scipion, (IV) Soissons, (V) Yacht. The HMW glutenin subunits are numbered according to Payne & Lawrence (1983).

ficulty. This suggestion has been generally accepted. In addition McIntosh, Hart & Gale (1990) gave a letter to each of the alleles coding for a (group of) HMW subunit(s).

In a study carried out by NIAB, Cambridge (Cooke, 1995), the HMW glutenin subunit composition of 746 currently grown wheat varieities from 19 countries all over the world was analyzed (Table 7.1). The varieties could be divided into 63 groups on the basis of the HMW glutenin subunits, four of which accounted for over a third of the varieties tested. Differences between various countries in distribution of gluten alleles and in level of gene diversity were noted. The *Glu-D1 f* allele (electrophoretic bands 2.2 + 12) appeared to be relatively common in Japanese varieties and absent in those from elsewhere. Novel subunits were reported on various occasions, for example in Italian wheats.

A similar study was done by Morgunov, Peña, Crossa & Rajaram (1993) at CIMMYT in Mexico. They found that *Glu-1* alleles are not associated with ecogeographical parameters in a worldwide context. The HMW glutenin composition turned out to be similar in cultivars from countries with different climatic conditions, e.g. Finland and Yugoslavia, or Italy and China.

Table 7.1. HMW glutenin alleles in a collection of 746 varieties; distribution in different countries (after Cooke, 1995)

Locus	Subunit (allele)		Country* AUS	BEL	CAN	CHI	CZE	DEN	FRA	GBR	GER	GRE	NET	ITA	IND	JAP	NZE	POL	POR	RUS	SPA	USA	Total nr.	Freq. (%)
Glu-A1	1	(a)	7	3	4	7	4	0	31	22	18	10	1	35	2	3	11	8	1	2	11	11	191	25.50
	2*	(b)	1	0	4	0	0	0	26	7	3	11	3	12	12	2	6	6	1	9	16	4	123	16.42
	N	(c)	4	7	5	7	15	10	125	68	43	14	8	27	28	12	12	23	1	1	14	8	435	58.08
Glu-B1	6 + 8	(d)	0	1	0	1	1	6	42	45	16	4	4	8	0	0	6	12	0	0	1	0	147	19.63
	7 + 8	(e)	4	2	6	3	5	1	53	7	5	4	0	10	19	16	3	5	1	1	9	8	162	21.63
	7	(a)	0	4	3	3	4	0	34	16	14	3	4	29	5	2	5	7	0	2	5	0	140	18.69
	7 + 9	(c)	2	1	3	3	7	2	43	10	25	3	4	14	4	2	11	10	1	9	11	13	178	23.77
	17 + 18	(i)	5	1	0	1	1	0	5	14	1	12	0	6	12	0	0	1	1	0	11	0	71	9.48
	14 + 15	(h)	0	0	0	0	0	0	2	5	3	4	0	0	0	0	2	1	0	0	0	2	19	2.54
	13 + 16	(f)	0	1	0	0	1	0	1	0	0	1	0	0	0	0	2	0	0	0	0	0	6	0.80
	20	(e)	1	0	1	2	0	1	1	0	0	0	0	7	1	0	0	1	0	0	1	0	16	2.14
	21	(j)	0	0	0	0	0	0	1	0	0	0	0	0	0	0	0	0	0	0	0	0	1	0.13
	22	(k)	0	0	0	1	0	0	0	0	0	4	0	0	0	0	0	0	0	0	3	0	8	1.07
Glu-D1	2 + 12	(a)	9	7	4	11	12	8	93	60	26	19	11	58	31	12	10	19	1	4	14	13	422	56.34
	3 + 12	(b)	0	0	0	0	0	0	9	3	0	0	0	0	0	1	0	0	0	0	0	0	13	1.74
	4 + 12	(c)	0	0	0	0	0	0	11	3	0	0	0	0	0	0	0	1	0	0	2	0	17	2.27
	5 + 10	(d)	3	3	9	3	7	2	69	31	38	16	1	16	11	0	19	17	2	8	25	9	289	38.58
	2.2 + 12	(f)	0	0	0	0	0	0	0	0	0	0	0	0	0	7	0	0	0	0	0	0	7	0.93
	2 + 10	(e)	0	0	0	0	0	0	0	0	0	0	0	0	0	0	0	0	0	0	0	1	1	0.13
Total			12	10	13	14	19	10	182	97	64	35	12	74	42	20	29	37	3	12	41	23	746	

* The countries have been abbreviated as AUS (Australia), BEL (Belgium), CAN (Canada), CHI (China), CZE (Czech Republic), DEN (Denmark), FRA (France), GBR (Great Britain), GER (Germany), GRE (Greece), NET (Netherlands), ITA (Italy), IND (India), JAP (Japan), NZE (New Zealand) POL (Poland), POR (Portugal), RUS (Russia), SPA (Spain), USA (United States of America)

Table 7.2. Quality scores assigned to individual HMW glutenin subunits or subunit pairs after Payne, Nightingale, Krattinger & Holt (1987)

Score	Chromosome		
	1A	1B	1D
4	–	–	5 + 10
3	1	17 + 18	–
3	2*	7 + 8	–
2	–	7 + 9	2 + 12
2	–	–	3 + 12
1	null	7	4 + 12
1	–	6 + 8	–

7.4.3. Relation between HMW subunits and bread-making quality

To his great merit, Peter I. Payne of Plant Breeding International (formerly Plant Breeding Institute), at Cambridge (UK), discovered a relationship between the presence of certain HMW glutenin subunits and the bread-making quality of wheat (Payne, Corfield, Holt & Blackman, 1981). This brought the use of SDS-PAGE within the interest of wheat breeders.

Certain high-molecular-weight subunits or pairs of subunits frequently occur in wheat varieties or selections with good bread-making characteristics. They are 1 and 2* on the *Glu-A1* locus, 17+18 and 7+8 on *Glu-B1* and 5+10 on *Glu-D1*. Other subunits or pairs of subunits mainly occur in wheats with inferior bread-making quality. They are 7 and 6 + 8 on *Glu-B1* and 4 + 12 on *Glu-D1*. Also, the absence of bands 1 and 2* (nullisome) on *Glu-A1* generally points to inferior bread-making quality. For more details, see Branlard & Dardevet (1985), Branlard & LeBlanc (1985), Johansson, Svensson & Heneen (1995), Moonen, Scheepstra & Graveland (1983), Odenbach & Mahgoub (1987), Payne, Corfield, Holt & Blackman (1981), Payne & Lawrence (1983).

Payne, Nightingale, Krattinger & Holt (1987) assigned quality scores to each of the commonly occurring HMW glutenin subunits. As the subunits 1 and 2* of chromosome 1A were shown to be associated with good bread-making qualities compared with the null allele, the former two were each given a score of 3 and the latter a score of 1.

The difference between loaf volumes associated with subunits 5 + 10 and 2+12, coded by chromosome 1D, is at least as great as that between subunit 1 and the null allele. However, there is another encoded subunit pair, 4 + 12, of chromosome 1D, whose quality association is inferior to 2 + 12. The subunits 5 + 10 are, therefore, given a score of 4. The subunits 2 + 12 were given a score of 2, and 4 + 12 a score of 1. By similar reasoning, scores have been given also to 3 + 12 and all encoded subunits of chromosome 1B shown in Table 7.2.

The quality score of a variety is simply calculated by summing the scores of the individual subunits it contains. The maximum score is 10 and the minimum is 3.

Some groups of scientists (e.g. Pogna, Mellini, Beretta & Dal Belin Peruffo, 1989) have compiled other scoring systems that correspond better to quality assessments in Italy and France.

A novel HMW glutenin subunit 21* was discovered by Johansson, Henriksson, Svensson & Heneen (1993) in Swedish wheats. This subunit was found to be allelic to subunit 1 and 2* encoded on chromosome 1A. Results obtained so far with Swedish grown wheats show that there is a positive correlation between subunit 21* and bread-making quality (Johansson, Svensson & Heneen, 1995).

7.4.4. Gliadins

Fifty percent of the storage proteins in wheat are gliadins. When applying acid polyacrylamide gel electrophoresis (APAGE), the gliadin fraction is divided into ω-, γ-, β-, and α-gliadins, based on the mobility of the subfractions.

Each cultivar may contain up to 50 gliadin subunits. The short arms of chromosome 1A, 1B and 1D contain the genetic information for the ω- and γ-gliadins as well as for the LMW subunits of glutenin (Section 7.4.1). Clusters of gliadin genes are called gliadin blocks (Metakovski, Novoselskaja, Kopus, Sobko & Sosinov, 1984; Sozinov & Poperelya, 1982). Additional gliadin loci, outside these blocks, have also been described. Studies on the influence of gliadins on bread-making quality are, therefore, more difficult then those on HMW glutenin subunits (Johansson, 1995).

Sozinov and Poperelya were able to find consistent associations between several gliadin blocks and Zeleny sedimentation volumes, which are used as bread-making quality parameters. Branlard & Dardevet (1985) and other investigators found quality differences between gliadins located on chromosomes 1A, 1B and 1D and on chromosomes 6A, 6B and 6D. However, it has been shown that the correlation between quality parameters and the gliadins located on chromosome group 1 might also be attributed to the LMW subunits of the glutenins. On these chromosomes, the genes encoding the LMW subunits of glutenin and those of the gliadins are very closely linked.

Van Lonkhuysen, Hamer & Schreuder (1992) demonstrated that 32 wheat lines with the same HMW glutenin-subunit composition (null, 7, and 2 + 12) but different gliadin composition varied greatly in bread-making quality. Loaf volumes ranged from 445 ml to 616 ml per 100 g of flour. Variation in the relative composition of four gliadins explained 82% of this variation.

7.4.5. Conclusion

The determination of glutenin and gliadin composition as a tool for the choice of parents and for selection in the offspring has been a major factor in quality breeding of wheat, although the results of different researchers are not consistent in all respects. We are aware that the information presented in sections 7.4.3 and 7.4.4 may soon be updated, as new results in this field of research succeed one another rapidly.

7.5. Cultivars giving sticky doughs

7.5.1. Wheat-rye substitution and translocation lines

Certain wheat varieties have the unpleasant property of producing sticky doughs, indicating that the dough sticks to machines or hands during kneading and successive processing steps. This makes it difficult for the baker to handle the dough. Stickiness is often caused by the fact that at least one of the parents is a wheat-rye substitution line, in which one wheat chromosome pair (1B) has been replaced by a rye chromosome pair (1R) – called a 1B/1R substitution. Sticky doughs may also occur in using varieties in which only a fragment of the rye chromosome has been transferred into the wheat chromosome. This is specifically the case when the short arm of the wheat chromosome (1BS) has been replaced by the short arm of the rye chromosome (1RS). Such progenies are referred to as 1BS/1RS translocation lines (Zeller, 1973; Zeller & Fischbeck, 1974; Zeller, Günzel & Fischbeck, 1982; Moonen, Scheepstra & Graveland, 1983).

Varieties with a wheat-rye substition or translocation have often been used as crossing parents in breeding programs. They have an improved resistance to diseases such as stem rust, leaf rust and stripe rust compared to 'normal' wheat lines, and also a yield advantage, even in the absence of diseases. Only later did it became apparent that dough made from flour of these varieties was difficult to handle in European bakeries due to its stickiness (Payne, Nightingale, Krattinger & Holt, 1987; Sreeramulu & Singh, 1994). In the 1970s, practically all countries of western Europe were confronted with sticky doughs. The problem arose first in Germany, where the Kranich, Benno and

Feldkrone varieties exhibited this inconvenient property. In France, the UK and the Netherlands, this was the case with Clement. Wheat lots containing these varieties were refused by the milling industry.

If wheat-rye substitution and translocation lines are involved in a breeding program, it is important to discard the part of the progeny that produces sticky dough as soon as possible. This characteristic can be recognized easily with the 'sticky dough test' developed by Bolling (1975). The test may be used by breeders as soon as enough seed is available (Section 8.4.1).

Despite this, varieties containing the 1B/1R chromosome can have good bread-making quality if their composition of HMW subunits is adjusted so that they would normally be regarded as 'over strong'. In varieties such as Rialto in the UK, France and Germany, the weakness of the 1B/1R chromosome is counter-balanced by the excessive strength of the HMW subunits.

7.5.2. Endogenic high α-amylase levels

There are two•types of α-amylase: the 'green amylases' and the 'malt amylases' (Section 4). Green amylases are found in the grains of immature wheat plants. During ripening, the activity of this enzyme rapidly decreases to a level of practically zero at harvest-ripeness. Malt amylases have a distinctly different chemical composition, but the same working mechanism, as green amylases. They hardly occur in sound, intact wheat grains. When ripe wheat grains start germinating, however, the activity of the malt-amylase enzymes increases rapidly. If this happens under natural conditions in the field, it causes *sprouting in the ear.*

In some varieties, the activity of the green α-amylases does not diminish during ripening. Sometimes, it even increases. The result is that ripe grains of these have a relatively high level of α-amylases, even when they have not sprouted. It makes the wheat unsuitable for bread baking. Some well-known examples of these are the English varieties Maris Huntsman, Fenman, Kinsman, Mardler and Norman. Each of these varieties is derived from the old Belgian variety Professeur Marchal, one frequently used by English wheat breeders during the 1960s, until this deleterious character was identified.

Acknowledgement

The author wishes to thank Dr Peter Payne for his critical comments on this chapter.

References

Bolling H (1975) Zur Problematik der 'nichtbackfähigen Weizen' – Eine Schnellmethode zu ihrer Erkennung'. Mühle u Mischfuttertechnik 112: 256–257.

Branlard G, Dardevet M (1985) Diversity of grain protein and bread wheat quality. (II) Correlation between high molecular weight subunits of glutenin and flour quality characteristics. J Cer Sci 3: 345–354.

Branlard G, LeBlanc A (1985) Les sous-unités gluténines de haut poids moléculaire des blés tendres et des blés durs cultivés en France. Agronomie 5: 467–477.

Cooke RJ (1995) Allelic variabiliy at the Glu-1 loci in wheat varieties. Plant Varieties and Seeds 8: 97–106.

Cookson MA (1975) In: Spicer A (ed), Bread: Social, nutritional and agricultural aspects of wheaten bread. London: Appl Sci Publ: 347–348.

Dhaliwal AS, McRitchie F (1990) Contribution of protein fractions to dough handling properties of wheat-rye translocation cultivars. J Cer Sci 12: 113–122.

Doekes GJ, Belderok B (1976) Kernel hardness and baking quality of wheat – A genetic analysis using chromosome substitution lines. Euphytica 25: 565–576.

Hamer R, Weegels PL, Marseille JP (1992) Prediction of the bread making quality of wheat: the use of HMW glutenin – A subunit based quality scoring system. J Cer Sci 15: 91–102.

McIntosh RA, Hart GE, Gale MD (1990) Catalogue of gene symbols for wheat. Cer Res Comm 18 suppl: 141–157.

Johansson E (1995) Wheat grain proteins: Accumulation and composition in breeding for improved bread-making quality. Dissertation Svalöv.

Johansson E, Henriksson P, Svensson G, Heneen WK (1993) Detection, chromosomal location and evaluation of the functional value of a novel high Mr glutenin subunit found in Swedish wheats. J Cer Sci 17: 237–245.

Johansson E, Svensson G, Heneen WK (1993) Quality evaluation of the high-molecular-weight glutenin subunits in Swedish wheat material. Proc 5th Int Workshop Gluten Proteins 7–9 June 1993. Detmold: Ass of Cereal Res: 568–575.

Johansson E, Svensson G, Heneen WK (1995) Composition of high-molecular-weight glutenin subunits in Swedish wheats. Acta Agric Scand; Sec B Soil and Plant Sci 45: 112–117.

Lukow OM, Payne PI, Tkachuk R (1989) The HMW glutenin subunit composition of Canadian wheat cultivars and their association with bread-making quality. J Sci Food Agric 46: 451–460.

Metakovski EV (1991) Gliadin allele identification in common wheat II. Catalogue of gliadin alleles in common wheat. J Genet and Breed 45: 325–344.

Metakovski EV, Novoselskaja AY, Kopus MM, Sobko TA, Sosinov AA (1984) Blocks of gliadin components in winter wheat detected by one-dimensional polyacrylamide gel electrophoresis. Theor Appl Genet 67: 559–568.

Moonen JHE, Scheepstra A, Graveland A (1983) The positive effects of the high molecular weight subunits 3 + 10 and 2* of glutenin on the bread-making quality of wheat cultivars. Euphytica 32: 735–742.

Moonen JHE, Zeven AC (1984) SDS-PAGE of the high-molecular-weight subunits of wheat glutenin and the characterization of 1R (1B) substitution and 1BL/1RS translocation lines. Euphytica 33: 3–8.

Morgunov AI, Peña RJ, Crossa J, Rajaram S (1993) Worldwide distribution of Glu-1 alleles in bread wheat. J Genet and Breed 47: 53–60.

Odenbach W, Mahgoub E (1987) Beziehungen zwischen der Backqualität und dem Vorkommen der Gene für die hochmolekularen Untereinheiten des Glutenins bei den deutschen Winter- und Sommerweizensorten. Getr Mehl u Brot 41: 195–198.

Payne PI (1987) Genetics of wheat storage proteins and the effect of allelic variation on bread-making quality. Ann Rev Plant Physiol 38: 141–153.

Payne PI, Corfield KG, Holt LM, Blackman JA (1981) Correlations between the inheritance of certain high-molecular weight subunits of glutenin and bread-making quality in progenies of six crosses of bread wheat. J Sci Food Agric 32: 51–60.

Payne PI, Lawrence GJ (1983) Catalogue of alleles for the complex gene loci, Glu-A1, Glu-B1 and Glu-D1 which code for high-molecular-weight subunits of glutenin in hexaploid wheats. Cer Res Comm 11: 29–35.

Payne PI, Nightingale MA, Krattinger AF, Holt LM (1987) The relationship between HMW glutenin subunit composition and the bread-making quality of British-grown wheat varieties. J Sci Food Agric 40: 51–65.

Peltonen J, Salopelto J, Rita H (1993) The optimal combination of HMW glutenin subunits coded at gene loci *Glu-A1* and *Glu-B1* for bread-making quality in Scandinavian wheats. Hereditas 118: 71–78.

Pogna NA, Mellini F, Beretta A, Dal Belin Peruffo A (1989) The high-molecular-weight glutenin subunits of common wheat cultivars grown in Italy. J Genet and Breed 43: 17–24.

Rimpau J, Röbbelen G (1978) Vererbung der Eigenschaft 'negatives Teigverhalten' von Clement und Maris Huntsman. Getr Mehl u Brot 32: 53–55.

Sozinov AA, Poperelya FA (1982) Polymorphism of prolamins and variability of grain quality. Qual Plant; Plant Foods Hum Nutr 31: 243–249.

Sreeramulu G, Singh NK (1994) High Mr glutenin subunits of Indian wheat cultivars: association of subunits 5+10 with the 1BL/1RS wheat-rye translocation. J Cer Sci 20: 217–225.

Svensson G (1981) Varietal and environmental effects on wheat milling quality. Agri Hort Genet 39, 103 pp.

Symes KJ (1965) The inheritance of grain hardness in wheat as measured by the particle size index. Austr J agric Res 16: 113–123.

Symes KJ (1969) Influence of a gene causing hardness on the milling and baking quality of two wheats. Austr J agric Res 20: 971–979.

Van Lonkhuysen HJ, Hamer RJ, Schreuder C (1992) Influence of specific gliadins on the bread-making quality of wheat. Cer Chem 69: 174–177.

Zeller FJ (1973) 1B/1R wheat-rye chromosome substitutions and translocations. Proc 4th Int Wheat Gen Symp. Agr exp Sta, Columbia, Mo, USA: 209–221.

Zeller FJ, Fischbeck GW (1974) Chromosomenadditionen, -substitutionen und -translokationen als Grundlagen für die Übertragung artfremden Erbmaterials in den Saatweizen (*Triticum aestivum* L.). Fortschr Pflanzenzüchtung 4: 5–55.

Zeller FJ, Günzel G, Fischbeck GW (1982) Veränderungen der Backeigenschaften des Weizens durch Weizen-Roggen-Chromosomentranslokation 1B/1R. Getr Mehl u Brot 36: 141–143.

8. Assay methods and instrumentation

8.1. Introduction

Breeders of bread wheat should be well aware of the baking requirements of new varieties, when they choose parents and select offspring of wheat crosses. The different elements of these requirements may be grouped according to the following characteristics (Irvine, 1975; Cookson, 1975):

Grain properties

- External aspects
- Protein content
- Grain hardness
- Alpha-amylase activity
- High molecular weight (HMW) glutenin subunits,

Milling quality

Dough properties

- Stickiness of dough
- Protein strength
- Baking quality.

This chapter briefly reviews the methods used by breeders to identify these characteristics. A suggestion on how to choose among these techniques is given at the end of the chapter.

8.2. Grain properties

8.2.1. External aspects

A wheat sample is spread out on the hand and assessed for grain colour, grain size and filling degree. The colour of the grains has to be typical for the variety in question. Gray kernels may be caused by a fungus infection at ripening or at harvest and may influence the colour of the milled products in a negative way. A light-brown, dull colour is often found in soft and mealy grains with a relatively low protein content. A dark-brown, shiny colour usually points to hard, vitreous grains with more protein. A sample taken from one and the same wheat batch may contain a mixture of both mealy and vitreous grains.

The milling industry is generally more interested in wheat lots with medium-size grains than in those with large grains. Medium-size grains have undergone a relatively short ripening period, resulting in a higher protein content. On the other hand, wheat grains should not be too small as that may be caused by premature ripening. Such grains have a somewhat boat-like shape. Their extraction rate is low as bran and endosperm cannot be separated properly. Moreover, the bran tends to shatter during milling, which will negatively affect the colour and the bread-making quality of the flour.

8.2.2. Protein content

Gluten determination. The dough properties of a flour are mainly determined by the *quantity* and the *quality* of grain proteins that are insoluble in a diluted salt solution, i.e. the *gluten*. To test protein content approximately 25 g of a wheat sample is ground in a small-scale disc mill (e.g. a Miag or Moulinex mill or a Glenn Creston disc mill). Ten grams of flour are kneaded for several minutes in a dish with 5 ml of a 2% NaCl solution until a stiff dough is obtained. After a resting period, the dough is taken from the dish and washed out with the fingers under a fast-dripping salt solution above a silk-gauze filter. Starch and water-soluble parts are washed out. The gluten ball remaining on the filter is then dried with the hands by rolling it with two fingers from one hand on the palm of the other hand. The hand palm must be dried at regular intervals; rolling stops when the gluten starts to become sticky. The weight of the gluten, multiplied by 10, is the *wet-gluten* content. A wet-gluten content of 20–27% (proportion of weight of the dry flour) is normal for bread making. The *dry-gluten* content may be determined by drying the wet-gluten for 24 h at room temperature or for shorter times at higher temperatures. A special machine has been developed to wash the gluten and carry out the entire analysis. Note that of the 'crude gluten' obtained by this method only 80% consists of proteins. It also contains considerable amounts of fatty acids and carbohydrates, as well as small quantities of cellulose and ash.

Protein determination by the Kjeldahl technique. For bread-making purposes, a protein level in the flour of at least 13% on a dry matter basis is required, whereas for biscuit making the level should not surpass 10–11%. Protein content is generally determined by the Kjeldahl technique. Approximately 1 g of ground wheat grain is transferred to a digestion flask. Concentrated sulphuric acid and small quantities of catalyst are added. The flask and contents are heated until white smoke is given off and the liquid becomes clear. The organic material in the flask is decomposed by the sulphuric acid, and the nitrogen in the protein is transformed into sulphate of ammonia. The fluid in the flask is then made alkaline by adding a strong lye, releasing anhydrous ammonia from the sulphate of ammonia:

$$(NH_4)_2SO_4 + 2\,NaOH \rightarrow Na_2SO_4 + 2NH_3 + 2H_2O.$$

To prevent losses of ammonia, the digestion flask is attached to a distillation device as soon as the lye has been added. The ammonia is displaced upon boiling and collected in an excess of boric acid. Eventually, the amount of nitrogen present is determined by titration. To calculate the amount of protein in the sample, the N content of the protein has to be known. This factor is 5.7 for wheat protein, corresponding to a N content of 17.55%. The sample is generally ground by hand. Subsequent steps, however, including the calculation of the results, may be carried out with a Kjeltec Autosampler of Tecator.

Protein determination according to the NIR technique. One could say that the introduction of protein content as a quality standard had to await the development of a method for analysing protein that could be performed cheaply and accurately by unskilled personnel with limited laboratory services. The technique that made this possible is called *Near Infra-red diffuse Reflectance Spectroscopy*, usually abbreviated to NIRS or simply NIR. Briefly, NIR is based on the following principle: monochromatic light (i.e. light of one single wavelength) shines on a grain sample; it is partially reflected; the intensity of the reflected light is measured by a photo-electric cell. By measuring the reflection intensity at various wavelengths, it is possible to eliminate the disturbing influence of other components than proteins. By using a wide variety of samples for the calibration, differences in origin of wheat, weather circumstances during growth and ripening, etc.,can be excluded. In practice determination is very simple. A sample is ground, transferred into a measuring cup and placed in the NIR equipment. After 20 s the data appear on a screen or are printed out. The whole determination takes about $2^1/_2$ minutes. When requested to report on a certain parameter of the sample, the computer software extracts the relevant data from its memory and transforms them (according to supplied calibration constants) into the required format, e.g. protein percentage. It is even possible to analyse more than one constituent in one operation. For example, protein, moisture, ash and grain hardness can be determined simultaneously.

8.2.3. Grain hardness

Large differences in grain hardness exist between wheat varieties (Section 2). Two categories are distinguished: varieties with hard grains, and varieties with soft grains. Hard-wheat varieties are pre-eminently suitable for bread making and soft-wheat varieties for biscuit making.

Pearling Index. A measured weight of grains is subjected to the abrasive action of a rotating carborundum disc. The loss in weight is determined. The test can be carried out in several ways. Each is a variation on the following commonly used procedure. Twenty grams of grains are ground for 70 s in a Strong Scott Barley Pearler at 1400 rpm. The proportion (%) of the pearled-off kernel weight to the original weight is the Pearling Index. The index varies between 20% for hard-wheat types to 55% for soft-wheat types. The results are well reproducible for samples of different moisture contents.

Grinding resistance. This involves the measurement of the time taken to fill a specific volume when wheat (20 g) is ground under standard conditions in a small disc-type mill or hammer mill. The grinding resistance can vary from about 30 s for hard grains to about 55 s for soft grains.

Particle Size Index (PSI) (Symes, 1965). At milling, the endosperm of wheat varieties with hard grains disintegrates for the greater part into single cells and clusters of *intact* endosperm cells. Due to the strong bond between starch granules and protein bodies, very few single starch granules and protein bodies are found in the flour. Soft-wheat grains fall apart in *fragments* of one or more cells and many single starch granules and protein bodies. Differences in grain hardness are thus expressed as differences in particle-size distribution of the flour. Twenty-five grams of wheat grains are ground into whole-meal flour with a disc-type laboratory mill or a small roller-mill. This ground material is then sifted for 10 minutes on a sieve with a mesh of 0.075 μm. The material that passes through the sieve is weighed. The proportion of this flour to original weight of grains is the *Particle Size Index* (PSI). It may vary between approximately 13% (for extremely hard grains) and 55% (for very soft grains). The test results are influenced by the moisture content of the grains and by the grinding technique.

Visual examination of the bran (Bell & Bingham, 1957; Bingham, 1962). A convenient single estimate of grain hardness was developed at the Plant Breeding Institute, Cambridge (UK), in the 1950s. The method was based on the tenacity with which the endosperm adheres to the bran. This character can be easily scored, and a given finding can be compared with the score of standard varieties. Grain samples are first brought to a moisture content of about 15% and then milled in a small laboratory roller-mill, e.g. a Brabender Quadrumat Junior mill. Then the bran is separated from the ground stock using a small hand sieve. In varieties classified as hard-milling, the endosperm breaks down into particles composed of whole cells, which readily separate from the skin, giving a bran with little adhering flour. In varieties classified as soft-milling, the endosperm tends to be broken open indiscriminately during milling, resulting in a bran that is more difficult to clean from starchy material. Bran samples can be scored from 1 (very soft wheat) to 5 (very hard wheat). The scoring is very reliable and unaffected by differences in protein content.

8.2.4. Alpha-amylase activity (Hagberg-Perten Falling Number)

The activities of amylolytic and proteolytic enzymes have to be very low in a commercial batch of wheat. Germination of wheat entails a considerable increase in α-amylase activity. A wheat lot with too many sprouted grains, and thus an amylase activity that is too high, is not suitable for bread, biscuit and/or pasta making.

The Hagberg-Perten Falling Number is a reliable measure of α-amylase activity. A flour suspension is heated in a boiling-water bath so that the starch will be gelatinized. Simultaneously, part of the starch is decomposed into sugars and dextrins by the amylases in the sample. The higher the amylase activity, the less viscous the remaining suspension will be.

Perten, a Swede, developed a Falling Number device that heats the flour suspension and measures its viscosity. The main parts of the device are a set of glass tubes with an inner diameter of 21 mm and a length of 220 mm, and a set of stirrers of specific dimensions and weight.

Seven grams of flour and 25 ml of water are introduced into a glass tube together with a stirrer. The tube is then placed in a boiling-water bath and a stopwatch is started. The suspension is stirred for exactly 1 minute at a frequency of two up-and-down motions per second.

The stirrer is brought in its upper-most position after 59 s and dropped at 60 s. As soon as the lowest position is reached, the stopwatch is stopped. The number of seconds between the moment that the tube has been introduced into the boiling-water bath and the time at which the stirrer has reached its lowest position is called the Hagberg-Perten Falling Number. Wheat varieties may be classified according to their suitability for bread and biscuit making by their falling number (Table 8.1).

Some wheat varieties have a relatively high α-amylse activity, even without any sprouting in the ear (Section 4). This is due to the fact that the activity of 'green' amylases, normally present in unripe seed, does not sufficiently decrease at ripening. Very often these varieties are derived from the Belgian variety Professeur Marchal. These wheats are also unsuitable for food-processing.

Table 8.1. The relation between the Falling Number of wheat and suitability for bread and biscuit making

Falling number	Suitability class
<< 120	High sprouting level, not suitable for bread – or biscuit making
120–180	Sprouted wheat, may be mixed with an unsprouted wheat lot
180–200	Low sprouted wheat
200–250	Unsprouted wheat
250–300	Unsprouted wheat, should be mixed with malt flour or sprouted wheat
>> 300	Unsprouted wheat, has to be mixed with malt flour or sprouted wheat

8.2.5. HMW glutenin subunits

Traditionally, storage proteins in wheat have been divided into two groups: *gliadins* and *glutenins*. They are usually defined by their molecular size in dissociating solvents. Gliadin molecules are relatively small; glutenins have large, heterogeneous molecules, built up from some 19 different subunits connected by disulphide bonds. The glutenin subunits fall into two unequal groups, the low-molecular-weight (LMW) subunits and the less frequent, high-molecular-weight (HMW) subunits.

By SDS-PAGE (sodium dodecyl sulphate, polyacrylamide gel-electrophoresis), storage wheat proteins can be split into their subunits, which can be made visible as a band pattern. Payne, Nightingale, Krattinger & Holt (1987) described a relationship between the presence of certain HMW glutenin subunits and the bread-making quality of wheat. This brought the use of SDS-PAGE to the attention of plant breeders (Section 7.4.3).

The analysis of HMW glutenin subunits can be done on either wheat grains or wheat flour. First, the grains have to be crushed with, for example, a hammer, resulting in some 40 mg of material per grain. It is even possible to analyse the embryo-less half of a grain and to keep the remaining part for sowing.

The technique as it has been described by Moonen, Scheepstra & Graveland (1982) is given here as an illustration. A crushed grain or 50 mg of flour is suspended in a centrifuge tube in 1 ml buffer of Tris-HCl (pH 6.8), 5% mercapto-ethanol and 2% SDS (sodium dodecyl sulphate). Mercapto-ethanol is a reducing agent with the ability to break the SS-bridges in the protein. SDS is a detergent that encloses the degradation products of glutenins and gliadins and prevents them associating.

The tube with contents is placed in a shaking device and shaken for two hours at room temperature. The tube is then transferred to a boiling-water bath – to enhance the reaction – and finally centrifuged for 5 minutes.

Fifty μl supernatant are applied to a vertical polyacrylamide slab-gel. The electrophoresis is carried out in a Pharmacia GE 2/4 system at room temperature. Two slab-gels are run at a time with 60 mA for 30 minutes. If necessary, the samples can also be analysed using a 7.5% or 10% gel, to separate subunits 2 and 2* or 9 and 10, respectively.

Proteins are fixed and stained with Coomassie Brilliant Blue solution at 60°C for one hour and destained with ethanol-acetic acid solution at 60°C overnight. After destaining, the protein subunits become visible as dark-coloured bands. These bands can be described and numbered, preferably according to Payne & Lawrence (1983) (Section 7.4.2).

8.3. Milling quality

The main factors determining milling quality of wheat are *ease of milling* and *yield of flour*. Milling quality should be determined for flour that is comparable to commercial bread and biscuit flours. This can only be achieved by the shearing and scraping action of one or more sets of mill rollers.

A small laboratory mill will give only a rough indication of the milling characteristics of a wheat sample. To obtain reliable results, a medium size test mill with a capacity of several kilograms per hour should be used. As this type of mill requires relatively large quantities of seed and as the analysis is rather time consuming, it is not very suitable for routine research on a breeding station. Breeding companies generally use one of the following laboratory mills for the making of bread and biscuit flour (LeBrun, 1986):

Brabender Quadrumat Junior mill. This is a very compact mill with four corrugated rollers of 70 mm diameter and 30 mm thickness (Figure 8.1). The rollers rotate in opposite directions at a constant speed. The upper two rollers are positioned diagonally, the lower two are placed underneath in a horizontal line. A round 240 μm mesh sieve is placed under the rollers. Flour and bran are collected in different drawers. Quantities of 100–1000 g of wheat can be milled at a time.

Brabender Sedimat mill. As far as size and construction are concerned, this mill is similar to the Quadrumat Junior mill described above. The main difference is that the Sedimat mill produces a very white flour with an extremely low ash content. As a consequence, the flour yield is very low, about

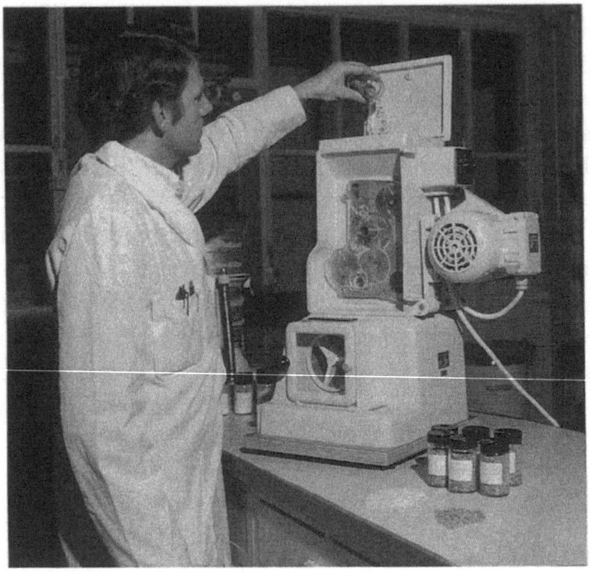

Figure 8.1. Brabender Quadrumat Junior laboratory mill.

18%. This mill has been designed especially for the Zeleny sedimentation test (Section 8.4.2).

Chopin-Dubois mill. This mill has both a break and reduction system. On the break side, it has three corrugated rollers with the same dimensions as those of the Quadrumat Junior mill. Two round sieves are placed underneath one another: a 150 μm sieve to separate the flour and a 800 μm sieve to separate semolina and bran. Semolina and bran are transferred to the reduction side of the unit, where the semolina is reduced to flour by a pair of smooth rollers and a round 170 μm sieve. The bran is not used for further analysis. The Chopin-Dubois mill can mill 100–1000 g of wheat at a time.

Brabender Quadrumat Senior mill. This is a middle size experimental mill with a capacity of 1–3 kg a time. It takes approximately 15 minutes to mill 1 kg of wheat. The mill actually consists of two Quadrumat Junior mills placed side by side: one with four corrugated rollers and corresponding sieves to break the grains; and one with four finely corrugated rollers and matching sieves for a further reduction of the material.

Bühler MLU 202 mill. This medium size experimental mill has three pairs of corrugated break rollers on its left-hand side and three pairs of smooth reduction rollers on its right-hand side. The rollers each have a diameter of 20 mm and a length of 12 cm. A plansifter is placed under each pair of rollers. (The flow diagram has been described in Section 3, Figure 3.4. The only difference is that here the mill has four instead of three pairs of smooth rollers.) The milled products (flour from the three break rollers, flour from the two reduction rollers, and shorts and bran) are collected separately. The five flour fractions are generally mixed. This Bühler mill has a milling capacity of 1–5 kg at a time. It takes 30 minutes to mill 1 kg of grain.

8.4. Dough properties

8.4.1. Stickiness of dough

The offspring of certain wheat varieties have the unpleasant property of producing sticky doughs, implying that the dough sticks to hands and machines during kneading and the successive processing steps, which makes it difficult to handle.

As described in Section 7, dough stickiness is often caused by at least one of the parents being a wheat-rye substitution line in which one chromosome pair (1B) has been replaced by a rye chromosome pair (1R), a so-called 1B/1R substitution line. Sticky doughs may also occur in varieties in which a fragment of the rye chromosome has been transferred into the wheat chromosome. This is specifically the case when the short arm of the wheat chromosome (1BS) has been replaced by the short arm of the rye chromosome (1RS). This type of lines is referred to as 1BS/1RS translocation lines.

If wheat-rye substitution and translocation lines are involved in a breeding program, it is important to discard progenies with sticky doughs as soon as possible. They can easily be recognized with a test developed by Bolling (1975). Approximately 200 g of wheat grains are milled to a whole-wheat flour in a laboratory hammer mill. Then 50 g of flour and 30 ml water are put into the beaker of a fast-rotating mixer and mixed into a stiff dough. Subsequently, the dough is taken from the beaker and shaped into a ball by hand. The stickiness of the ball is assessed by the feel of pressing the dough several times in the hand. This very simple assessment is based on only one criterion: sticky or not sticky. Rimpeau and Röbbelen (1978) elaborated a classification into five groups. Dhaliwal and McRitchie (1990) extended the test by not only assessing the stickiness on the hands but also on the inside of the mixer, on a bench surface and on a paper towel. Every sub-test results in a score, the sum of which is a stickiness score, which varies between zero and 20.

8.4.2. Protein strength

Saunders' chewing test. At the turn of the century, a Canadian wheat breeder, Charles Saunders, produced the hard spring wheat cultivar Marquis. This very successful wheat consolidated Canada's reputation in Europe as a supplier of the highest quality milling wheat for bread flours. By the late 1920s, over 80% of the spring wheat area in Canada and USA was planted with this variety alone. Saunders' screening test for quality was a very simple one. He used to chew a handful of kernels from each line. Based on the hardness between his teeth he selected for milling quality; the gluten, which he had in his mouth after chewing the wheat, was tested between his fingers for strength and elasticity.

Wheat-meal fermentation test (dough-ball test; Pelshenke test). The wheat-meal fermentation test is generally carried out according to its original description by Pelshenke (1933). A certain amount of grains is ground to a coarse meal by a laboratory disc- or cone-type mill. With water and yeast, 10 g of this meal is kneaded to a dough by hand. The dough is split in two parts and each part is rolled into a ball. Each ball is then placed in a beaker filled with water of 32 °C. The CO_2 formed by the yeast inflates the dough and the ball rises to the water surface after some 10 minutes. After a certain amount of time the dough is no longer able to retain the gas and starts to crack and fall apart. The time between the placing of the dough in the water and its falling apart is measured and referred to as the *Pelshenke value* or *P value*. The P value depends on both the gas-forming and gas-retaining capacity of the dough. As bread-making quality is largely dependent on these two factors, Pelshenke assumed that the P value would indicate the bread-making ability of the wheat concerned. P values can vary from 40 minutes (poor bread-making quality) to more than 250 minutes (excellent bread-making properties). However, the predictive value of intermediate P values is often very poor.

The Zeleny sedimentation test (Zeleny, 1947). By measuring sedimentation, it is possible to combine information on protein quantity and quality in one value. The sedimentation test is based on the principle that high-quality wheat proteins swell in a lactic acid solution more than poor-quality proteins do. The Zeleny protocol requires flour of specific fineness, which is usually produced by a specific mills. The Brabender Sedimat mill was especially designed for this purpose. It is a laboratory roller-mill producing a low ash content and a very low flour yield of about 18%. In practice the test is executed as follows: 3.2 g of flour is suspended in 50 ml water in a glass cylinder with a scale from 0–100 ml. The cylinder is shaken a few times to moisten the flour particles

Figure 8.2. Zeleny sedimentation test; the level of the sediment is a measure of the baking quality of the analyzed sample.

properly. Then 25 ml of a watery solution of lactic acid and isopropyl alcohol is added and the cylinder is shaken again. The protein of the flour starts to swell. Finally, the cylinder is left standing upright for 5 minutes in order to allow the suspended protein to precipitate. Then the volume of the sediment is read off the cylinder scale, giving the *sedimentation value* (Figure 8.2). The higher the gluten content in the flour and the better the bread-making properties of the gluten are, the more sediment will be formed. Thus, the sedimentation value is a measure of both the quantity and quality of the gluten. An advantage of the sedimentation test is that it can be mechanized to a large extent by using a shaking device. By doing so, any subjective element is excluded. Moreover, the test is simple and fast and does not involve expensive machinery or chemicals (except the Brabender Sedimat mill). The test is generally carried out in series of eight samples, which takes some 30 minutes (excluding the milling and weighing of samples and the cleaning of the glassware).

The SDS sedimentation test. The Zeleny sedimentation test and the SDS sedimentation test work on the same principle but differ in details. The Zeleny test was developed in the USA in the late 1940s and has been adopted in many European countries. The SDS sedimentation test was devised at the Flour Milling and Baking Research Association, Chorleywood (UK), some 30 years

Table 8.2. Relation between Zeleny sedimentation values and suitability for bread and biscuit making of flour

Sedimentation value	Classification
< 20	Low protein content, suitable for biscuit making
20–30	Medium protein content, medium bread-making quality
30–40	High protein content, medium bread-making quality
> 40	Very high protein content, very good bread-making quality

later, to avoid the necessity of a special flour produced by the Sedimat mill. It has been widely used in the UK, but has not been taken up to any significant extent elsewhere in Europe, except for the former Soviet Union. Although it can be performed with flour, the SDS test normally uses wheat ground on a disc-type mill with a 1 mm sieve. Six grams of ground wheat are added to 50 ml water in a 100 ml stoppered measuring cylinder, and the meal is dispersed by vigorously shaking. The contents are re-shaken twice at 2 minutes intervals. Following the last shaking, 50 ml of a SDS-lactic acid solution is added and mixed by inverting the cylinder. Inversion is repeated at intervals of 2, 4 and 6 minutes. After the last inversion the content of the cylinder is allowed to settle for 20 minutes before the sedimentation volume is read. The procedure allows for four samples to be assessed at once and four more to be started up during the settlement step of the previous four samples. An equation predicting Zeleny volumes from SDS volumes was derived. The rough rule of thumb is *Zeleny = SDS – 20.*

The mixograph (Belderok, 1977). As early as 1933, Swanson and Working published the description of a 'recording dough mixer' suitable for the analysis of the quality of wheat flour. Initially, this mixograph was used quite successfully for quality breeding in USA and Australia. It found its way to Europe much later. There are two versions: the macro-mixograph, with a capacity of 30–35 g flour, and the micro-mixograph with a capacity of 10 g flour; the latter version is especially used by breeders (Figure 8.3). The heart of the mixograph is a regularly rotating pin-kneading machine (1 and 2 in Figure 8.3). Ten grams of flour are kneaded for 7 minutes with a fixed amount of water (usually 5.5–6 ml). The kneading time is registered on the horizontal axis and the power exerted on the dough on the vertical axis of graph paper passing through a recording device. In the beginning, the mixograph curve increases to a maximum and then decreases (Figure 8.4). Several parameters may be read from a mixogram:

Figure 8.3. Micro-mixograph of Swanson and Working for the analysis of dough characteristics; the kneading system has been tilted; kneading basin (1), kneading pens (2) and recorder (3).

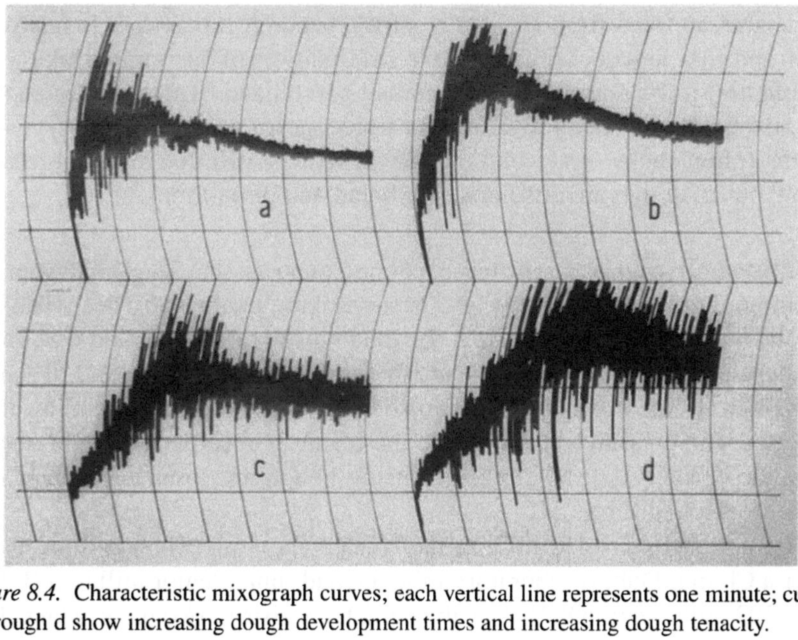

Figure 8.4. Characteristic mixograph curves; each vertical line represents one minute; curves a through d show increasing dough development times and increasing dough tenacity.

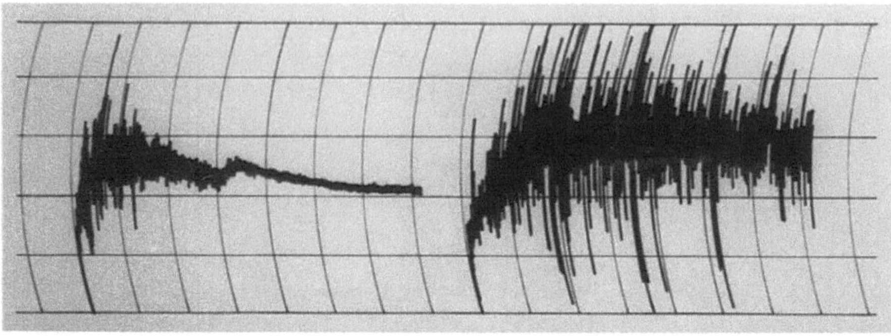

Figure 8.5. Characteristic mixograph curves of wheat giving a weak and sticky dough of the Clement type (left) and wheat giving a stiff and short dough of the Felix type (right). Both types are undesirable.

- the height of the curve, mainly dependent on the protein content of the sample and the amount of water added;
- the developing time of the dough, i.e. the time span between starting the mixing process and reaching the maximum level of the curve;
- the dough relaxation, i.e. the difference in height of the curve at its maximum and 3 minutes later;
- the tenacity of the dough, measured as the bandwidth after a fixed time.

All heights are read from the centre of the band. A mixograph is relatively cheap and easy to work with. Moreover, the analysis of one sample takes only ten minutes, if the sample has been milled beforehand. Another advantage is that mixograph curves allow a breeder to recognize samples with very weak doughs (often sticky) and samples with extremely stiff doughs (Figure 8.5). Both types of dough are unsuitable for bread and biscuit making.

The Chopin alveograph. In francophone countries the Chopin alveograph is commonly used to determine the bread-making quality of wheat. The original device was developed in 1921 by the Frenchman Chopin and was called an extensimeter; in subsequent years the device was modified and called an alveograph. In the alveograph, a thin slice of dough is inflated with air until it breaks. The pressure profile inside the dough is recorded on graph paper. Thus, the changes that take place in rising dough are simulated, albeit in a much accelerated form.

The first steps involve milling the wheat sample to produce flour, using either a Chopin-Dubois laboratory mill, a Quadrumat Senior mill or a Bühler MLU 202 mill (Section 8.3). The alveograph consists of two parts: a kneading unit and the actual alveograph. The kneading basin has a horizontal slot in its side that may be open or closed with a small plate. The kneading arm can rotate in two directions: if it turns to the right with the slot closed, the unit works as a kneader; if it turns to the left with the slot open, an 18 mm thick dough is pressed out through the slot.

The procedure is as follows: 125 ml of a salt solution (concentration 25 g/l) is added to 250 g of the flour to be analyzed. After 6 minutes kneading, the kneading arm is turned in the opposite direction and the slot is opened so that a slice of dough is pressed out of the mixing bowl. Five round pieces of 50 mm diameter are cut from the dough slice. After a resting period of 20 minutes, the dough pieces are, one by one, placed horizontally in the alveograph. Compressed air is blown underneath the dough, causing it to balloon. Eventually, the 'balloon' starts to leak and collapses. The ultimate size of the balloon depends on the elasticity of the dough. The relation between pressure and elasticity is registered as a curve, an *alveogram*. The curve increases steeply to its maximum, then it decreases rather strongly at first and somewhat more gradually afterwards, finally to drop sharply as the dough collapses (Figure 8.6).

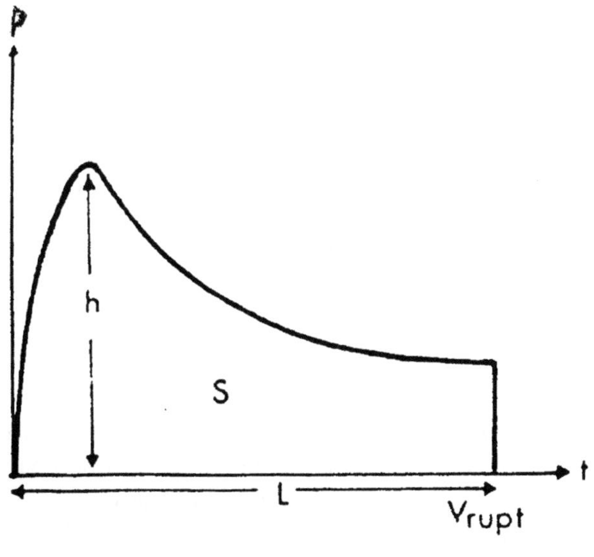

Figure 8.6. Representative alveogram.

Five curves, all starting at the same position, will thus be registered on the recording paper. The average of these five curves is used to characterize a sample. The following measurements may be taken from this average alveogram:

P: the dough resistance to the pressure (stiffness), measured by the 'over-pressure' or highest pressure achieved during the test = the length of h (in mm) multiplied by 1.1 (correctionfactor of the manometer);

L: the dought elasticity, measured by the length of the alveogram to the point of rupture (in mm);

S: the area under the curve, measured with the help of a planimeter (in cm^2);

W: the 'W of Chopin' is the amount of energy required to inflate the dough bubble to bursting point. It is derived from the area S under the curve: $W = 6.54 \times S \times 10^3$ ergs.
W is generally taken to be a measure of protein strength. The classification of wheat varieties in several countries is based on the 'W of Chopin' (see Table 8.3 and Figure 8.7).

P/L: is an indicator of the balance between P (dough stiffness) and L (dough elasticity).

Table 8.3. Relation between the 'W of Chopin' and the suitability for bread and biscuit making of flour

W of Chopin	Classification
< 50	Not suitable for making bread and biscuits
50–100	Biscuit-making quality
100–150	Medium bread-making quality
150–250	May be used as improver wheat, 'blé améliorant'.
250–350	High quality wheat with a high protein level and excellent bread-making quality, 'blé de force'.

8.4.3. Baking quality

Micro-baking tests for bread. The standardization of baking trials is not an easy job. First of all, the grains of the sample to be analysed must be milled into a baking flour. At micro level, this can only be done with one of the laboratory roller-mills described in Section 8.3. Then, a dough with constant technological properties must be made. This means that for each sample the amount of water to be added to the flour must be determined apart. An experienced test baker can do this by touch. The kneading time and mixing intensity of the dough, the duration of the several resting periods, etc., have to be properly standardized. The whole test takes several hours and is relatively expensive. Breeding companies, do not, therefore, perform micro-baking tests themselves but contract this out to specialized laboratories

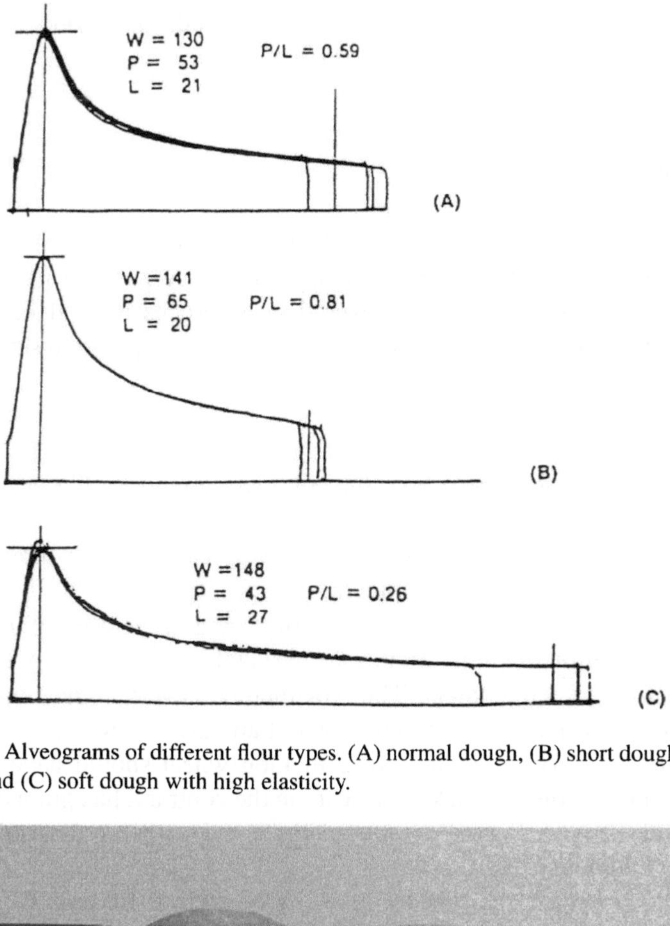

Figure 8.7. Alveograms of different flour types. (A) normal dough, (B) short dough with poor elasticity and (C) soft dough with high elasticity.

Figure 8.8. Micro-baking test for bread.

and institutions. A selection of references on micro-baking tests based on 10–100 g of flour is given at the end of this chapter.

Micro-tests for biscuit, cake and pasta making. For the sake of completeness, it should be mentioned here that standard test methods for making biscuits, cakes and pastas are given in literature as well. They are often based on 40–100 g of flour. As a start, readers may consult the Approved Methods of the American Association of Cereal Chemists.

8.5. Evaluation

Each of the assay methods just described determines one, or exceptionally two, parameter(s) of milling and baking quality. A positive result on one parameter is no guarantee for an overall good processing value. Breeder's material is therefore generally submitted to a series of quality tests.

8.5.1. Early stage of breeding

During the first stage of breeding, the complete offspring of a cross may be assessed for its processing value:

- *External quality*: this property is relatively easy to determine and does not require any special equipment.
- *Protein content*: extremely high or low values generally refer to abnormal growth or ripening conditions. Several techniques may be used to determine protein content. The NIR method is very convenient because it combines analyses of protein content and grain hardness.
- *Grain hardness*: as a rule, bread-making wheat should have hard, vitreous grains; biscuit-making wheat, on the contrary, has grains with soft, opaque kernels. Several simple methods exist for the determination of grain hardness.
- *Protein strength*: the breeder may choose from three tests to determine gluten quality: Zeleny sedimentation, SDS sedimentation or a mixograph test. All three are quick and simple methods and suitable for large numbers of samples.

8.5.2. Additional assessments

For specific reasons certain analyses may be done on a limited number of samples.

- *Alpha-amylase activity*. If the α-amylase activity is too low, the miller can readily increase it through the addition of malt flour or fungal amylases, but if it is too high there is no way of reducing it. Thus, α-amylase level in wheat has become a major quality factor. The analysis of the Falling Number is only useful if one, or both of the parents of a cross has high natural α-amylase activity, as in the case of descendents from the variety Professeur Marchal.

- *SDS-polyacrylamide gel-electrophoresis.* With this technique it is possible to determine the HMW glutenin subunits. A guide for wheat breeders who wish to develop varieties with improved bread-making quality is therefore to cross genotypes that have complementary good-quality subunits and to select progeny with an increased quality score.

8.5.3. Final breeding stages

In the final stage of quality breeding, the milling and baking characteristics of a restricted number of promising lines have to be assessed on an industrial scale, before introducing them on the commercial circuit. It is common practice for a breeding company to co-operate with the milling and baking industry, which can give a professional opinion on the practical value of a new variety.

References

1. *General publications*

ICC-Standards Methods of the International Association for Cereal Chemistry (ICC). Detmold, Verlag Moritz Schäfer (also available in a French and a German edition).

AACC (1990) Approved Methods (2 vol) 8th edn. St Paul, MN: American Association of Cereal Chemists.

Arbeitsgemeinschaft Getreideforschung (1987) Standard Methoden für Getreide, Mehl und Brot, 6th edn. Detmold: Verlag Moritz Schäfer.

Mauzé C, Richard M, Scotti G (1972) Guide pratique. Controle de la qualité des blés. Institut Technique des Céréales et des Fourragères, 177 pp.

2. *Ten-gram baking tests*

AACC (1990) Approved methods, Method 10–10B. Optimized-straight bread-baking method.

Meppelink EK (1981) Einsatzmöglichkeit des Mikrobackversuchs in der Weizenzüchtung. Getr Mehl u Brot 35: 107–109.

Shrogren MD, Finney KD (1984) Bread-making test for 10 g of flour. Cer Chem 61: 418–423.

3. *One hundred-gram baking test*

AACC (1990) Approved methods, Method 10-09. Basic straight-dough bread-baking method; Long fermentation.

AACC (1990) Approved methods, Method 10–10B. Optimized-straight bread-making method.

Finney KF (1984) An optimized, straight-dough, bread-making method after 44 years. Cer Chem 61: 20–27.

4. *Specific references*

Belderok B (1977) Der Mixograph als Hilfe bei der Weizenzüchtung auf Backqualität. Getr Mehl u Brot 31: 93–95.

Bell GDH, Bingham J (1957) Grain quality; a genetic and plant breeding character. Agric Rev 3: 10–22.

Bingham J (1962) The contribution of plant breeding to grain quality in wheat. In Recent advances in processing cereals, Monograph 16. London: Society Chemical Industry: 34–54.
Bolling H (1975) Zur Problematik der 'nichtbackfähigen Weizen' – eine Schnellmethode zur ihrer Erkennung. Die Mühle und Mischfuttertechnik 112: 256–257.
Cookson MA (1975) In Spicer A (ed), Bread: social, nutritional and agricultural aspects of wheaten bread. London: Appl Sci Publ, pp 347–348.
Dhaliwal AS, McRitchie F (1990) Contribution of protein fractions to dough handling properties of wheat-rye translocation cultivars. J Cereal Sci 12: 113–122.
Irvine GN (1975) Milling and baking quality. In Spicer A (ed), Bread: social, nutritional and agricultural aspects of wheaten bread. London: Appl Sci Publ, pp 115–126.
LeBrun J (1986) Mouture d'essai de blé tendre en laboratoire. Paris: Bureau Interprofessionnel d'Etudes Analytiques, BIPEA, Paris, 26 pp.
Moonen JHE, Scheepstra A, Graveland A (1982) Use of the SDS-sedimentation test and SDS-polyacrylamidegel electrophoresis for screening breeder's samples of wheat for bread-making quality. Euphytica 31: 677–690.
Payne PI, Corfield KG, Holt LM, Blackman JA (1981) Correlations between the inheritance of certain high-molecular weight subunits of glutenin and bread-making quality in progenies of six crosses of bread wheat. J Sci Food Agric 32: 51–60.
Payne PI, Lawrence GJ (1983) Catalogue of alleles for the complex gene loci Glu-A1, Glu-B1 and Glu-D1 which code for high-molecular-weight subunits of glutenin in hexaploid wheat. Cer Res Comm 11: 29–35.
Payne PI, Nightingale MA, Krattinger AF, Holt LM (1987) The relationship between HMW glutenin subunit composition and the bread-making quality of British-grown wheat varieties. J Sci Food Agric 40: 51–65.
Pelshenke P (1933) A short method for the determination of gluten quality in wheat. Cer Chem 10: 90–96.
Rimpau J, Röbbelen G (1978) Vererbung der Eigenschaft 'negatives Teigverhalten' von Clement und Maris Huntsman. Getr Mehl u Brot 32: 53–55.
Svensson G (1981) Varietal and environmental effects on wheat milling quality. Agri Hortique Genetica 39: 3–103.
Symes KJ (1965) The inheritance of grain hardness in wheat as measured by the particle size index. Aust J Agric Res 16: 113–123.
Zeleny L (1947) A simple sedimentation test for estimating the breadbaking and gluten qualities of wheat flour. Cer Chem 24: 465–475.

PART TWO

Breeding for bread-making quality in Europe

J. MESDAG[1] and DA. DONNER[2]
[1] *Former cereal breeder at*
the Foundation for Agricultural Plant Breeding
(SVP, now part of Plant Research International, Wageningen)
[2] *Cereal breeder and former secretary of*
The Netherlands Grain Centre (NGC, Wageningen)

General introduction to the Nordic countries

1. Climate

The climate of northern and northwestern Europe, which includes the Nordic countries (NC), is greatly influenced by the Gulf Stream. The climate is strongly maritime but with trends towards continental in eastern parts of the region.

The maritime climate of *Denmark* is marked by mild winters, cool summers and a fairly evenly distributed rainfall. The differences between regions are small. The vegetation period (the period with daily mean temperatures above +5 °C) varies between 200 and 220 days.

The climate of *Finland* is influenced both by the Atlantic and by the great continental landmass to its east; the maritime influence is strongest along the Baltic coastline. In agricultural regions annual rainfall is about 600 mm.

The climate of *Norway* is strongly influenced by the Gulf Stream. The high mountain ranges, however, cause great variations and regions may be found with a continental and others with a maritime climate, although the latter is prevalent. Typical is also the great variation in annual rainfall from one region to another. Along the coast, as far north as Trondheim, the vegetation period is 200 days; it decreases to the east and the north.

Sweden shows great climatic variation. The climate in Scania, for example, is about the same as in Denmark. Variations during summer are fairly small. The winter temperatures and the duration of winter varies considerable within the country, however. In eastern Sweden the annual rainfall is 500–600 mm, with an uneven distribution, which often results in severe early summer droughts. In western Sweden conditions are somewhat more favourable. The vegetation period varies from 210–220 days (Åkerberg, 1965).

2. Harvest and yield of wheat

The harvest of wheat in the Nordic countries currently totals more than 5.5 Mt annually, of which the Danish wheat crop alone comprises more than 3 Mt (Table NC.1). Denmark accounts for nearly one-half of the area under wheat (more than 400 000 ha, and even 650 000 ha in 1997) in the Nordic countries. Wheat production of the Nordic countries more than doubled during the period 1975–1995, due in particular to the expansion in Danish wheat

Table NC.1. Average yield (kg/ha) and production (1000 t) of winter and spring wheat in Nordic countries

		Denmark 1988–1990	Finland 1988–1990	Norway 1988–1990	Sweden 1986–1990
Yield	ww	7 180	3 720	4 560	5 880
	sw	5 030	3 130	3 860	4 330
Production	ww	3 024	81	25	1 399
	sw	62	392	145	316
	total	3 086	473	170	1 715

Source: Salovaara & Fjell (1995).

production, which has increased four-fold since 1970 and which has benefited from Denmark's membership of the EU (Salovaara & Fjell, 1995).

3. Use of the wheat harvest

Less than one-quarter of the total wheat harvest of the Nordic countries is used for human consumption. The main use of wheat is as an animal feed, either directly on the farm or as a component of industrially-produced animal feed. In the past, Denmark and Sweden have exported wheat on a regular basis, and occasionally Finland, whereas the volume of the Norwegian crop has never met domestic need (Salovaara & Fjell, 1995) (Table NC.2).

Table NC.2. Wheat export and import (1000 mt) from and to Nordic countries

	Denmark	Finland	Norway		Sweden
	1987–1990	1988–1990	1990	1991	1988–1990
Export	65	25[a]	0	0	472
Import	116	87[c]	159	130	56[bc]

[a] Export under international food-aid programme.
[b] Approximate value.
[c] Mainly US or Canadian wheat, including durum wheat for pasta production. Source: Salovaara & Fjell (1995).

4. Quality of Nordic wheat

There is much variation in quality due to location and year. The protein content of Nordic winter wheat seldom exceeds 12–13% (dry matter basis), whereas spring wheat may have an average protein content of 13–15%. Farmers are encouraged to produce high-protein wheat by using a split nitrogen application. A protein pricing system is applied in Sweden, Finland and Norway. The premium to the farmer for higher protein also makes spring-wheat growing somewhat more competitive to winter wheat. In the domestic wheat trade in Sweden and Norway, spring wheat goes under the term 'quality wheat', which relates to its higher protein content, among other properties. Practically all wheat cultivars in the Nordic countries are hard-kernel types and hence have good milling properties. Electrophoretic analyses have shown that the wheat varieties grown in northern Europe possess HMW subunits characterized by high quality scores in Payne's classification system.

5. Consumption of wheat

The per capita consumption figures for wheat as food in Denmark, Finland and Sweden are among the lowest recorded in Europe and the OECD[1] countries. Typically, Norway has a higher wheat consumption and lower rye consumption than its neighbouring countries. However, relatively low wheat consumption in Denmark, Finland and Sweden is compensated for in part by the relatively high consumption of rye, which is used not only for making crisp bread and whole rye bread but also for various 'mixed' or brown breads containing some rye flour (Table NC.3).

Table NC.3. Annual human consumption (kg per capita) of wheat, rye, barley and oats in Nordic countries

	Denmark	Finland	Norway	Sweden
wheat	56.0	58.4	80.8	68.6
rye	20.2	19.8	9.3	15.1
barley	0.1	2.0	0.4	0.4
oats	4.8	5.3	2.8	3.6
wheat + rye	76.2	78.2	90.1	83.7

Sources: OECD Food balance sheets (1991); Salovaara & Fjell (1995).

[1] OECD = Organisation for Economic Cooperation and Development.

6. Wheat flour usage

Most wheat flour is used by bakeries for making various types of bread. Alongside this industrial use, one-quarter to one-third of all wheat flour produced in Finland, Norway and Sweden is sold for household use. In Denmark the proportion is lower: namely 16% as compared with 35%, 30% and 25% for Finland, Norway and Sweden, respectively. Different types of flour products and flour mixes for home bakers are marketed. Most home baking flour is lower in protein content and quality than flour delivered to commercial bakers. Home baking is more common in the sparsely populated northern parts of Norway, Sweden and Finland than in the southern parts, where the consumers have more access to bakeries. In spite of this, home baking seems to have survived even in urban areas.

J.M.

References

Åkerberg E (1965) Plant husbandry in the Nordic Countries in relation to edaphic and climatic factors. Acta Agricultura Scandinavica Suppl 12: 7–23.

Salovaara HA, Fjell KM (1995) Wheat usage in northern Europe. In: Faridi H, Faubion JM (eds), Wheat end uses around the world. St Paul (MN): AACC: 149–166.

Norway

1. Introduction

Norwegian (N) agriculture has been shaped by its geographic position, far to the north, with short cool summers and long winters with extreme conditions. Early ripening in annual crops harvested for seed, and winterhardiness in perennial crops, are absolute requirements for varieties to be grown and these are the primary aims of plant breeders (Wexelsen, 1965). With increased importance of domestic wheat production during the last two decades, additional breeding objectives are: yield, disease and lodging resistance and bread-making quality.

Spring-sown varieties dominate in wheat. The values in Table N.1. demonstrate the decrease of wheat area between the 1930s and the 1970s and its gradual increase thereafter. Grain yields increased from 2000 kg/ha in the 1930s to a maximum of 5420 kg/ha in 1993.

On the basis of variety trials done in south east Norway from 1899 until 1960, Strand (1964) calculated the differences in yielding capacity of the varieties used at the beginning and the end of this period. He found a difference of 725 kg/ha for winter wheat, of which 52% could be attributed to genetic differences and the rest to improved husbandry. The corresponding figures for spring wheat were 425 kg/ha and 53%.

Likewise, on the basis of a great number of variety trials in the country, the improvement in grain yield during the period 1960–1992 was assessed

Table N.1. Wheat area (1000 ha), grain yield (t/ha) and total production (1000 t) in Norway during the 20th century

Year(s)	Area	Yield	Production
1924–1925	9	1.54	13
1934–1938	28	2.01	56
1948–1952	28	2.06	58
1961–1965	8	2.57	19
1969–1971	4	3.13	11
1979–1981	15	4.26	63
1989–1991	46	4.32	203
1995–1997	65	4.42	288

Sources: IIA and FAO statistics.

(Strand, 1994). For winter wheat the annual increase in recorded grain yield was 88 kg/ha and for spring wheat 75 kg/ha; genetic improvement of cultivars contributed 57% and 47% to this increase, respectively.

2. General remarks on wheat breeding

Plant breeding work in Norway began very early in the 20th century. Plant breeders were greatly influenced by work done at Svalöf (Sweden). Initially the work centered on selection in local varieties.

Hybridization in self-fertilizing crops was begun around 1920. Even though early ripening was already an absolute necessity, the introduction of combine harvesting in the 1950s strengthened the demands for this characteristic (Wexelsen, 1965). It is true to say that a short growing period is of crucial importance in this country; yielding ability is of secondary importance. Thus, the challenge of breeding for improved baking quality is primarily one of combining this characteristic with a restricted length of the growing period.

Traditionally, Norwegian wheat quality was not considered to be sufficient for bread making. Therefore imported high quality wheat was added to the home-grown wheat before milling. To increase the proportion of Norwegian wheat in the flour, which was about 30% in 1987, there was a need for new varieties with higher protein content and better protein quality. Although in the Norwegian wheat-breeding programmes the improvement of both protein content and quality was the aim, most effort has been invested in improving the second characteristic (Uhlen & Mosleth, 1987). The proportion of Norwegian wheat in the flour was 60% in 1997.

3. Breeding institutes

Traditionally all plant breeding in Norway was publicly funded, and cereal breeding was the responsibility of the Department of Crop Science (formerly Farm Crops Institute) at the Agricultural University of Norway, and the State Experiment Stations Møystad near Hamar and Voll (later transferred to Kvithamar) near Trondheim. Between 1955 and 1975 additional funds were provided by the Norwegian Agricultural Research Council, and in 1975 the cereal breeding activities were merged into a research programme with funding from public and semipublic organisations. In 1993 all cereal breeding was transferred to a private company, Norsk Kornforedling AS, located at the experiment station Bjorke near Hamar, but part of the breeding activities are still carried out on a contract basis by the public research institutions (Ringlund, 1998).

4. Winter wheat

The area of winter wheat was restricted in Norway until 1980, but has increased since then. Accordingly, the breeding work in winter wheat has been intensified.

Essential characteristics in wheats for cultivation in the Oppland/Hedmark region of Norway are good overwintering capacity, stiff straw, satisfactory grain quality and an average yield of 3 t/ha. In trials done between 1920 and 1930, a new variety, named Heid, met these requirements except for its weak straw. According to official government laboratory tests, baking strength was good: this wheat needed only a small addition of improving flours. It seems likely that the variety Heid is a selection from the Russian quality line Stavropol-193 (Glaerum, 1939). From the offspring of a cross of Stavropol-193 with the Finnish variety Labors Elite 05, the varieties Sigyn and Sigyn II were selected. This work was done at Møystad (Wexelsen, 1965). The variety Sigyn apparently inherited the good flour quality, which is also typical for Heid. Owing to its very stiff straw, Sigyn exceeded Heid in grain yield (Bjaanes, 1942, 1953a). In a comparison of the baking quality of a number of Norwegian varieties with that of Gluten, the baking quality of Sigyn and Heid was above average, but not equal to that of Gluten (Bjaanes, 1953b).

The main winter-wheat varieties grown during the period 1959–1981 are presented in Table N.2. The great influence of Swedish varieties is notable.

Table N.2. The main home-bred winter-wheat varieties grown in Norway between 1959 and 1997

Variety	Pedigree	Year of release
Sigyn II	Stavropol/Labors Elite 05	1950
Skjaldar	Sigyn/Fram II//Redcoat/3/Trond	1976
Rida	Sigyn/Fram II//Redcoat/3/Trond	1976
Kalle	Yorkstar/Trond//Mo67-38/3/ Kavkaz/4/Odin/3/Vakka//Fram II/Sigyn	1990

Sources: Skinnes (1982) and Ringlund (1998).

During the period 1956–1973, a total of 157 trials with winter cereals was carried out in the southeastern part of the country. A summary of the results for the main winter-wheat varieties is presented in Table N.3a and N.3b. It is evident that the agricultural value of the winter-wheat varieties improved from the years 1956–1973 to 1974–1981 in terms of grain yield, straw length, lodging resistance and earliness of ripening (Table N.3a). It is also interesting to note the high grain yield of the rye variety Kungsrug II in these trials. The Zeleny sedimentation value, as far as it had been determined, was on

Table N.3a. Winter-wheat varieties (and one winter-rye variety for comparison) in trials in southeastern Norway during 1956–1973

Variety	Grain yield (rel.)[a]	Zeleny value[b]	Straw length (cm)	Lodging (%)	Date of ripening in August
Gluten	82	–	103	35	14
Virtus	96	–	105	51	14
Odin	90	27	101	20	16
Sigun 11	80	–	107	51	13
Trond	100	32	95	31	11
Mo 64-43	98	26	97	12	10
Kungsrug II (*rye*)	124	9	116	29	12

[a] Grain yield 100% = 4200 kg/ha.
[b] Apparently not determined for some varieties.
Source: Skinnes (1982).

Table N.3b. Winter-wheat varieties in trials in eastern Norway during 1974–1981

Variety	Grain yield (rel.)[a]	Zeleny value	Straw length (cm)	Lodging %	Date of ripening in August
Trond	100[a]	34	89	39	10
Skjaldar	108	34	83	39	9
Rida	106	33	83	18	8

[a] Grain yield: 100% = 4290 kg/ha.
Source: Skinnes (1982).

a reasonable level for winter wheat. The improvements are still clearer in the two younger varieties, Skjaldar and Rida (Table N.3b). These varieties were grown on a considerable scale in Norway between 1977 and 1981: the maximum percentage of Skjaldar was 100 and that of Rida 40 (Skinnes, 1982).

A new winter-wheat variety, named Kalle, bred by the Department of Crop Science of the Agricultural University of Norway at Ås was released in 1990. It is suitable for areas with short growing periods and severe winters. It was derived from the cross Moystad 7044/Kavkaz/Moystad 7066. The grain has a soft endosperm with a protein quality that is superior to that of other varieties released in Norway. The grain yield is higher than that of Rida, but lower than that of the Swedish variety Folke (Uhlen, Ringlund & Skinnes, 1992).

Table N.3c. Winter-wheat varieties in trials in eastern Norway during 1993–1997

Variety	Country of origin	Year of release in Norway	Lodging (%)	Grain yield (rel.)[a]	Protein in grain (%)	Spec. SDS
Folke	Sweden	1987	64	98	10.6	5.49
Kalle	Norway	1990	54	84	11.0	5.03
Portal	Germany	1993	31	95	11.5	6.36
Rudolf	Sweden	1993	37	100	10.7	5.60
Mjolner	Sweden	1996	18	102	10.7	5.95
Bjorke	Sweden	1996	14	93	11.4	5.78
Terra	Denmark	1997	14	97	10.5	5.91

[a] Grain yield: 100% = 6330 kg/ha.
Source: Åssveen et al. (1998).

It is obvious that the mean grain yield in these trials (Table N.3c) is considerably higher than in those of 1956–1973 and 1974–1981. The straw stiffness of the more recently released winter-wheat varieties tested in eastern Norway during 1993–1997 was clearly improved (Table N.3c). It is interesting to note the great variation in Specific SDS Sedimentation values of those varieties. During this period the area of the Norwegian variety Kalle deminished from 30% in 1993 and 1994 towards a few percents in 1997 (Åssveen et al., 1998).

5. Spring wheat

5.1. Breeding of spring wheat at Moystad

In the Oppland region during 1925–1941, a number of Norwegian spring-wheat varieties was compared with Finnish and Swedish varieties (Glaerum, 1941). Grain yields of Froya and Ås were equal during 17 years of trials; their baking quality was satisfactory. Another variety, Froya II, gave a very good flour for making bread without any addition of foreign high quality wheat. Grain yield of the variety Froya II was 200–300 kg/ha higher than that of Froya and Ås, although they ripened practically at the same time

A series of crosses between the home-bred cultivars Fram I or Fram II and the foreign cultivars Pika, Marquis, Sopu or Diamant II was made by Bjaanes in 1936. Disease resistant individual plants were selected from F_2 progenies. Subsequent families were screened up to F_7 for agronomic value and baking quality, using the Pelshenke test. One outstanding heterogenic F_2-family was investigated in more detail, leading to the conclusion that high quality was not

Table N.4. Results of trials with spring-wheat varieties at Vollebekk during 1960–1974

Variety	Year of release	Grain yield (rel.)	Zeleny sediment value	Straw length (cm)	Lodging (%)	Duration development (days)
Diamant II	1938	88	39	92	52	115
Ås	1945	92	36	95	55	113
Norrona	1952	95	37	90	53	111
Drott	1954	95	39	92	18	119
Svenno	1954	93	54	84	28	114
Nora	1959	95	39	89	50	112
Rollo	1963	100[a]	49	80	12	110
Moystad	1966	98	46	87	25	112
Runar	1972	110	47	80	3	110
Reno	1975	112	47	78	0	113

[a] Grain yield: 100% = 3700 kg/ha.
Source: Strand (1975).

inherited recessively. This is in contradiction with the findings of Pelshenke in Germany (Bjaanes, 1951).

Selection in the offspring of the cross Fram II/Sopu resulted in the varieties Norrona and Nora, released in 1952 and 1959, respectively. In these varieties a valuable combination of earliness and yielding ability was created (Wexelsen, 1965).

The later variety Trym was selected from the progeny of the cross Fylgia/Huron (from Sweden and Canada, respectively); it was released in 1951. This variety had nearly the same straw stiffness as Diamant II, but exceeded it in yield by 6%; the difference in baking quality was small (Bjaanes, 1953b).

Table N.4. lists 10 main spring-wheat varieties grown successively in Norway in order of their year of release. Note how a number of important agricultural characteristics has been improved: grain yield; Zeleny sedimentation value; straw length; percentage of lodging; and, more or less, the duration of the growing period. The figures illustrate the struggle wheat breeders have to combine as many essential characteristics as possible (Strand, 1975).

5.2. Breeding of spring wheat at the Agricultural University of Norway

At the Department of Crop Science, selection for mildew resistance began as early as 1920. Crosses were made with a resistant line found in a local variety. Three varieties originated from this work: Fram II, released in 1939; Snogg II, released in 1940, and Ås II, released in 1945.

Table N.5. Spring-wheat variety trials in eastern Norway 1992–1997

Variety	Year of release	Breeder[a]	Lodging (%)	Dor-mancy	Grain yield[b]	Protein in grain	Specific SDS sediment
Tjalve	1987	Sv-W	4	6	95	13.4	5.53
Bastian	1989	NK	2	9	90	13.2	5.89
Polkka	1992	Sv-W	2	4	94	13.7	4.86
Brakar	1995	NK	14	11	95	13.0	5.59
Avle	1996	Sv-W	4	10	100	13.5	5.44
Vinjett		Sv-W	7	12	108	12.8	5.28

[a] Sv-W = Svalöf-Weibull (Sweden); NK = Norsk Kornforedling (Norway).
[b] Grain yield: 100% = 5240 kg/ha.
Source: Åssveen et al. (1998).

During 1945–1965 the wheat production in Norway was drastically reduced. The old varieties were not suited for combine harvesting. A new breeding programme was initiated by Strand in 1959 in which sprouting resistance, resistance to shattering and lodging, together with disease resistance, earliness and yield improvement were the main breeding objectives. This programme resulted in the varieties Rollo, Runar and Reno, released in 1963, 1972 and 1975, respectively. Semidwarf wheats from the CIMMYT Wheat Program were introduced in 1968, and the varieties Bastian and Brakar, both semidwarfs were released in 1989 and 1995, respectively (Ringlund, 1998).

Tjalve was the main variety during 1990–1997, with 70% of the spring wheat area in 1991 and about 50% in each of the years 1993–1997. The Norwegian variety Bastian was grown on 40% of the spring-wheat area in 1993 and on about 20% in each of the years 1994–1997; the other Norwegian variety, Brakar, was grown on a few percents of the area in 1997 (Åssveen et al., 1998). Thus, the position of home bred spring-wheat varieties in Norway during the 1990s is of importance, but the competition of foreign, especially Swedish, varieties is heavy.

It is obvious that lodging was on a low level in these trials and that the Specific SDS sedimentation value in general was high for these varieties (Table N.5).

6. Investigations on glutenin composition

A total of 13 spring-wheat varieties were investigated for their glutenin subunit composition (Uhlen, 1990a). Among these cultivars was one land variety, some varieties selected from a land variety, and varieties selected from crossing populations, all released between 1926 and 1975. The material is listed in Table N.6.

Uhlen concluded that the 13 spring-wheat varieties had a limited allelic variation on the *Glu-1* loci. Note that the *Glu-A1* subunit 2∗ was already present in the Norwegian land variety Borsum and (subsequently) in its derivates. This is in contrast with the absence of this subunit in land varieties of several other north and west European countries. It may be assumed that the variation in the Norwegian land varieties was limited to the subunits 2∗, 7+8, 7+9 and 2+12. The subunits 5+10 were introduced into the Norwegian spring-wheat varieties Norrona and Nora through the Finnish variety Sopu, which in its turn acquired the subunits through the Canadian cultivar Marquis.

Table N.6. HMW-subunit composition of Norwegian spring-wheat varieties

Variety	Pedigree	Year of release	HMW-glutenin subunit composition		
			Glu-A1	Glu-B1	Glu-D1
Borsum	Land variety		2*	7+8/7+9	2+12
Ås	Sel.from a land var.	1926	2*	7+9	2+12
Froya	Sel. from Borsum	1933	2*	7+8/7+9	2+12
Fram I	Jo3/MO07	1936	2*	7+8	2+12
Fram II	Jo3/MO07	1940	2*	7+8	2+12
Snogg II	Jo3/Sibirian/Ås	1940	2*	7+8/7+9	2+12
Trym	Fylgia/Huron	1951	2*	7+9	2+12
Norrona	Fram II/Sopu	1952	2*	7+8	5+10
Nora	Fram II/Sopu	1959	2*	7+8	5+10
Rollo	Kärn/Norrona	1963	2*	7+8	5+10
Moystad	Mo 043-40/Kärn II	1966	2*	6+8	2+12/5+10
Runar	Els/^{7}*Rollo	1972	2*	7+8	5+10
Reno	Tammi/Kärn II/Els	1975	1	7+9	5+10

Source: Uhlen (1990a).

A total of 212 breeding lines, obtained from the spring-wheat breeding programme in 1984–1986, were investigated for the effect of specific HMW glutenin subunits on bread-making quality. These lines had not been subjected to any selection for bread-making quality (Uhlen, 1990a). The results of these investigations are presented in Table N.7.

According to Uhlen, the greater variation in *Glu-1* alleles found in the breeding lines as compared with the varieties of spring wheat reflects the more extensive use of foreign germplasm in crosses during later decades. In particular, germplasm from CIMMYT was used in the Norwegian spring-

Table N.7. Frequencies (%) of the alleles at the loci *Glu-A1, Glu-B1* and *Glu-D1* for 13 varieties and 212 pure breeding lines of spring wheat

		13 varieties	212 breeding lines
Glu-A1	2*	92	74
	1	8	24
Glu-B1	6+8	8	12
	7+8	46	44
	7+9	23	12
	7+8/7+9	23	–
	7	–	2
	13+16	–	25
	17+18	–	2
	20	–	2
Glu-D1	2+12	54	24
	5+10	38	76
	2+12/5+10	8	–

Source: Uhlen (1990a).

wheat breeding programme. It is likely that the subunits 17+18 were transferred from this material. The presence of the subunits 13+16 could be attributed to a distinct CIMMYT line that was used as a crossing parent. In fact, the frequency of these subunits 13+16 in a world collection of 300 varieties that was investigated by Payne & Lawrence (1983) was very low. Uhlen presumed that the prevalence of subunits 5+10 in the breeding lines reflects the extent to which the varieties Runar and Reno contribute to the pedigrees of subsequent lines.

Approximately 25–27% of the variance in Zeleny sedimentation volume in this material could be related to the composition of HMW glutenin subunits. Although this estimate was lower than that reported for UK- and Spanish-grown wheat varieties, Uhlen considered SDS-PAGE as a useful screening test for bread-making quality in the Norwegian wheat breeding programme.

By relating several HMW subunits to Zeleny sedimentation value and extensogram values, it appeared that the quality of the lines with subunits 5+10 surpassed that of the lines with subunits 2+12; this finding is in agreement with that of many other investigators.

Among the HMW subunits encoded by the 1B chromosome, the subunits 13+16 were strongly related to good quality. Among the other *Glu-B1* subunits present in this material, both 6+8 and 7+9 seem to provide better quality than 7+8, at least in combination with the *Glu-D1* subunits 5+10. Further studies were intended to investigate whether there is a different interaction with the *Glu-D1* subunits 2+12.

Uhlen (1990b) examined the quantitative variation of the HMW glutenin subunits present in Norwegian wheat material. Electrophoretic patterns of HMW glutenin subunits for 111 spring-wheat breeding lines were analysed quantitatively by densitometer scanning. Significant differences in quantity between subunit bands were found, but only with a slight effect on the variation in Zeleny sedimentation volume. From these results Uhlen concluded that selection for bread-making quality according to the electrophoretic patterns of HMW glutenin subunits can be made qualitatively without taking the quantity of the subunits into consideration.

A set of 97 of the 212 breeding lines analysed for HMW glutenin subunit composition was analysed further for their gliadin composition using acid polyacrylamide gel electrophoresis (Mosleth & Uhlen, 1990). Of the 97 lines, 51 were grown in 1985 (85B2) and the other 46 lines were grown in 1986 (86B2). The data were analysed by the multivariate data analysis Partial Least Squares (PLS) regression to study relations between protein composition and bread-making quality. The PLS regression analyses were first carried out using the protein content and quantitative expressions (by use of densitometer scanning) of the electrophoregrams of HMW glutenin subunits and gliadins as the x-variables. Zeleny sedimentation volume or extensogram parameters were used as the y-variable. From these analyses, six quality-related gliadin bands were identified. Three ω-gliadin bands and one γ-gliadin band, corresponding to the gliadin block Gld1B8 (Gli-1Bk according to Metakovsky, 1991) were positively associated with the quality tests. Two other gliadin bands, one ω- and one γ-gliadin band, encoded by chromosome 1B, were negatively associated with the quality tests.

New PLS regression analyses were carried out using the protein content and qualitative expressions of the protein composition (presence or absence

Table N.8. The proportion (%) of the variance in Zeleny sedimentation volume that could be predicted by the three different sets of x-variables for breeding lines grown in 1985 and 1986

x-variables	Tested lines:	
	85B2 (n=51)	86B2 (n=46)
protein content	17	54
protein content + HMW glutenin units	39	64
protein content + gliadin bands	47	73

Source: Mosleth & Uhlen (1991).

of the different polypeptides) as x-variables and Zeleny sedimentation volume as the y-variable. The x-variables protein content, presence or absence of the different HMW glutenin subunits, and presence or absence of the six gliadin bands was added subsequently to the PLS analyses. The results are shown in Table N.8. for the lines grown in each year separately. The PLS analyses were evaluated by cross validation (Mosleth & Uhlen, 1991).

The figures in this table make clear that information on the protein content, the HMW glutenin subunits, as well as the six gliadin bands, contributed to the prediction of the quality test. Mosleth & Uhlen suggested that the difference in results between the two sets of lines could be due to the larger variation in the protein content in the '86B2' material than that in the '85B2' material. They commented that it is not known wether the quality associations were due to the gliadin bands themselves or to the Low Molecular Weight (LMW) glutenin subunits genetically linked with the gliadins.

In another set of 97 spring wheat lines breeding lines grown in 1990, the LMW glutenin subunits were investigated by SDS-PAGE (Mosleth et al., 1993). Quality-related LMW glutenin subunits were identified by multivariate analysis of the electrophoregrams; protein quality was evaluated by the Zeleny sedimentation test; and gliadin composition of the material was analysed by APAGE. The results indicated that the LMW glutenin subunits with the strongest correlation to quality were linked to the gliadin block Gli-1Bk, which was found to be related to flour quality. Mosleth et al.'s conclusion from this investigation (1993) as well as from an earlier investigation (Mosleth & Uhlen, 1990) was that the quality relations might be due to these specific gliadins themselves or to the LMW glutenin subunits genetically linked to them. Due to their ability to form intermolecular disulphide bonds, the LMW glutenin subunits most likely play an important role in the formation of the gluten network.

Research on gluten proteins in Norway has been continued at the Agricultural University of Norway and at the Norwegian Food Research Institute, both in Ås. Studies on the composition of LMW glutenin subunits in Norwegian varieties and breeding lines (Flæte, in preparation) have been carried out, and their influence on gluten quality is under investigation. Research is also initiated to study endosperm hardness and the effect of endosperm hardness and gluten quality of wheat for different end-use products. The aim is to provide further knowledge of the production and use of Norwegian quality wheat (Uhlen, 1998).

J.M.

Acknowledgments

The author wishes to thank Professor Kåre Ringlund and Dr Anne Kjersti Uhlen from the Agricultural University of Norway (Ås) for their valuable comments on this chapter and their help by providing additional information on wheat breeding in Norway.

References

Åssveen M, Linnerud H, Weiseth L (1998) Sorter och sortsprövning. In Abrahamsen U (ed), Jord- og plantekultur 1998. Landbruksforlaget: 37–43.

Bjaanes M (1942) Breeding of autumn wheat at Moistad Exp Farm. PBA 18: 823.

Bjaanes M (1951) Studies in spring wheat breeding. PBA 22: 1064.

Bjaanes M (1946, 1947) Trials with autumn wheat. Meld Stat Forsoksg Moistad. PBA 24: 1813, 1814.

Glaerum O (1939) Variety trials with autumn wheat and breeding of winter wheat. Med Statens Forsoks Moistad. PBA 18: 163.

Glaerum O (1941) Trial with spring wheat varieties. PBA 18: 862.

Metakovski EV (1991) Gliadin allele identification in common wheat II. Catalogue of gliadin alleles in common wheat. J Gen & Br 45: 325–344.

Mosleth E, Uhlen AK (1990) Identification of quality-related gliadins and prediction of bread-making quality of wheat from the electrophoretic patterns of gliadins and HMW subunits of glutenin. J Agr Sc 4: 27–45.

Mosleth E, Uhlen AK (1991) Associations between the composition of gliadins and HMW glutenin subunits and the gluten quality of wheat (*T. aestivum* L.). In Bushuk W, Tkachuk R (eds), Gluten Proteins, St Paul, MN: American Association of Cereal Chemists.

Mosleth E, Longva Å, Uhlen AK, Magnus EM (1993) Identification of quality-related Low Molecular Weight glutenins in bread wheats. In: Gluten Proteins (1993) Publ by Assoc of Cereal Research, Detmold: pp 539–548.

Payne PI, Lawrence GJ (1983) Catalogue of alleles for the complex gene loci, *Glu-A1*, *Glu-B1*, and *Glu-D1* which code for high-molecular-weight subunits of glutenin in hexaploid wheats. Cereal Research Communication 11: 29–35.

Ringlund K (1998) personal communication, Ås, Wageningen.

Skinnes H (1982) Variety trial with winter cereals in Eastern Norway during the period 1956–81. Meldinger fra Norges Landbrukshogskole 62 (18): 27.

Strand E (1964) Contributions by new varieties to the increase in yield level of small grains in South East Norway. Meldinger fra Norges Landbrukshogskole 43 (13), 16 pp.

Strand E (1975) Breeding and testing of spring wheat cultivars 1960–74. Meldinger fra Norges Landbrukshogskole 54 (22): 26.

Uhlen AK (1990a) The composition of HMW glutenin subunits in Norwegian wheats and their relation to bread-making quality. Norw J Agric Sc 4: 1–17.

Uhlen AK (1990b) Quantitative analyses of HMW glutenin subunits present in Norwegian wheats. Norw J Agric Sc 4: 19–26.

Uhlen AK, Mosleth E (1987) Influence of HMW subunits of glutenin on baking quality in wheat (I) Relationship between HMW glutenin subunits composition and baking quality in Norwegian breeding material. In Lasztity R, Békés F (eds), Proc 3rd Intern Workshop on Gluten Proteins, Budapest: pp 543–547.

Uhlen AK, Ringlund K, Skinnes H (1992) Kalle: a new early winter-wheat variety. Norsk Landbruksforsking 6: 275–283.
Wexelsen H (1965) Plant breeding in Norway. In Scandinavian Plant Breeding, publ. by the Organizing Committee for the Eucarpia Congress 1965, Acta Agriculturæ Scandinavica (suppl 12) Stockholm 1965: pp 72–96.

Sweden

1. Introduction

Climatic conditions for crop cultivation in Sweden (S) vary greatly. Cereal is grown mainly in the southern and central regions, on about 40% of the total area of arable crops. Wheat has never been a major crop in terms of cultivated area. Spring barley and spring oats each cover one-third of the total cereal area, whereas the autumn-sown bread cereals, i.e. wheat and rye, together cover about 30%. This latter percentage was more or less stable from 1920 until 1955, but among the bread cereals the culture of rye has gradually decreased in favour of wheat. The total area of wheat grown in Sweden increased considerably during the first half of the 20th century, whereas in the second half this area has remained more or less stable (Table S.1). The area under rye increased again after 1975, when breeding for sprouting resistance resulted in a better adaptation to the climate (Gunnarsson, 1986). This was of importance as the Swedish population consumes much rye, in various forms.

Table S.1. Wheat acreage (1000 ha) and production (1000 t) in Sweden between 1925 and 1997 and the distribution of winter wheat (ww) and spring wheat (sw) for a number of years

Year(s)	Area	Prod.	Year(s)	Area[a]			
				ww		sw	
1925–1926	138	281	1920	118	(7.2)	27	(1.6)
1934–1938	290	696	1932	214	(14.2)	64	(4.3)
1948–1952	323	677	1951	147	(10.9)	177	(13.1)
1961–1965	264	909	1965	228	(15.4)	60	(4.0)
1969–1971	259	958	1980	243		54	
1979–1981	252	1088	1985	190		98	
1989–1991	294	1825	91–95	232		36	
1995–1997	311	1880	1996	292		42	

[a] The figure between brackets represents the percentage of total cereal area.
Sources: IIA and FAO statistics, Broekhuizen (1969), Svensson (1987) and Yearbook of Agricultural Statistics 1997.

At the end of the 19th century, winter wheat was dominant and the area of spring wheat was very restricted. But the area under spring wheat increased gradually between 1900 and 1930. A further increase in spring wheat cultivation during the 1930s was due mainly to the availability of improved varieties

(Åkerman, 1954). With increased production of spring wheat, the insufficient baking quality of winter wheat varieties could be compensated for, more or less. Nowadays, the proportion of spring wheat grown varies, depending on the weather while drilling winter wheat; on average it is only 10–15% of the total wheat area.

As mentioned in the Chapter 'General introduction to the Nordic countries', under 'Consumption of wheat', the per capita consumption of wheat in Sweden is among the lowest in Europe and the OECD countries. However, this is compensated for in part by the relatively high consumption of rye. Under 'Harvest and yield of wheat' of the same chapter it was indicated that Sweden exports considerable amounts of wheat annually. However that wheat export has fallen during the 1990s as more wheat is being used for feed.

A general shift in appreciation from rye to wheat in bread consumption, together with a shift from home to industrial baking, in combination with a continuing development of industrial baking techniques, induced changes in screening methods for baking quality in wheat breeding (Olered & Hummel-Gumælis, 1975; Olered & Johansson, 1986). Farmers have been encouraged to grow high quality varieties by the an official price-scale system that has been in force since 1931. Moreover, an annual survey of harvest quality was introduced and continued for many years: see Table S.2 (Olered & Johansson, 1986).

Table S.2. Average values for bread-making quality of winter and spring wheat according to the survey of the Swedish bread grain crop in 1974 and 1978 compared to the bread-making quality of Kosack (ww) and Dragon (sw), tested between 1993 and 1997

	Protein content		Loaf volume		Sediment value
year	1974	1978	1974	1978	1974
winter wheat	11.7	12.6	736	719	26
spring wheat	13.0	13.3	942	950	52
years	1993–1997		1993–1997		
Kosack (ww)	11.3		769		
Dragon (sw)	13.6		970		

Sources: Olered & Hummel-Gumælis (1975); Olered & Johansson (1986); Larsson et al. (1997).

2. General review of wheat breeding for yield and quality

Selection in local varieties was already practised before 1880, some 25 years before Johannsen published (in 1903) a paper on the scientific background of selecting individual plants and testing their progenies separately (MacKey, 1963). The initial selections resulted in improved yields but decreased baking quality. The import of large quantities of high quality wheat was therefore necessary to compensate for the inferior quality of home grown wheat. The introduction of high yielding varieties bred in Germany and Great Britain did not contribute to an improvement of the baking quality. Moreover, those varieties were poorly adapted to Swedish conditions.

Combination breeding, involving crosses of indigenous and foreign varieties, was introduced around 1910. The first recorded was Squarehead / Swedish land wheat, aimed at combining stiff straw and high yields with winter-hardiness and quality, respectively. By successive recombination many varieties were obtained over a period of 60 years, of which the winter-wheat varieties Skandia IIIB, Odin and Diana have been widely grown (MacKey, 1963). Iduna (1911) and Standard (1921) are notable products of the program at Weibullsholm.

The Hungarian variety Bankuti 178 was introduced into discontinuous winter wheat backcrosses with the aim of improving bread-making quality. The same goal was pursued in spring wheat by incorporating the North American variety Marquis, reknown worldwide for its outstanding baking quality.

Baking tests played an important role in selection for bread-making quality; loaf volume, especially, was a criterion often used. Indirect tests to estimate baking quality in early stages of selection were undoubtedly employed, but the results have never been published. Noteworthy is the important role which Berg attributed to kernel-hardness and milling quality.

3. Breeding institutes

Two breeding institutions have been involved in Swedish wheat breeding since 1870: the private enterprise Weibull Plant Breeding at Landskrona and the Swedish Seed Association at Svalöf, working on a cooperative basis already in 1886. Both institutes had their branch stations at different locations in the country, and in view of great climatic differences the breeding and testing work at these locations is of major importance. The two amalgamated in 1993, continuing their breeding activities for a great number of agricultural and horticultural crops under the name Svalöf-Weibull AB.

4. Winter wheat

4.1. Winter-wheat breeding at Weibullsholm

At Weibullsholm, where winter-wheat breeding started in 1904, Berg introduced the Hungarian variety Bankuti 178[1] in 1928. Bankuti 178 was used as a parent in crosses with several winter strains because of its excellent milling and baking quality, and its outstanding resistance to various leaf diseases. Berg had initiated breeding to improve bread-making quality. F_2-plants were selected and backcrossed with Äring and Ankar II. After a thorough testing programme of these backcross progenies one selection emerged as the very best; it was released in 1943 under the name Eroica.

The improved baking quality of Eroica and its milling properties in particular, were a noteworthy advance on existing varieties. Eroica became very popular, covering some 90% of the winter-wheat area in southern Sweden, as well as in Denmark during several years. It was also grown in northern Germany.

Following this first result, a range of varieties, derived from crosses of Eroica and closely related strains, were marketed. Among these, the variety Banco, released in 1953, was outstanding. It showed a clear transgression in grain yield and baking quality (Table S.5). Fajersson (1994) suggested that it was probably the presence of Bankuti 178 from two sides of Banco's pedigree that resulted in the clear transgression in baking quality, as well as in kernel yield and protein content, compared with Eroica, which has Bankuti 178 from only one side in its pedigree.

Through the offspring of Bankuti 178 a hard endosperm was introduced in Swedish wheat varieties, resulting in a higher extraction rate during the milling process. Berg started systematic selection on kernel hardness.

Fajersson, Berg's successor in 1946, proceeded along the lines set out by his predecessor. He produced the great variety Starke, released in 1959, derived way back from Squarehead and Swedish land wheat, again with the strong influence of Bankuti 178 via Eroica and via a sister line of Banco. Starke strongly dominated Swedish and Danish winter-wheat production for a decade and was also grown in Poland and Germany (Fajersson, 1964). Starke II, based on a single plant selection from Starke, was an improvement in respect of winter-hardiness, straw stiffness and grain size. The programme of intercrossing more or less closely related derivatives of Starke (some of which were introduced on the market, e.g. Holme) yielded the successful variety Folke, released in 1982 (Table S.6).

[1] Bankuti 178 was presumed to be derived from a spontaneous cross between Marquis and a Hungarian winter wheat (Berg, 1940, cited by Fajersson, 1994).

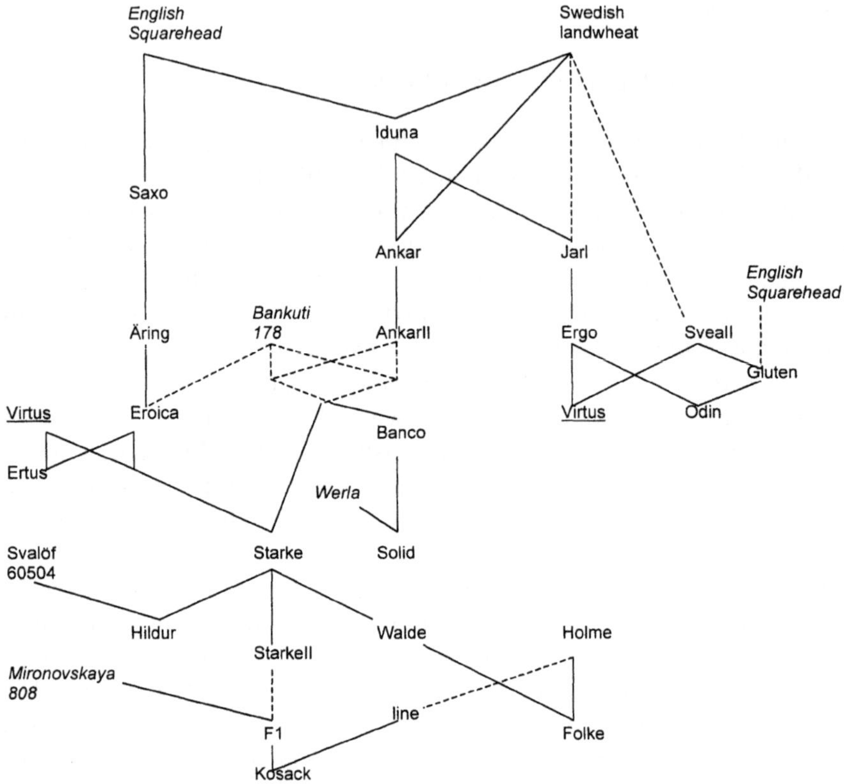

Figure S. 1. Simplified pedigree of Kosack.
Foreign varieties are printed in italics. Underlined variety names are derived from parents whose origin is shown elsewhere in the same Figure. Unbroken lines represent a direct descendence, broken lines represent an indirect descendence.

With Mironovskaya–808, an additional source of baking quality was introduced into the Weibull winter-wheat breeding programme. From cross-progenies that included mildew-resistant material based on Starke II and Holme, the variety Kosack, released in 1984, was selected on the evidence of outyielding its parents as well as Folke. Kosack has a hard endosperm and is easy to mill; its loaf volume is slightly better than that of Folke (Table S.6). Soon after its release, Kosack became very popular; it covered 56% of the area under winter wheat in 1987 and 88% in 1994. A simplified pedigree of Kosack is presented in Figure S.1.

Table S.3. Loaf volume of main winter-wheat varieties grown in southern Sweden during the 1940s and 1950s

Variety	Year of release	Period of trial	Protein content (%)	Loaf volume	
				ml	rel.
Sammetsvete[a]		1931–1950	11.9	677	100
Skandia IIIB	1939	1944–1950	10.2	608	90
Ergo I;II	1934;1949	1935–1950	10.7	551	81

[a] Sammetsvete was used as a standard variety in baking trials at Svalöf; this variety was one of the best Swedish varieties as far as baking quality was concerned.
Source: Åkerman (1954).

4.2. Winter-wheat breeding at Svalöf

The Swedish Seed Association at Svalöf has been involved in winter-wheat breeding since 1886 (Andersson, 1963). It was supported by scientific research in plant breeding methodology. A cereal laboratory was founded in 1930 and systematic investigations into the baking quality of wheat varieties have been carried out since then. Loaf volume was the most important criterion for quality and the Swedes were among the first to use this criterion for investigating the quality of wheat (Table S.3.); in many other countries indirect tests were commonly used. The variety Odin, released in 1950, was the first outstanding result, and became rather widely accepted in cultivation in the period 1950–1965 (Fajersson, 1963). Solid, released in 1973, was a further step forward, both in yield and quality. Hildur, like Solid, based on Banco, was a definite improvement in quality over Holme, but not in grain yield. Data on yield and quality of these varieties are presented in Tables S.4 and S.5.

Table S.4. Baking quality of the main winter-wheat varieties grown in mid-Sweden during the 1940s and 1950s

	Period of trial	Protein content (%)	Loaf volume	
			ml	rel[a]
Odin	1946–1950	10.2	609	90
Aros	1948–1950	10.1	617	91

[a] The loaf volume is related to that of Sammetsvete (100%) (see Table S.3.).
Source: Åkerman (1954).

Table S.5. Productivity and baking value of Swedish winter-wheat varieties

Variety	Year of release	Productivity kg/ha	Productivity rel.[a]	Loaf volume ml	Loaf volume rel.[a]	Protein content (%)
Iduna	1912	4250	100			
Standard	1921	4813	113			
Eroica	1943	5125	121		100	
Banco	1953	5375	127		105	
Starke	1959	5875	138		101	
Source: Fajersson (1963, 1964)						
Ergo	1934		100		100	
Odin	1950		106		106	
Source: Fajersson (1963)						
Starke II	1970		100	751		12.1
Solid	1973		103	757		12.6
Source: Olsson (1973)						
Holme	1973	6115	100	722	100	12.3
Hildur	1978	6060	99	758	105	12.2
Source: Olsson & Kristiansson (1979)						

[a] Relative values for productivity and loaf volume are related to varieties within the same block.

4.3. Yield and quality of Swedish winter-wheat cultivars

Data on yield and quality of varieties released from 1912 to 1978 are presented in Table S.5. An increase in grain productivity of almost 40% over some 45 years is apparent. From the comparison of the data of Odin and Ergo it appears that both productivity and loaf volume of Odin were better than for Ergo, which was one of Odin's parents. Compared to Starke II, Solid was an improvement both in yield and protein content. It is interesting that not only the kernel yield of Solid is slightly higher than that of Starke II but also the protein content of its kernel is somewhat higher, namely 0.5%. Data in Table S.6. refer to quality only. Apparently, once baking quality had been brought to a satisfactory level, breeders were able to maintain that level while raising the yielding capacity considerably.

Table S.6. Baking trials with samples from official trials (A, B) and from Weibullsholm (C)

	Variety	Year of release	No. of trials	Period of trial	Loaf volume ml	Loaf volume rel.[a]	Protein content (%)
A	Starke II	1970	11	1977–1979	728	100	
	Folke	1982	11	1977–1979	749	103	
	Holme	1973	17	1977–1981	781	100	
	Folke	1982	17	1977–1981	777	99	
Source: Fajersson & Svensson (1982b)							
B	Holme	1973		1981–1984	755	100	11.0
	Folke	1982		1981–1984	760	101	10.7
	Kosack	1984		1981–1984	780	103	10.6
Source: Fajersson & Svensson (1985)							
C	Starke II	1970	3		925	100	11.8
	Kosack	1984	3		870	94	11.4
	Holme	1973	9		935	100	12.1
	Kosack	1984	9		895	96	11.9
Source: Fajersson & Svensson (1985)							

[a] Relative values for productivity and loaf volume are related to varieties within the same block.

5. Spring wheat

5.1. Pedigree of spring-wheat cultivars

As for winter wheat, foreign varieties with good bread-making qualities were used in the breeding programmes of spring wheat by both Swedish cereal breeding institutes. The spring wheat line Cahn 1833 was one such wheat used as a basis for breeding. It was introduced at Weibull in 1927 with a wheat collection of mainly Russian and North American origin, which had been gathered by Cahn. Although the exact origin of Cahn 1833 is not known, it has been described by Berg as a Marquis type of wheat and it was noted for its very good gluten quality (Fajersson, 1950). Other foreign varieties used by Swedish breeders were Marquis from the USA, Svalöfs Kolben (selected from Heines Kolben, in its turn selected from the French land variety Saumur de Mars), Roter Schlanstedter from Germany, Kolibri from Germany and Sicco (of moderate baking quality) from the Netherlands.

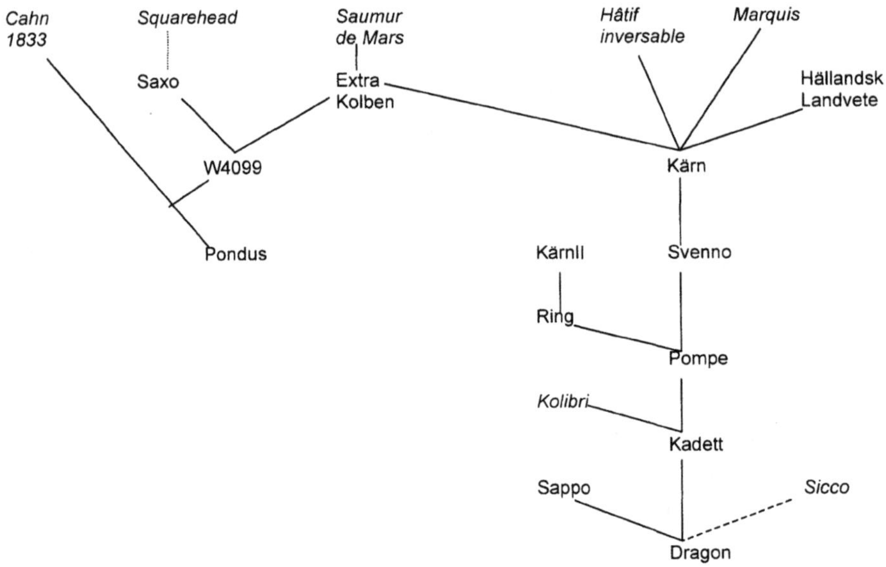

Figure S. 2. Simplified pedigree of the spring wheat Dragon (foreign varieties are printed in italics).

Other varieties contributed to improving yield, straw stiffness, disease resistance and early maturity. It is remarkable that a number of these introductions occured so early in the history of plant breeding. As was the case with winter wheat, breeders succeeded in improving the bread-making quality of Swedish spring-wheat varieties, while, at the same time, considerably increasing grain yield. To illustrate the background of Swedish spring wheats, the pedigree of the variety Dragon, released in 1990, is presented in Figure S.2.

5.2. Yield and quality of spring-wheat cultivars

The primary aim of Berg's breeding of spring wheat at Weibullsholm was to create varieties combining a better straw stiffness and higher yields with the existing high quality level. One approach adopted early on by Berg to attain this goal was the transfer of better agronomic characteristics to spring wheat by crosses with winter wheat (Lundin et al., 1970). But Berg also built up new spring-wheat material by introducing French (Hâtif inversable), German (Extra Kolben) and Canadian (Marquis) sources in crosses with Swedish land wheat. From this four-way cross varieties such as Kärn and Svenno were selected (Svensson, personal communication, 1998).

Åkerman (1951) stressed that the breeding of spring wheat at Svalöf was not primarily based on Swedish land varieties, although it was common know-

Table S.7. Results of breeding spring-wheat varieties in Sweden

Variety	Year of release	Resistance to lodging	Grain yield (rel.)	Loaf volume (rel.)
Halland LV		3	76	90
Svalöfs Kolben	1909	4.5	74	103
Atle	1936	6	91	93
Diamant II	1938	6	91	93
Kärn II	1947	7	100	99
Svenno	1954	7.5	100	98
Ring	1961	8	102	98
Pompe	1969	8.5	107	

Sources: Fajersson (1963) and Lundin et al. (1970).

Table S.8. Baking trials under application of optimal bromate with grain samples from official trials, baked in different laboratories

Variety	No. of trials	Protein in flour (%)	Period of trials	No. of trials	Loaf volume	
					ml	rel.
Drabant	7	12.5	1977–1981	22	968	100
Kadett		12.6			906	94
Pompe	6	13.1	1977–1981	22	991	100
Kadett		12.5			906	91

Source: Fajersson & Svensson (1982a).

ledge that their baking quality was good. Rather, the early ripening spring-wheat variety Svalöfs Kolben was primarily used as a basis; it has a very good baking quality. Svalöfs Kolben was selected from Heines Kolben, which in its turn was selected from Saumur de Mars, a selection from a French land variety (Åkerman, 1931; Wienhues & Giessen, 1957). Thus one could say that the quality of Svalöfs Kolben is of French origin.

Yield per hectare has risen by some 40% compared to the Halland land variety, which was cultivated until about 1910; Svalöfs Kolben was not any better in yield, but had a much better quality. Varieties successfully marketed since 1936 combined increased yield levels with gradually improved quality (Table S.7).

The variety Kadett, released in 1981, was very interesting because it out-yielded Drabant and Pompe by some 10% (!). However, loaf volume in standard baking tests was smaller for Kadett than for the two other varieties (Table S.8). This smaller loaf volume was not expected, micro-tests (Pelshenke & Zeleny) indicated that Kadett has a better protein quality than Pompe and

Drabant. Moreover the mixograph indicated that Kadett has strong rheological characteristics, obviously too strong for the Swedish baking industry. (Anticipating on paragraph 6.2: Kadett has the HMW subunits 5+10 on *Glu-D1*).

In a number of trials it was demonstrated that intensive dough mixing favoured the loaf volume of Kadett more than that of the two other varieties. In a trial with very intensive dough mixing, the loaf volume of Kadett was higher than that of the other two varieties: Pompe 100%, Drabant 101% and Kadett 108%. In a few large baking factories, where intensive dough handling practices are used, the results with Kadett are good; this is in agreement with the previous trials.

In this way, a new aspect was introduced into the testing procedure for wheat varieties or lines, namely the intensity of dough mixing, which should be in agreement with the recipe employed by bakers and the baking industry: either a low, an intermediate or an intensive mixing procedure. For breeders who export their varieties to different countries this aspect may be of interest (Fajersson & Svensson, 1982a).

Another aspect is that in Sweden, Finland and Norway a considerable proportion of wheat flour is sold for household use. As mentioned in the Chapter on the Nordic Countries (Section 6), for Sweden this amounts to 25% of the wheat flour production. It is reasonable to suppose that the kneading intensity in household mixers is lower than in industrial processing, so that it may be assumed that a variety like Kadett will be handled in a sub-optimally way. Therefore, in agreement with the Swedish milling and baking industry, wheat breeders in Sweden returned to supplying spring wheats demanding only moderate kneeding intensity. Dragon has been the dominant variety since 1991. This variety has the HMW subunits 2+12 (Svensson, personal communication, 1998).

A good 10 years after these trials, Johansson (1994b) once more paid attention to the special behaviour of Kadett during dough mixing, particularly in connection with its special HMW glutenin composition (see Section 6.1 below).

Generally speaking, it may be stated that while the grain yield in the varieties developed at Svalöf was successfully raised their baking quality lagged behind. This may be seen in the data presented in Table S.9.

The variety Prins, released in 1965, was compared with its parents Diamant II and Kärn II and a landwheat by Persson et al. (1986). Prins proved to possess outstanding bread-making quality as expressed in loaf volume but its protein content was comparatively low (Table S.10). In practice, Prins could not compete with Ring, a variety released four years before.

Table S.9. Kernel yield and baking quality of the main spring-wheat varieties released at Svalöf during the first half of the 20th century

Variety	Year of release	Rel. grain yield (rel.)	Protein content (%)	Loaf volume	
				ml	rel. (%)
Halland LV	< 1900	?	13.1	672	84
Sv.Kolben	1909	100	13.2	800	100
Extra Kolben II	1926	115	12.6	714	89
Diamant II	1938	117	12.6	699	87
Progress	1942	123	12.6	712	89

Source: Åkerman (1951).

Table S.10. Grain yield and baking quality of the spring-wheat variety Prins compared with its parents

Variety	Grain yield		Protein content (%)		Loaf volume	
	kg/ha	rel. (%)	kernel	flour	ml	rel. (%)
Land wheat	3 390	100	14.9	11.9	746	100
Diamant II	3 640	107	14.8	11.9	791	106
Kärn II	4 000	118	13.7	11.2	815	109
Prins	4 170	123	13.4	10.9	852	114

Source: Persson et al. (1986).

The Swedish varieties Atle and Svenno were grown in the UK to a considerable extent in the 1950s and early 1960s. This was the case for Sappo in the 1970s.

5.3. Yield and protein content of recent spring-wheat cultivars

A spring-wheat variety with higher protein content in the kernel was released by Weibull in 1992; it was named Dacke (CI5484/Pompe//Trippel/3/W17269/4/W19151). Data in Table S.11 show that the protein content in the kernel of Dacke is higher than that of Dragon and Drabant. It is 1.1 percentage point higher than that of Drabant, whereas their kernel yield is about the same. The gluten content of Dacke (37%) is much higher than that of Drabant (31%). The loaf volume of the three varieties is very high, Dacke being the highest with 103%. A sister line of Dacke, named Sport, was released in 1993. Sport has a protein content which is 2.5% higher than that of Drabant, but the grain yield is 20% lower. The higher protein content in the endosperm seems to be mainly the result of a more efficient transfer of protein from the straw to the kernel (Svensson, 1989). Sport is grown to some extent; it is of special interest to ecological farmers.

Table S.11. The content of raw protein in the kernel of some recent spring-wheat varieties (the mean of 72 official trials), the kernel yield (the mean of 86 official trials) and the loaf volume

Variety	Year of release	Protein content (%)	Kernel yield		Loaf volume	
			kg/ha	rel. (%)	ml	rel. (%)
Drabant	1972	13.0	5 450	100	995	100
Dragon	1990	13.3	5 940	107	992	100
Dacke	1992	14.1	5 540	102	1 023	103
Sport	1993	15.5	4 400	81	1 170	117

Source: Svensson (1992a).

This introduction of spring-wheat varieties with a higher protein content in the kernel was not the first in Sweden, as already in the early 1970s some spring-wheat varieties with a genetic potential for higher protein content – up to 1% more – were available (Snabbe, Amy, Sonnet). However, the yield of these varieties was somewhat lower and farmers showed little interest in them.

To conclude, opportunities for genetically increasing protein content in spring wheat exist, but their effect is restricted in comparison with environmental effects (Johansson, 1988).

6. Fundamental research on protein quality

6.1. HMW glutenin composition of Swedish wheats

The influence of HMW glutenin subunits on dough characteristics and bread-making quality of wheat grown under Swedish conditions was investigated by Johansson et al. (1993, 1994, 1995a). Subunits that appear to be of special interest for Swedish bread wheat are subunit 21∗ on chromosome 1A, the subunit pair 14+15 on chromosome 1B and the sub unitpair 5+10 on chromosome 1D. The newly described subunit 21∗ has not been encountered in other European wheats.

The HMW glutenin subunit composition of a restricted number of widely grown – both older and recent – Swedish varieties of winter and spring wheat, together with a great number of modern breeding lines was investigated by Johansson et al. (1995b). Cultivars released before 1960 were not included in the investigation as they often contain a mixture of subunits within one cultivar. Therefore the variation in baking quality of the material under study was relatively restricted, which may be of importance for the interpretation of the results. A summary of the results is presented in Table S.12.

Variation in HMW glutenin subunits within the three groups of material indicates some interesting differences:

Table S.12. Frequencies of HMW glutenin alleles and *Glu-1* scores of Swedish winter- and spring-wheat cultivars and breeding lines

Locus	HMW glut. subunits	*Glu-1* score	Older cultivars	Current cultivars	Breeding lines
Winter types (*number*)			$n = 9$	$n = 5$	$n = 112$
Glu-A1	0	1	44	20	16
	1	3	11	0	4
	2*	3	44	80	80
Glu-B1	6+8	1	67	80	61
	7	1	0	0	3
	7+8	3	0	0	6
	7+9	2	33	20	30
	13+16	3	0	0	0
	14+15	?	0	0	0
	17+18	3	0	0	0
Glu-D1	2+12	2	89	80	93
	4+12	1	0	0	1
	5+10	4	11	20	6
Spring types (*number*)			$n = 5$	$n = 5$	$n = 110$
Glu-A1	0	1	0	0	2
	1	3	0	20	12
	2*	3	100	80	83
	21*	?	0	0	3
Glu-B1	6+8	1	40	0	5
	7	1	0	0	1
	7+8	3	20	0	3
	7+9	2	40	60	53
	13+16	3	0	0	2
	14+15	?	0	40	34
	17+18	3	0	0	2
Glu-D1	2+12	2	100	80	72
	4+12	1	0	0	0
	5+10	4	0	20	28

Source: Johansson et al. (1995b).

- in the winter-wheat material, neither the subunit 21∗ nor the subunits 14+15 were found in any genotypes of either older and recent cultivars or breeding lines; the subunits 5+10 were found in one older variety (Virgo) and in one more recent variety (Portal, of German origin);

- in the spring-wheat material, the subunit 21∗ was only found in three of the breeding lines; the subunits 14+15 were identified in two more recent varieties (Dacke and Sport) and in 38 breeding lines;
- the subunits 5+10 were not identified in the older varieties, but were in the more recent variety Kadett and in 31 of the breeding lines.

These findings indicate that:

- variation in subunit composition within the Swedish winter-wheat material is limited;
- variation within the more recent spring-wheat varieties is larger than within the older ones;
- variation within the spring-wheat breeding lines is even larger than within the spring-wheat varieties.

Johansson et al. concluded that the increased variation in the breeding lines should provide breeders with greater opportunities for incorporating new subunit combinations in bread wheats that would achieve the higher gluten strength requested by the milling and baking industry. But they added: 'However, the best subunit combination is dependent on the protein concentration of the cultivar, the rest of the genetic background, the yearly weather fluctuations and other factors. The influence of weather fluctuations is of considerable significance in Sweden, because of varying conditions during the growing season. More research has to be done in this area before a satisfactory recommendation of a HMW subunit combination can be made'.

6.2. HMW glutenin subunits related to quality characteristics

There are some striking differences in the evaluation of HMW glutenin subunits of Swedish (Johansson et al., 1995b) and Norwegian (Uhlen, 1990) bread-wheat varieties as compared to the findings of Payne et al. for wheats grown in the UK.

The *Glu-A1* subunit 2∗, to which Payne assigned a quality score 3, showed only a slight but insignificant correlation with Zeleny sedimentation in Johansson's studies. For the subunit 1, also evaluated with score 3 by Payne, Johansson found a negative (but insiginificant) correlation with gluten strength. Moonen et al. (1983), as well as Pogna et al. (1989) and Branlard et al. (1992), gave a lower score to subunit 1 than to subunit 2∗. On the other hand, Johansson found a high positive correlation between the newly described subunit 21∗ and sedimentation value.

The *Glu-B1* subunits 7+8, 7+9 and 6+8, evaluated by Payne with a score 3, 2 and 1, respectively, were not significantly correlated to sedimentation value in the Swedish studies. Johansson et al. (1993) found a positive correlation between subunit 14+15 (not described by Payne) and gluten strength. The evaluation of subunit 13+16 by Payne (score 3) more or less agrees with that of Uhlen, who found a positive correlation with sedimentation value in Norwegian material. For the subunit 17+18 (score 3 by Payne), however, Uhlen found a negative correlation with quality parameters in Norway.

The evaluation of the *Glu-D1* subunits is under discussion in Sweden: the most negative correlation with gluten strength was found for subunit 2+12 (score 2 by Payne) and the most positive correlation for subunit 5+10 (score 4 by Payne).

Johansson et al. (1994) pointed out that most of the Swedish varieties with subunit 5+10 require high kneeding intensities to produce optimal loaf volumes. This is in agreement with the findings of Fajersson & Svensson (1982a) (see Section 5.1) for the variety Kadett (subunit compostion 1,7+9, 5+10).

Johansson et al. suggested that the introduction of cultivars with subunits such as 5+10 may be an answer to bakers' demands for a higher flour quality.

A possible explanation for the different evaluation of Johansson and Uhlen compared to that of Payne et al. may be the lack of variation in baking quality in the Swedish and Norwegian material. All varieties in the studies of Johansson and Uhlen were relatively young, and thus of improved quality, whereas Payne et al. studied a larger group of varieties covering more years and a wider range in quality. This leads to more clear-cut judgments about the value of the different subunits. Moreover, the results may be influenced by the conditions under which the material was grown, such as day length, which affects growth rythm, and consequently, grain filling of the wheat.

6.3. HMW glutenin subunits and D-zone omega gliadins: the relation with quality characteristics

Johansson (1996) compared three electrophoretical methods for analysing D-zone *omega* gliadins: A-PAGE of gliadins; SDS-PAGE of gliadins; and SDS-PAGE of total proteins. In the latter method, the gel showed the HMW subunits of glutenin and a number of distinguishable D-zone *omega* gliadins; both protein patterns can be designated in their own specific way as described in literature. A broad collection of spring and winter wheats was investigated with this method (see Table S.13).

The conclusion to be drawn from these data is that the proportion of variation in Zeleny sedimentation volume explained by the HMW subunits of glutenin (a) is rather different for the three groups of material; it is generally low, especially for the group of 118 advanced winter wheat lines (SWB). The proportion of the variation in Zeleny sedimentation volume explained by the HMW subunits and the D-zone omega gliadins, together, (b) is still rather low and the additional explanation for group SSB is lower than for group SWB.

Table S.13. The proportion of the variation in Zeleny sedimentation volume explained by the HMW subunits of glutenin (X), or by the HMW subunits and D-zone omega gliadins together (Y), or by these two factors along with the protein concentration (Z)

Group of material tested	Proportion of variation (%) explained by:		
	X	Y	Z
181 SWCB entries[a]	10	21	38
280 SSB entries[b]	14	18	47
118 SWB entries[c]	4	18	37
Total 579 entries	9	19	41

[a] Spring and winter wheat cultivars and breeding lines of various origin.
[b] Swedish spring-wheat breeding lines from the 1994 field trials of Svalöf-Weibull AB at Landskrona.
[c] Swedish winter-wheat breeding lines from the 1994 field trials of Svalöf-Weibull AB at Landskrona.
Source: Johansson (1996).

For the total of 579 entries, one can state that the percentage of variation in Zeleny sedimentation volume explained by the HMW subunits (a) is approximately equal to that explained by the other protein group (b). Johansson concluded that both protein patterns may be used as parameters in breeding for bread-making quality.

It is unfortunate that these comparisons were made for sedimentation volume and not for the *specific* sedimentation volume, which was used in a number of other comparisons in the same publication.

Positive relationships were found between some specific gliadin bands (d7 and d2d4) and the specific Zeleny volume. The literature cites the suggestion that the variation in Zeleny sedimentation volume could be due either to the D-zone *omega* gliadins themselves or the LMW glutenin subunits associated with them. Following the former suggestion and the results of her own investigations, Johansson concludes that it is important for the breeder to know the composition of either the *omega* gliadins or the LMW glutenin subunits, since a significant part of the variation in bread-making quality is determined by these proteins.

J.M.

Acknowledgments

The author wishes to thank Dr Gunnar Svensson (Sveriges Lantbruksuniversitet, Alnarp) for his valuable comments on the manuscript of this chapter and for providing additional information on wheat breeding in Sweden.

References

Andersson G (1963) The Swedish Seed Association, its history and organization. In Åkerberg E, Hagberg A, Olsson G, Tedin O (eds), Recent Plant Breeding Research. Stockholm: Almqvist & Wiksell; New York, London: John Wiley & Sons, pp 9–23.

Åkerman A (1931) Über die Backfähigkeit Schwedischer Weizensorten und die Möglichkeit ihrer Verbesserung durch Züchtung. Der Züchter 3: 1–8.

Åkerman A (1948) The breeding of wheat. In Åkerman A, Tedin O, Fröier K (eds), Svalöf 1886–1946; History and present problems. Lund, pp 72–97.

Åkerman A (1951) Fünfzig Jahre Weizenzüchtung in Schweden. Z f Pflanzenz 29: 346–365.

Åkerman A (1954) Qualitätszüchtung bei Weizen. In Bericht über die Tagung in Detmold am 9. und 10. Februar 1954: Arbeitsgemeinschaft Getreideforschung E.V., Detmold, pp 205–220.

Berg SO (1947) Is the degree of grittiness of wheat flour mainly a varietal character? Cereal Chem 24: 274–282.

Branlard G, Pierre J, Rousset M (1992) Selection indices for quality evaluation in wheat breeding. Theor and Appl Gen 84: 57–64.

Carlson J (1970) The distribution of the cereal varieties in Sweden during the last 25 years. Agri Hort Gen 28: 49–54.

Fajersson F (1950) Weibulls original Pondusvarvete. Agri Hort Gen 8: 119–152.

Fajersson F (1963) Methods and achievements in Swedish wheat breeding. In MacKey J (ed), Proceedings of the Second International Wheat Genetics Symposium, Lund, pp 11–26.

Fajersson F (1964) Weibulls Starke Winterweizen – eine Sorte für Schweden, Dänemark und gewisse Gebiete Deutschlands und Polens. Agri Hort Gen 22: 5–28.

Fajersson F (1994) Transgressionszüchtung für gute Qualitäts- und Anbaueigenschaften. Die Mühle und Mischfuttertechnik 131 (38): 115–116.

Fajersson F, Svensson G (1982a) Weibulls Kadett – eine neue Sommerweizensorte für Süd-Schweden, die BRD und Dänemark. Agri Hort Gen 40: 29–50.

Fajersson F, Svensson G (1982b) Weibulls Folke Höstvete. Agri Hort Gen 40: 5–28.

Fajersson F, Svensson G (1985) Weibulls Kosack Höstvete. Agri Hort Gen 43: 5–28.

Gunnarsson E (1986) Rye breeding at Svalöf. In Olsson G (ed), Svalöf 1886–1986, Research and Results in Plant Breeding. LTs förlag-Stockholm, pp 56–61.

Johansson E, Henriksson P, Svensson G, Heneen WK (1993) Detection, chromosomal location and evaluation of the functional value of a novel high M. glutenin subunit found in Swedish wheats. J of Cereal Sc 17: 237–245.

Johansson E, Svensson G, Heneen WK (1994) Quality evaluation of the high-molecular-weight glutenin subunits in Swedish wheat material. In Proc 5th International Workshop on Gluten Proteins, Detmold, 7–9 June 1993, Association of Cereal Research, Detmold, pp 568–575.

Johansson E, Svensson G (1995a) Contribution of the High Molecular Weight Glutenin Subunit 21∗ to Breadmaking Quality of Swedish Wheats. Cereal Chem 72: 287–290.

Johansson E, Svensson G, Heneen WK (1995b) Composition of High-Molecular-Weight Glutenin Subunits in Swedish Wheats. Acta Agric Scand (sect B) 45: 112–117.

Johansson E (1996) Quality evaluation of D-zone omega gliadins in wheat. Pl Breeding 115: 57–62.

Johansson H (1988) A Swedish perspective on plant breeding and baking properties of wheat. In Asp NG (ed), Cereal Science and Technology in Sweden; Proc Intern Symp, June 1988, Lund University, Sweden, pp 70–81.

Larsson S, Magnet B, Hagman J (1997) Stråsäd, trindsäd, oljeväxter, potatis: sortval 1998. Uppsala 1997.

Lundin P, Ewertson G, Svensson G (1970) Plant Breeding at Weibullsholm. Agri Hort Gen 28: 12–48.

MacKey J (1963) Autogamous plant breeding based on already highbred material. In Åkerberg E, Hagberg A, Olsson G, Tedin O (eds), Recent Plant Breeding Research. Stockholm: Almqvist & Wiksell; New York, London: John Wiley & Sons, pp 73–88.

Moonen JHE, Scheepstra A, Graveland A (1983) The positive effects of the high molecular weight subunits 3+10 and 2∗ of glutenin on the bread-making quality of wheat cultivars. Euphytica 32: 735–742.

Olered R, Hummel-Gumælis T (1975) Report on the quality inventory of the Swedish bread grain crop in 1974. Sveriges Utsädeför Tidskr 85: 175–189.

Olered R, Johansson H (1986) Determination of the technological quality of bread grain. In Olsson G (ed), Svalöf 1886–1986: Research and Results in Plant Breeding. LTs förlag-Stockholm, pp 157–164.

Olsson G (1973) Svalöfs Solid winter wheat. Sveriges Utsädesför Tidskr 83: 201–208.

Olsson G, Kristiansson B (1979) Svalöf Hildur winter wheat. Sveriges Utsädesför Tidskr 89: 67–73.

Persson G, Mattson B, Kristiansson B (1986) Breeding self fertilizers. In Olsson G (ed), Svalöf 1886–1986: Research and Results in Plant Breeding. LTs förlag-Stockholm, pp 62–69.

Pogna NE, Mellini F, Bretta A, Dal Belin Peruffo A (1989) The high molecular weight glutenin subunits of common wheat cultivars grown in Italy. J Gen Br 43(1): 17–24.

Svensson G (1987) Situation der Weizenzüchtung in Skandinavien. Getreide, Mehl und Brot 41: 291–293.

Svensson G (1989) Breeding for improved baking quality. Sveriges Utsädesför Tidskr 99: 117–120.

Svensson G (1992a) Dacke – högproteinvete som möter industrins krav. Weibulls Årsbok 1992, p 7.

Svensson G (1992b) Tryggve – en ny höstvetesort med stabilt falltal. Weibulls Årsbok 1992, p 8.

Svensson G (1998) Personal communication, Lund, Wageningen.

Uhlen AK (1990) The composition of HMW glutenin subunits in Norwegian wheats and their relation to bread-making quality. Norw J Agric Sc 4: 1–17.

Wienhues F, Giessen JE (1957) Die Abstammung Europäischer Weizensorten. Z Pfl Z 37: 217–230.

Yearbook of Agricultural Statistics (Sweden), 1997.

Finland

1. Introduction

Being situated almost entirely north of 60°N with half of the arable land even north of 62°N, Finland (FIN) needs crop varieties that develop rapidly during the long hours of daylight during the short, cool summers. Winter wheat needs a high degree of winter hardiness (Kivi, 1968). The total area under arable crops in Finland increased from 845 000 ha in 1875 (Ignatius, 1878) to 1 250 000 ha in 1968 (Kivi, 1968). In the same period the proportion of cereals decreased from 62 to 44.5%. The proportions of winter and spring wheat were 2.4 and 7.2% respectively, in 1968 and 0.6 and 6.7% in 1986 (Table FIN.1).

In 1875 the production of wheat was not sufficient to meet the needs for consumption of the 1.9 million inhabitants. Wheat flour was one of the main products imported into Finland from Russia, and was even the most important in years with a bad harvest (Ignatius, 1878). The area under wheat was and

Table FIN.1. Areas (1000 ha) sown with cereals, total wheat, winter wheat and spring wheat in Finland between 1875 and 1995

Year(s)	Cereals	Wheat	Winter	Spring
1875[a]	523			
1921-1925[b]	787	15	9	6
1931-1935[b]	859	40	18	22
1950	889	189	13	176
1960[b]	995	181	37	144
1968[c]	1077	240	60	180
1970	1197	176	56	119
1980	1171	124	27	97
1986[d]	1210	166	15	151
1990	1250	181	38	153
1995	983	101	13	88

Sources: Yearbook of Farm statistics, [a] Ignatius (1878), [b] Broekhuizen (1969), [c] Kivi (1968) and [d] Mukula & Rantanen (1989).

still is small, but increased over the years at the expense of rye, as was the case in virtually all European countries. The largest area of wheat (286 000

Table FIN.2. Area (1000 ha), grain yield (t/ha) and production (Mt) of wheat in Finland between 1925 and 1997

Year(s)	Area	Yield	Production
1924–1925	15	1.55	23
1934–1938	78	1.82	142
1948–1952	171	1.54	263
1961–1965	259	1.73	448
1969–1971	184	2.42	445
1979–1981	110	2.39	267
1989–1991	150	3.49	521
1997	124	3.86	477

Sources: IIA and FAO statistics.

ha) was grown in 1963; from then on it has steadily decreased, with large fluctuations from year to year. In times past, many farmers, even in Lapland, used to grow small plots of bread grain, mainly of spring wheat, for domestic use. This area is not included in any statistics. Nowadays, wheat cultivation is largely concentrated in the south (Kivi, 1968): winter and spring wheat between 60°N and 61 30°N and spring wheat also in a region along the Botnic Gulf at 63°N.

Rye bread is very popular in Finland, but in spite of this the area under rye gradually decreased from 65 000 ha in the early 1960s to 27 000 and 22 800 ha respectively, in 1986 and 1997.

Evidently the average yield per hectare has increased considerably over the years (Table FIN.2), but it has not reached the level of southern Sweden (Peltonen-Saino & Karjalainen, 1991). Because of the shorter growing season in Finland it seems unlikely that this will ever change. The production of wheat suitable for bread making varies a great deal, due to great fluctuations in climatic conditions. Over the period 1970–1983 the annual national requirement for bread-making wheat was covered for 181–18% (!) with a mean of 108% (Kettunen, 1986).

The aim of wheat cultivation was defined in 1988 as self-sufficiency for the nearly 5 million inhabitants. It was estimated that for complete self-sufficiency 200 000 ha of wheat would be required, of which the share of spring wheat would have to be 65–80% (Juuti, 1988). In addition to productivity, bread-making quality is of major importance in Finnish wheat breeding, as legislation limits the use of imported high quality wheat by flour mills (Peltonen, 1991).

2. General remarks on wheat growing in Finland

2.1. Winter wheat

Because of the need for extreme winter-hardiness and adaptation to the short growing season, winter-wheat breeders confined themselves initially to the selection of single-plant-based strains from land varieties with good baking quality. Resistance to frost was transferred to Finnish winter wheats from old varieties and was also inherited from certain Russian varieties, such as Charkov and Ukrainka (Kivi, 1968). During the 1960s, the need increased for varieties with modern straw characteristics and resistance to sprouting and disease. In response, plant breeders had to broaden the genetic background of their material with foreign varieties. (Kivi, 1968; Varis, 1975). The main aim of winter-wheat breeding was to improve stability of performance by selecting cultivars with versatile overwintering ability, wide adaptation, stiff straw and resistance to the most important diseases (Juuti, 1980b). In addition, selection focused on two different uses:

- an early maturing, stiff-strawed bread wheat of good baking quality;
- a high yielding, stiff-strawed wheat for feeding purposes.

2.2. Spring wheat

Early on, breeding spring wheat consisted of single-plant selection from land race populations and from impure spring-wheat varieties that had spread to Finland around the turn of the century. During the first half of the 1910s, crosses were made to establish a broader basis for selection. Early domestic wheats were crossed with foreign varieties, such as the Canadian variety Marquis and the Australian Aurore, sources of yielding ability, straw characteristics, quality and disease resistance (Juuti, 1988). In the 1930s, purposeful research into the quality of spring wheat was undertaken; breeding work focused on the quantity and characteristics of the protein (Pesola & Veijola, 1954, cited by Juuti, 1988). Varieties such as Tammi II and Kimmo from Hankkija, and Sopu, Hopea and Apu from Jokioinen were typical of the results of this endeavour. During the 1970s and 1980s, quality breeding was largely based on domestic varieties bred during earlier periods. This work resulted in the varieties Ruso, Tähti, Ulla and Tapio. Foreign cultivars with earliness, high protein content, resistance to sprouting in the ear and yielding ability were selected from the global collection of breeding material for use as crossing parents (Juuti, 1988).

3. Plant breeding institutes in Finland

The Institute of Plant Breeding at *Jokioinen*, north-west of Helsinki, is part of the Agricultural Research Centre of the Ministry of Agriculture. The Plant Breeding Institute of *Hankkija* is a private institute; its activities are financed by a farmers cooperative organization, the Wholesale Society Hankkija, which began its breeding work in 1913. The working programme includes the breeding of all agricultural crops which are important in Finland. The Hankkija and Jokioinen Breeding Institutes merged in 1994 to become *Boreal Plant Breeding*, settled at Jokioinen. This is a state-owned establishment, doing research in cereals, potato, pea, spring turnip, rape, meadow fescue, timothy and clovers.

4. Winter wheat

4.1. The area under winter wheat

In Finland, the cultivation of winter wheat is concentrated on clay soils of the southern regions. The area cultivated has always been irregular but it began to increase steadily from 1920 onwards (Broekhuizen, 1969; Kivi, 1968). Since 1970, however, the area has decreased, and in 1995 it was only 13 000 ha, representing 13% of the total wheat area. The main winter-wheat varieties grown in Finland since 1940 are listed in Table FIN.3. It is remarkable that

Table FIN.3. Area (ha) and main varieties of winter wheat in Finland between 1940 and 1996

Year(s)	Area	Varieties
1940–1950		Varma
1955	18 000	Varma (57%), Virtus (9%), Vakka (5%), Olympia (10%)
1965	50 600[a]	Vakka (40%), Antti (14%), Varma (19%), Elo (13%)
1968	60 000	Vakka, Elo, Linna
1974	40 000	Vakka (60%), Linna, Nisu, Elo
1979	11 000	Vakka (50%), Aura, Nisu, Linna
1986	15 000[b]	Aura (91%), Vakka (9%)[c]
1991	38 100	Aura (46%), Hankkijan Ilves (31%)
1996	25 200	Aura (41%), Tryggve (18%)

Sources: Yearbook of Farm statistics, [a] Broekhuizen (1969), [b] Hankkija Plant Breeding Institute (1987) and [c] Svensson (1987).

some varieties have been grown for many years: for example, Vakka for more than 30 years (1953–1987) and Aura for 20 years (since 1979).

4.2. Methods to determine baking quality at Hankkija

For quality tests, protein content, Pelshenke value and Zeleny sedimentation value are used on a routine basis. Baking tests are carried out on the more advanced material, under supervision of the Grain Research Committee (Varis, 1975).

4.3. Yield and quality features of winter-wheat varieties

From Table FIN.4, it is evident that one effect of 20 years of breeding work has been yield improvement. It is striking that the 'old' variety Vakka has the highest value for loaf volume (Table FIN.5). Conversely, the differences in extensibility between the varieties are very restricted. There are varietal differences in resistance to extension: Elo and Linna have especially low values.

When the *Glu-1* quality score in Table FIN.7 is compared with the quality characteristics of the five varieties in Table FIN.5, the relation between Pelshenke value, resistance to extension and quality score is obvious.

Table FIN.6 presents a summary of results of winter-wheat trials in a number of years preceding 1988; the mean kernel yield of 457 g/m^2 is high for Finland.

From the point of view of the combination of yield and quality, Aura is the only genotype of special interest, since its yield and sedimentation value

Table FIN.4. Winter wheat at Anttila and Nikkilä Stations[a] in 1970–1974

Variety	Year	Grain yield (rel.)[b]	Protein content %	Pelshenke value	Zeleny number	Falling number
Vakka	1953	100	15.5	145	26	323
Elo	1963	104	15.1	58	38	288
Jyvä	1965	108	16.2	102	26	338
Linna	1965	117	15.6	48	34	335
Nisu	1966	108	15.3	122	27	352
Aura	1976	118	15.4	75	46	354

[a] The station of Anttila is situated at 60°42′N and that of Nikkilä at 61°55′N.
[b] Grain yield 100% = 3710 kg/ha.
Source: Varis (1975).

Table FIN.5. Some quality characteristics of winter wheat in baking tests of the Grain Research Committee

Variety	Loaf[a] volume	Protein content (%)	Pelshenke value	Resistance to extension	Extensi-bility
Vakka	100	13.9	162	163	201
Elo	87	13.2	55	100	210
Jyvä	94	14.4	122	172	203
Linna	89	13.9	41	83	230
Nisu	94	14.2	138	177	185

[a] Loaf volume 100 = 650 cc.
Source: Varis (1975).

Table FIN.6. Summary of winter-wheat trials in a number of years preceding 1988

Variety	Year of re-lease	Kernel yield (rel.)	Protein content (%)	Zeleny sedim. value	Falling number	Plant height (cm)	Matu-rity (days)	Over-winte-ring
Vakka	1953	92	11.3	14	261	98	340	85
Linna	1965	98	10.7	24	254	112	348	83
Nisu	1966	95	10.0	19	310	101	345	87
Aura	1976	104	10.4	27	272	99	344	88
Skjaldar	1976	96	10.4	26	292			
Ilves	1984	98	11.1	16	286	94	342	90
Pitko	1985	98	11.5	19	238	99	342	85
Mir-808	1984	84	11.3	37	270	90	341	76
Otso	1989	103	10.5	16	283	104	344	85
Trial-mean		100[a]	10.8	25.8	271	103	345	86

[a] Trial mean for kernel yield = 457 g/m^2.
The number of trials included differed for varieties from 2 for Nisu to 30 for Vakka.
Source: Rekunen (1989).

are higher than average; the protein content in this trial is rather low. The proportion of overwintering plants of the Russian cultivar Mironovskaya-808 is notably low compared to the other varieties, which may explain its low yield. Also its sedimentation value is high, which is unusual for a winter wheat; the older variety Vakka has the lowest value. The difference of 8 days in number of days to maturity between the earliest and the latest variety will be of importance in the continual efforts to breed varieties that can be harvested early.

Table FIN.7. Winter-wheat varieties bred in Finland (until 1985) and indications for their bread-making quality

Variety	Breeding station[a]	Year of release	Pedigree	HMW subunits 1A	1B	1D	*Glu-1* score
Rusopäävehnä	Ha	1921	mixed population of Pudel wheat[b]				
Jalostettu Villavehnä	Ha	1921	sel.from Uusimaa (=South Finnish local wheat)	2*	7+9	2+12	7
Sukkula I	Ha	1922	sel from South Finnish local wheat	2*	7+9	2+12	7
Sukkula II	Ha	1928	selection from Sukkula I	2*	7+9	2+12	7
Varma	Ha	1933	Svea/local variety Orimattila	1	7+9	2+12	7
Pohjola	Jo	1933	selection from Uusimaa	2*	7+9	2+12	7
Sampo	Jo	1933	Thule II/landrace	1	20	2+12	6[c]
Panu	Ha	1936	Svea/local variety Noso	2*	20	2+12	6[c]
Olympia	Jo	1941	selection from Uusimaa	1	7+9	2+12	7
Vakka	Jo	1953	Varma/Kehra	2*	7	5+10	8
Antti	Ha	1955	F_1 (Ta 02325/Ta 02278)//Ukrainka	1	7+9	5+10	9
Elo	Ha	1963	Charkov//Lokalahti/Svea/3/Varma	1	7+9	2+12	7
Linna	Ha	1965	Panu/line from No (l.v.)//Virtus	2*	7+9	2+12	7
Jyvä	Jo	1965	selection from Vakka[d]	2*	7+9	5+10	9
Nisu	Jo	1966	Varma/Kehra	2*	7+9	5+10	9
Aura	Jo	1976	Ertus/Vakkaa	2*	7+9	2+12	7
Ilves	Ha	1984	Hja b 356/Vakka	2*	7	5+10	8
Pitko	Jo	1985	Ta 05901/Vakka	1	7	5+10	8

[a] Ha = Hankkija; Jo = Jokioinen.
[b] Pudel wheat = Swedish selection from English squarehead (Zeven & Zeven, 1976).
[c] Subunit 20 has been given score 1 according to Ruiz et al. (1995).
[d] After Zeven & Zeven (1976).
Source: Kivi (1968, 1980) and Sonntag et al. (1986).

From the presentation of the HMW subunits for the 17 winter-wheat varieties in Table FIN.7, the following conclusions can be drawn:

- The variation of subunits on the 1A locus is restricted in so far that the nullisome is not found in this material. Varieties number 2, 3 and 4 in the Table have the 2* allele at the 1A locus. As these varieties are selected directly or indirectly from local Finnish wheats, it may be assumed that the 2* allele was present in these local wheats.
- The variation of subunits on the 1B locus is also limited. Two varieties, however, have the subunit 20, of which very few was known at the moment of the publication (1986); Ruiz et al. (1995) later assigned these a quality score of 1.

- Subunits 5+10 on the 1D locus do not occur among the listed varieties released before 1953. The influence on the *Glu-1* score of the subunits 5+10 compared with that of the subunits 2+12 is evident: of the six varieties in Table FIN.7 possessing subunits 5+10, three have a quality score of 8 and three a score of 9; all other varieties have a score of 7.

A comparison can be made between the varieties bred at each of the two breeding stations. Comparison of the mean score for the nine varieties from Hankkija (7.2) and for the eight varieties from Jokioinen (7.6) shows that the difference is small. Actually, the average values are generally high for winter wheat.

4.4. Some conclusions on winter-wheat breeding in Finland

Plant breeders in Finland have very successfully achieved their aim of combining good baking quality with overall agronomical value of winter-wheat varieties. The quality, measured as the *Glu-1* score, of older winter-wheat varieties was already rather high, and this has been further improved in varieties that have been released since 1953.

5. Spring wheat

5.1. The area under spring wheat

Spring wheat has been grown on unrecorded areas for hundreds of years (Grotenfelt, 1922, cited by Juuti, 1988). From 1930 onwards, there was a significant increase in the amount grown. According to Pesola (1950, cited by Juuti, 1988), this was due to the release of Svalöf's Diamant, to the milling protection of domestic wheat and to weather conditions favouring spring-wheat cultivation. The annual variation of the area under spring wheat was and still is considerable, due to great annual fluctuations in weather. Self-sufficiency in bread wheat was reached during the 1960s, whereupon efforts were made to limit the cultivation. Subsequently, the demand for domestic spring-wheat varieties with good baking quality increased (Kivi, 1970). The most important spring-wheat varieties grown in Finland since the 1930s are listed in Table FIN.8.

5.2. Aims and results of spring-wheat breeding in Finland

At Tammisto (Hankkija), great emphasis has been given to the creation of early ripening, high yielding varieties of good baking quality. Huttunen (1950) reported that the spring-wheat line A-3455 (later known as Terä) gave a high yield of high baking quality.

Table FIN.8. Areas (1000 ha) and main varieties of spring wheat in Finland

Year(s)	Area	Predominant region	Main varieties
1930–1955		southern	Diamant (from Sweden)
		central, northern	Kimmo, Tammi
1960	44	southern	Diamant, Svenno (from Sweden), Kärn II
		central	Tammi
		northern	Apu
1962	261	southern	Svenno, Norrona (from Norway)
		northern	Apu
1969	136		Svenno, Ruso, Apu, Touko
1974	160		Ruso (50%), Apu, Tähti
1980	88		Ruso (50%)
		southern	Tähti, Touko
1986	150		Ruso, Tapio (together 50%)
1991	152		Reno (28%), Satu (17%), Heta (9%)
1996	87		Tjalve (57%), Reno (18%)

Sources: Broekhuizen 1969; Yearbook of Farm Statistics, 1997.

In 1960, Huttunen reported on efforts to replace the Swedish variety Svenno, which was then one of several rather popular varieties in southern Finland; unfortunately it ripened too late. Another popular variety at that time was Norrona, from Norway; the low protein content of its grain was a drawback, however, reason for the milling industry to oppose its cultivation.

The aim of breeding spring wheat was described by Kivi (1970) as the development of varieties whose baking properties should at least be similar to those of the older, early-ripening Finnish varieties, such as Tammi and Kimmo. Additional aims were resistance to sprouting in the ear and characteristics to increase the capability of utilizing high fertilizer inputs, for example, straw stiffness and early maturity. A number of varieties, listed in Tables FIN.9 and FIN.11, meeting these aims were released during 1967–1980, among them Tapio and Tähti. Tähti combined high yield with higher than average protein content, which to some extent contradicts the observation that these features are often negatively interdependant. The quality of its protein is good, according to its Pelshenke value, its Zeleny sedimentation value, its resistance to extension and its extensibility. However, its loaf volume of 102% is somewhat lower than is to be expected from these indirect quality mesurements. The variety seems to be an example of successful efforts to break the interdependence between yielding ability and protein content.

Table FIN.9. Spring wheat at the Tammisto and Anttila Stations; means of the years 1970–1974

Variety	Breeder	Year of release	Grain yield (rel.)	Protein content (%)	Pelsh. value	Zeleny number	Falling number
Ruso	Ha[a]	1967	100[b]	15.2	163	43	231
Apu	Jo	1949	89	16.1	89	43	244
Touko	Jo	1950	94	16.2	101	40	283
Svenno	Wb	1953	95	15.8	126	58	242
Veka	Ha	1970	102	15.2	167	62	229
Tähti	Jo	1972	103	15.9	149	55	308
Ulla	Ha	1975	89	17.1	174	67	252

[a] Ha = Hankkija; Jo = Jokioinen; Wb = Weibull.
[b] 100 = 4280 kg/ha (Ruso was the standard variety).
Sources: Hovinen & Varis (1975a, 1975b).

Table FIN.10. The quality of five spring-wheat varieties; means of the years 1970–1974

Variety	Loaf volume (rel.)[a]	Protein content (%)	Pelshenke value	Resistance to extension	Extensibility
Ruso	100	14.2	164	222	205
Apu	111	15.6	90	98	234
Svenno	112	15.6	126	161	221
Tähti	102	15.3	150	214	225
Ulla	122	16.9	181	221	242

[a] 100% = 659 ml; baking test with additives and 80 minutes of raising time.
Sources: Hovinen & Varis (1975a, 1975b).

As a declining trend in the protein content of spring wheat grown in Finland became apparent (Salovaara, 1986), these indications for varietal differences in protein content of the harvested grain were of major importance.

Salovaara (1986) and Peltonen (1991) observed that varieties such as Kadett and Tapio became popular because of their increased yielding ability (Table FIN.11), but their protein content was just average. The situation became more complicated again because of the release of the cultivars Polkka and Laari, which have a high protein content but low protein quality. To allevi-

Table FIN.11. Spring-wheat trial of the Hankkija Plant Breeding Institute at Anntila, 1988

Variety	Year of release	Kernel yield (rel.)	Protein content (%)	Zeleny sedim.	Falling number	Plant height (cm)	Maturity (days)
Drabant		103	11.6	27	162	90	109
Ruso	1967	97	12.2	30	173	94	103
Tähti	1972	89	12.7	43	206	97	109
Runar	1972	103	12.5	44	209	88	103
Ulla	1975	93	13.6	57	178	87	98
Reno	1975	99	12.4	38	227	86	103
Tapio	1980	105	11.9	39	184	90	105
Luja	1981	96	13.0	52	194	84	101
Kadett	1981	109	11.7	45	171	89	106
Heta	1988	93	14.0	43	214	88	100
Polkka	1989	108	12.8	52	186	86	104
Manu	1994	102	13.5	62	201	90	101
Trial mean		100[a]	12.7	41	190	87	103

[a] trial mean for kernel yield = 397 g/m^2; the number of trials varied between 55 (for Ruso) and 9 (for Polkka).
Source: Rekunen (1989).

ate the situation, quality criteria as specified by industrial baking technology and segregation into quality classes should be introduced. No official quality classification system for wheat is in force in Finland (Juuti, 1998).

5.3. Yield and quality features of spring-wheat varieties

The data in Table FIN.9 show a clear increase in yield between the varieties Apu and Touko, released in 1949 and 1950, respectively, and Ruso, Tähti and Veka, released some years later. The same is true for Pelshenke value. For yield, Ulla is an evident exception, but its protein content, Pelshenke value, Zeleny number and loaf volume are strikingly better. Ulla is an example of breeding for a relatively short growing period together with good straw stiffness. However, because of its lower grain yield it did not become important in practice.

Unfortunately, Table FIN.10 does not present the quality figures of Veka and Touko. The loaf volume of the oldest and of the most recent variety (Apu and Ulla, respectively) are the highest. The loaf volume of Ulla is certainly enhanced by its very high protein content. It is interesting enough to note that for the variety Apu there is a relation between its lower Pelshenke value, its

Table FIN.12. Spring-wheat varieties bred in Finland until 1995

Variety	Br. Year	Pedigree	HMW subunits *Glu-1* 1A	1B	1D	score
Ruskea	Ha 1919	selection from LV from Halland (Sw)	2*	7+8	2+12	8
Tammi	Ha 1922	local wheat from western Finland	?			
Pika I	Ha 1927	Ruskea (early local strain)	2*	7+8	2+12	8
Pika II	Ha 1934	Canadian local wheat/N.Karelian local wheat	N	7+9	2+12	5
Sopu	Jo 1935	Marquis / Ruskea	2*	7+8	5+10	10
Hopea	Jo 1936	Marquis / Ruskea	2*	7+8	5+10	10
Tammi II	Ha 1939	McIntosh/line 01214	2*	7+9	5+10	7
			N	7+9	5+10	5[a]
Kimmo	Ha 1941	sel. from population from Pisarev in Moscow	1	7+9	5+10	9
Apu	Jo 1949	Garnet / Pika	N	7+8	2+12	6
Touko	Jo 1950	Diamant / Hope	N	7+8	5+10	8
Kiuru	Jo 1951	Aurora / Sopu	1	7+8	5+10	10
			N	7+8	5+10	8[a]
Terä	Ha 1952	Hopea/3/Aurora//Finnish local var./Pärl	1	7+9	5+10	9
Ruso	Ha 1967	Reward/Pika//unknown pollinator	1	7+9	5+10	9
Veka	Ha 1970	Kärn / Tammi	2*	6+8	5+10	8
Tähti	Jo 1972	Kärn//Aurora/Pika	1	7+9	5+10	9
Ulla	Ha 1975	Tammi/Ta 4431	2*	7+9	5+10	9
			N	7+9	5+10	7[a]
Taava	Ha 1978	mutant from Ruso	1	7+9	5+10	9
Tapio	Ha 1980	Hja 33929 / Kolibri	N	7+9	5+10	7
Luja	Jo 1981	Svenno//Hopea/Tammi	2*	7+8	5+10	10
Heta	Ha 1988	Hja a 1105 / Hja a 1099	2*	6+8	5+10	8
Manu	Ha 1994	Ruso/Runar	2*	7+8	5+10	10
Mahti	Ha 1995	Cebeco 1036/Hja 20519	1	7+8	5+10	10

[a] The varieties Tammi II, Kiuru and Ulla showed allelic variation at *Glu-A1*; for this reason two *Glu-1* scores have been calculated.
Sources: Sontag et al. (1986) and Sontag-Strohm (1997).

lower resistance to extension and its *Glu-1* quality score according to Payne (see Table FIN.12).

From the point of view of yield and quality, there are two interesting genotypes, Polkka and Manu, whose kernel yield, protein content and Zeleny sedimentation value are higher than the mean value. These two cultivars are also interesting when they are compared to Drabant and Kadett, especially for their higher protein content and sedimentation value. Heta has the highest protein content and is otherwise very similar to Ulla. The difference between varieties in the number of days until maturity is more than 10%, which is a large difference.

Table FIN.12 has been compiled from the results of two investigations, one by Sontag, Salovaara & Payne (1986) and one by Sontag-Strohm (1997). In the first investigation the HMW-subunit composition was determined for 18 varieties; this study was based on the analysis of six grains per cultivar only and was therefore considered by the authors, to be very preliminary. In the second investigation, three recent spring-wheat varieties were added to those of the first investigation and 20 single seeds were analysed. Thus HMW subunits are available for 21 varieties; three of them showed allelic variation at *Glu-A1*, i.e.Tammi II, Kiuru and Ulla, so that each of these three varieties has two values of the *Glu-1* quality score.

The average quality score for the 21 spring-wheat varieties is 8.5, which is very high compared to average scores for spring-wheat varieties in many other countries (Sontag, Salovaara & Payne, 1986). If the allelic variation at *Glu-A1* for three varieties is taken into account, the mean score should be recalculated, which at 8.2, is still very high.

The very old varieties Ruskea and Pika I, released in 1919 and 1927, respectively, both have the 2* allele at the *Glu-A1* locus and the 7+8 alleles at the *Glu-B1* locus; in the quality score these alleles are awarded the high value of three and thus contribute to a high total value. The pedigrees of these varieties make clear that the alleles 2* and 7+8 must have been present in old Finnish land varieties. Later varieties have North American and Russian ancestors, along with Finnish varieties, in their pedigrees. Around the middle of the 20th century the Australian variety Aurora was also introduced as a parent in some crosses.

It is remarkable that baking quality, according to the quality score, is high from the earliest (1919) to the most recent (1995) variety mentioned in this list. This observation agrees with the results of a study on the progress in breeding for yield and for quality traits since the 1930s, where 10 spring-wheat cultivars bred in Finland were investigated. The 10 cultivars were released during 1939–1990. No trend between baking quality and first year of marketing was recorded, whereas the increase in grain yield arising from genetic improvement was 20% (Peltonen-Sainio & Peltonen, 1994).

5.4. Some conclusions on spring-wheat breeding in Finland

For the 21 spring-wheat varieties from which a *Glu-1* quality score could be calculated (Table FIN.12), the first 11 were released during 1919–1952, and the last 10 during 1967–1995. The mean quality score for the first 11 varieties is 8.2, and for the last nine 8.9. The conclusion can be drawn that Finnish spring-wheat breeders have been able to breed varieties for a good

75 years with improved overall husbandry value, while maintaining or even improving slightly baking quality, as measured in the *Glu-1* quality score.

The mean *Glu-1* quality score for the 15 Finnish winter- and 21 spring-wheat varieties is 7.6 (Table FIN.7) and 8.5 (Table FIN.12), respectively.

6. Methods for fractionating HMW and LMW glutenins and gliadins and their use in breeding and selection

6.1. Some considerations on the application of HMW glutenin subunits in breeding

Comparison of the *Glu-1* quality score of 35 Finnish wheat varieties with that of foreign varieties (Sonntag et al., 1986), showed that the average score of Finnish varieties (8.0) was much higher than the mean scores of cultivars grown in Great Britain or in the Federal Republic of Germany (5.2 and 5.8, respectively) and comparable to the average score of cultivars grown in Australia (8.0).

Peltonen, Salopelto & Rita (1993) investigated the optimal combination of HMW glutenin subunits at the *Glu-A1* and *Glu-B1* loci associated with bread-making quality in spring wheat. A total of 43 cultivars and lines, all of which had the subunit combination 5+10 at the *Glu-D1* locus in common, were studied; the material was grown in 1989 and 1990 at the Hankkija Plant Breeding Institute. Regression analysis of the electrophoretic data of subunits at the *Glu-A1* and *Glu-B1* loci and indirect bread-making quality parameters was carried out. This analysis showed that the presence of subunit pair 7+8 (*Glu-B1*), with a high *Glu-1* score, may lead to overstrong gluten. The combination of subunit 2∗ (*Glu-A1*), with a high *Glu-1* score, and the subunit pair 6+8 (*Glu-B1*), with a low *Glu-1* score, was associated with increased bread-making quality.

These results were rather different from those from Swedish investigations, in which the newly discovered allele 21∗ (*Glu-A1*) and the alleles 14+15 (*Glu-B1*), both in combination with the alleles 5+10 (*Glu-D1*), were considered as the optimal glutenin composition (Johansson et al. 1993, 1994b, 1995a).

Yet, some older Finnish spring-wheat varieties (Ruskea and Pika) have the subunits 2∗ (*Glu-A1*) and 7+8 (*Glu-B1*) in combination with 2+12 (*Glu-D1*). The old varieties Sopu and Hopea and the modern varieties Luja and Manu have the same subunit composition, even in combination with 5+10, from which one would expect still more 'over-strong gluten'.

Peltonen, Salopelto & Rita concluded that for a more accurate use of SDS-PAGE electrophoresis as a selection tool for high yielding and high-quality wheat genotypes, it would be beneficial to clarify the relationships between different HMW glutenin subunits and the yield characteristics of wheat.

6.2. A method to separate HMW and LMW glutenin subunits

Sontag-Strohm (1996) has recently described a new method for the electrophoretic fractionation of both high-molecular-weight and low-molecular-weight glutenin subunits. In this method gliadins are removed in an electrophoretic step with reversed polarity, which precedes the main electrophoretic procedure. In the latter procedure the HMW glutenins and LMW glutenins are fractionated in two separated groups of bands. Sontag-Strohm considered this new method to be just as informative and reliable for the analysis of glutenin subunits of wheat as the two-step one-dimensional SDS-PAGE, and easier to perform.

6.3. Investigations on the gliadin and LMW glutenin composition of Finnish spring-wheat material

Sontag-Strohm (1997) studied the gliadin low- and high-molecular-weight subunit alleles at *Gli-1*, *Glu-3* and *Glu-1* loci in 21 Finnish spring-wheat cultivars and advanced breeding lines. The HMW subunit composition of the spring-wheat cultivars is listed in Table FIN.12. The aim of her study was to describe the alleles at *Gli-1* and *Glu-3* loci in the spring-wheat varieties bred in Finland using 2D and 1D separations. The gliadin and glutenin alleles in advanced spring-wheat breeding lines were also studied using 1D separations.

The linked *Gli-1*, *Glu-3* allelic pairs were described according to the catalogue published by Metakovsky (1991), using one letter for each locus. The number of allelic pairs found in both groups of wheat material was restricted. It is interesting that the *Glu-1*, *Gli-3* combination found in the oldest spring-wheat varieties is also present in some varieties of a later date. Even so, the introduction of thus far unknown alleles by crosses of Finnish and foreign varieties is apparent.

The 86 advanced breeding lines contained 23 new combinations of *Gli-1*, *Glu-3* alleles compared to the cultivars studied in 1990. This is seen to reflect the use of varieties from other countries (e.g. Sweden, Germany, Great Britain and the Netherlands) as crossing parents.

In the same study, the effect of selection for gluten quality on the distribution of gliadin and glutenin alleles was monitored for three years. The selection was based on data of total protein content, Zeleny sedimentation volume, Pelshenke test number, wet gluten content and HMW glutenin subunit alleles. After three successive years of selection, 15 lines remained in which one-third of the new combinations of alleles was still present.

The relationship between baking quality and the high- and low-molecular weight (HMW and LMW) glutenin subunit and gliadin alleles on group 1 chromosomes in 13 cultivars and 62 advanced breeding lines of spring wheat grown in Finland was studied by Sontag-Strohm & Juuti (1997). The 75 spring wheats contained 18 HMW and 16 LMW glutenin compositions, as well as in all 50 different glutenin subunit combinations. Most of the spring wheats in this investigation (80%) contained HMW subunit pair 5+10 (*Glu-D1*), which was associated with higher Pelshenke number than allele subunit pair 2+12 (*Glu-D1*).

Variation in the *Gli-B1*, *Glu-B3* composition was only associated with differences in Pelshenke number for wheats with the HMW subunit pair 5+10 in the background. Thus the effect of these *Gli-B1*, *Glu-B3* alleles is only considered as a refinement of the influence of the HMW subunit pair 5+10.

Comparing results of investigations in Finland with those in Sweden, it is of interest to note that in the Finnish results the difference between the *Glu-1* alleles 5+10 and 2+12 is of major importance for the Pelshanke number, whereas this difference is of secondary importance for Swedish grown wheats (Johansson, 1995b).

7. Interaction between kernel yield and protein content

Joy & Peltonen (1993) studied the effect of the enzyme nitrate reductase (NR) and photosynthetic rate on grain yield and quality characteristics at different growth stages. They analysed the data from three lines selected in a winter-wheat population from Virtus/Pohjola. Combining the results of the three lines, Joy & Peltonen concluded that a long-term capacity for high NR activity and high photosynthetic rate has a positive influence on grain and protein yield, and possibly on protein quantity. This requires an adequate uptake of nitrogen by the plant combined with the ability to relocate this nitrogen to the seeds.

The challenge for a breeder is whether the high NR activity and the high photosynthetic rate found in this study at an early stage of development (13)in one line and at a later stage (47) of development[1] in another can be combined in one single genotype with high long-term NR activity and high photosynthetic rate.

[1] Growth stages according to Zadoks et al. (1974); growth stage 13 = three leaves of the seedling are unfolded; growth stage 47 = the sheeth of the flag leaf is opening.

8. Stability of quality traits

Considering the marginal conditions for wheat growing and the high risk factors endangering wheat production in Finland, it is not surprising that the stability of quality traits intrigues Finnish breeders and investigators.

Peltonen-Sainio & Peltonen (1993) investigated the stability of some characteristics in spring-wheat cultivars, which were grown at two locations for seven years. They found no significant genotype × environment (G × E) interaction for hectolitre weight, Zeleny sedimentation value and falling number. A significant G × E interaction was found, however, for thousand kernel weight, protein concentration and protein yield.

Another study (Peltonen et al., 1990), including 21 years of field experiments with spring wheat, showed that early varieties were more stable than late varieties. Extreme conditions of rainfall and temperature (very low and very high) affected the gluten content in a negative sense. Surprisingly, there was no significant correlation between these climatic factors and grain yield. The yield to quality ratio was mainly affected by excessive rains before heading and high temperatures during grain filling.

In a two-year field experiment with four spring-wheat varieties grown at two locations, Peltonen (1992) studied the effect of genotype and environment on yield and quality of bread wheat. The main factor influencing the quantity to quality ratio appeared to be the duration of the reproductive period, which was strongly regulated by the environment. Genotypes with a longer reproductive period had a poorer bread-making quality (measured as wet gluten content, Zeleny and farinograph values). The quantity to quality ratio was affected more by a genotype × year interaction than by genotype × location interaction. Peltonen suggested that not only the HMW glutenin composition, but also the amount of S-rich prolamins (LMW glutenin subunits, α-, β- and γ-gliadins) could play an important role in reducing the negative correlation between yielding ability and grain protein content.

J.M.

Acknowledgment

The author wishes to thank Dr T. Juuti (Boreal, Jokioinen) for his help in reviewing this chapter and for providing several literature references on wheat breeding in Finland.

References

Broekhuizen S (1969) Atlas of the cereal-growing areas in Europe. Agro-ecological Atlas of Cereal Growing in Europe, vol II, 157 pp. Wageningen: Pudoc; Amsterdam, London, New York: Elsevier Publishing Company.

Hovinen S, Varis E (1975a) Spring wheat. The Hankija Pl Br St 1975: 34–39.

Hovinen S, Varis E (1975b) Hankkija's Ulla Spring Wheat. The Hankija Pl Br St 1975: 142–143.

Huttunen E (1950) Spring wheat. Siemenjulkaisu 1950: 29–38.

Huttunen E (1960) Spring wheat. Siemenjulkaisu 1960: 36–44.

Ignatius KEF (1878) Le Grand-Duché de Finlande. Notice Statistique. Helsingfors: 140 pp.

Johansson E, Henriksson P, Svensson G, Heneen WK (1993) Detection, chromosomal location and evaluation of the functional value of a novel high Mr glutenin subunit found in Swedish wheats. J of Cereal Sc 17: 237–245.

Johansson E, Svensson G, Heneen WK (1994) Quality evaluation of the high molecular weight glutenin subunits in Swedish wheat material. In Proc 5th International Workshop on Gluten Proteins, Detmold, 7–9 June 1993, Association of Cereal Research, Detmold: pp 568–575.

Johansson E, Svensson G (1995a) Contribution of the High Molecular Weight Glutenin Subunit 21∗ to Breadmaking Quality of Swedish Wheats. Cereal Chem 72: 287–290.

Johansson E, Svensson G, Heneen WK (1995b) Composition of High Molecular Weight Glutenin Subunits in Swedish Wheats. Acta Agric Scand (sect B) 45: 112–117.

Joy P, Peltonen J (1993) Breeding implications of the negative correlation between yield and protein concentration in a winter-wheat population. Acta Agriculturae Scandinavica; Sect B: Soil Pl Sci 43: 82–88.

Juuti T (1980a) Bread cereals – Spring wheat. The Hankkija Pl Br Inst 1980: 28–32, 38.

Juuti T (1980b) Bread cereals – Winter wheat. The Hankkija Pl Br Inst 1980: 33–35, 39.

Juuti T (1988) Targets in the breeding of spring wheat. J Agric Sc Finland 60: 281–291.

Juuti T (1998) Personal communication.

Kettunen L (1986) Self-sufficiency of agriculture in Finland in 1970–1983. J Agric Sc Finland 58: 143–150.

Kivi E (1965) Spring wheat. Siemenjulkaisu 1965: 31–39, 54.

Kivi E (1968) Main features of agricultural Plant Breeding. Peat & Plant News 1(4): 45–53.

Kivi E (1970) Spring wheat. Siemenjulkaisu 1970: 35–41.

Kivi E (1980) List of varieties released by the Hankkija Plant Breeding Institute from 1919–1980. The Hankkija Pl Br Inst: 158–161.

Metakovsky EV (1991) Gliadin allele identification in common wheat (II) Catalogue of gliadin alleles in common wheat. J Genet Br 45: 325–344.

Mukula J, Rantanen O (1989) Climatic risks to the yield and quality of field crops in Finland. V. Springwheat 1969–1986. Annales Agriculturae Finniae 28: 21–28.

Peltonen J, Karvonen T, Kivi E (1990) The effect of climatic factors on production of spring wheat quantity to quality ratio in Southern Finland. J Agric Sc Finland 62: 227–236.

Peltonen J (1991) Testing cultivar differences of wheat flour in a Breeding Program with small-scale 'Home Bakery'. Vorträge Pflanzenzücht 20: 333.

Peltonen J (1992) Influence of environment and genotype on spring wheat yield and bread-making quality under Finnish conditions. Acta Agric Scand; Sect B: Soil Pl Sci 42: 111–117.

Peltonen J, Salopelto J, Rita H (1993) The optimal combination of HMW glutenin subunits coded at gene loci *Glu-A1* and *Glu-B1* for bread-making quality in Scandinavian wheats. Hereditas 118: 71–78.

Peltonen-Sainio P, Peltonen J (1993) Stability of quality traits in spring cereals cultivated under the growing conditions of Southern Finland. Acta Agric Scand; Sect B: Soil Pl Sci 43: 45–52.

Peltonen-Sainio P, Peltonen J (1994) Progress since the 1930s in breeding for yield, its components and quality traits of spring wheat in Finland. Plant Breeding 113: 177–186.

Peltonen-Sainio P, Karjalainen R (1991) Genetic yield improvement of cereal varieties in northern agriculture since 1920. Acta Agric Scand 41: 267–273.

Rekunen M (1989) Trial results 1988. Tiedote 37 of the Hankkija Plant Breeding Institute, p 15.

Ruiz M, Rodriguez-Quijano M, Carillo JM (1995) Relaciones entre la subunidades de gluteninas de alto peso molecular (HMW), distintas puntuaciones *Glu-1Y* varios parametros de calidad en trigo harinero. Invest Agr (prod y prot veg) 10 (1): 15–29.

Salovaara H (1986) Wheat and flour quality related to baking performance in industrial French breadprocesses. Acta Agric Scand 36: 387–398.

Sontag T, Salovaara H, Payne PI (1986) The high-molecular-weight glutenin subunit compositions of wheat varieties bred in Finland. J Agric Sc Finland 58: 151–156.

Sontag-Strohm T (1996) Electrophoretic fractionation of wheat glutenin subunits after removal of gliadins by reversed SDS-PAGE; a rapid communication. J Cereal Sc 24: 87–89.

Sontag-Strohm T (1997) Gliadin and glutenin subunit alleles on Group 1 chromosomes in Finnish Spring Wheats. Acta Agric Scand; Sect B: Soil Pl Sci 47: 98–105.

Sontag-Strohm T, Juuti T (1997) Association between gliadin and glutenin subunit alleles and baking quality in spring-wheat cultivars and breeding lines grown in Finland. Acta Agric Scand; Section B: Soil Pl Sci 47: 106–111.

Suomela H (1970) Wheat uses and quality tests of wheat. Siemenjulkaisu 1970: 165–171.

Svensson G (1987) Situation der Weizenzüchtung in Skandinavien. Getreide, Mehl und Brot 41: 291–293.

Teemu Veijola (1960) The problem of the baking properties of wheat. Siemenjulkaisu 1960: 202–214.

Varis E (1960) Winter wheat. Siemenjulkaisu 1960: 25–36.

Varis E (1965) Winter wheat. Siemenjulkaisu 1965: 27–30.

Varis E (1970) Winter wheat. Siemenjulkaisu 1970: 27–34.

Varis E (1975) Winter wheat. Siemenjulkaisu 1975: 30–33, 39.

Westermarck N (1956) Die Finnische Landwirtschaft. Gesellschaft Pellervo, Helsinki, 88 pp.

Yearbook of Farm Statistics (1997) Information Centre of Ministry of Agriculture and Forestry, Helsinki.

Zadoks JC, Chang TT, Konzak CF (1974) A decimal code for the growth stages of cereals. Eucarpia Bulletin 7 (10 pp).

Zeven AC, Zeven-Hissink NCh (1976) Genealogies of 14 000 Wheat Varieties. The Netherlands Cereal Centre – NGC, Wageningen. The International Maize and Wheat Improvement Centre, Mexico.

Denmark

1. Introduction

The total area of agricultural land in Denmark (DK) in 1996 of 2.7 million hectares comprised 1.545 million hectares of cereals, of which 43.6% was under winter wheat and only 0.3% under spring wheat. The balance, almost 50%, was made by barley and a small area of winter rye and spring oats.

Table DK.1 demonstrates the increasing importance of wheat, coinciding with a decrease in barley growing during 1974–1996. Whereas barley, and especially spring barley, had been the main cereal crop for many years, the importance of wheat increased under the influence of Denmark's membership of the European Union. Within the total area of barley, the proportion of winter barley also increased, especially during the 1990s.

Wheat in Denmark is mainly autumn sown. The area under spring wheat is very small: only 2% of the total wheat harvest was produced by spring wheat during 1988–1990 (Salovaara & Fjell, 1995). In 1996 this proportion was even lower. Figures on wheat growing in different periods or years are presented in Table DK.2, where the very large increase in area and production since the 1980s is notable.

1.1. The use of the grain of Danish wheat

In the 1990s the main part (approximately 60%) of the Danish wheat crop has been used for domestic animal feed. The Danish milling industry uses between 350 000 and 400 000 t for human consumption of which on the average 85% is Danish grown and 15% is imported wheat. About 30% of the crop is exported as bread wheat or as animal feed to different countries.

Table DK.1. Cereal area (1000 ha) and relative importance (%) of wheat and barley between 1974 and 1997 in Denmark

Year(s)	Cereals	Wheat	Barley
1974–1977	1 769	6.5	84.4
1984–1987	1 595	22.4	67.7
1991–1994	1 514	38.1	54.0
1996	1 545	43.6	49.4

Sources: Eurostat.

Table DK.2. Area (1000 ha), grain yield (t/ha) and wheat production (1000 t) in Denmark since 1925

Year(s)	Area	Yield	Production
1924–1925	70	3.03	212
1934–1938	126	3.04	383
1948–1952	78	3.63	285
1961–1965	130	4.13	535
1969–1971	111	4.59	509
1979–1981	135	5.14	692
1989–1991	499	7.25	3 616
1995–1997	657	7.18	4 716

Source: IIA and FAO statistics.

1.2. Early interest in baking quality of Danish-grown wheats

Towards the end of the 1920s, Jörgensen tested wheat varieties grown in Denmark in baking trials on behalf of the Danish Home-Grown Wheat Comittee. The varieties were compared with Manitoba wheat and with Swedish varieties, such as Extra Kolben. It appeared that important differences in baking quality between Danish varieties were available (Åkerman, 1931). Jörgensen was one of the first scientists to carry out experiments to improve baking trials in use at the time. These baking trials were carried out with and without addition of potassium bromate. In addition, the amount of yeast to be added and the duration of the fermentation time for optimal CO_2 production was investigated (Jörgensen, 1931).

2. General review of wheat breeding for yield and quality

Table DK.3 lists the number of winter-wheat varieties admitted to the Official List over a number of years; clearly, the number of varieties from Danish breeders is far lower than the number from foreign breeders.

2.1. Winter wheat

Fourteen Danish winter wheat varieties have been released since 1986. Some information of these varieties is summarized in Table DK.4. The baking quality of Danish winter-wheat varieties varies strongly: the older varieties Anja and Kraka and the more recent varieties Lakona, Ure and Terra have intermediate scores for sedimentation value and loaf volume. Sevin Sejet and

Table DK.3. The number of winter- and spring-wheat varieties admitted to the Danish Official List of Varieties since 1986

Year	Winter wheat		Spring wheat	
	total	home-bred	total	home-bred
1984	14	2	3	1
1986	14	2	3	1
1988	14	2	3	1
1990	21	3	3	1
1992	36	4	3	0
1994	36	4	5	0
1996	33	4	3	0
1997	35	3	5	0

Source: Officiel Sortsliste.

Wase have low scores for these characteristics. For the sake of comparison: the mean scores for sedimentation value and loaf volume for foreign winter-wheat varieties on the Danish list (1993) were 5 and 4, with a variation from 2 to 9 and 1 to 7 respectively. Winter wheat varieties bred in Denmark were mainly important between 1984 and 1989, due to the varieties Kraka and Anja. Based on the areas of certified seeds, these varieties were grown on 61% (44–80%) of the area under winter wheat during this period. Thereafter, the importance of both varieties decreased, with varieties bred in Germany, Sweden, Great Britain and the Netherlands becoming increasingly important. Due mainly to sowing of the variety Terra, the proportion of Danish varieties grown in the country increased to about 19% during 1995 and 1996, but this was halved in following years (Source: Annual Reports of the Danish Plant Directorate).

2.2. Spring wheat

The Danish spring-wheat variety that featured on the Recommended List between 1982 and 1993 was Vitus Sejet, from the breeding firm Landbrugets Kornforaedling (LK), later named Sejet Planteforaedling. During the test period 1979–1981:

- the grain yield of Vitus Sejet was 105% compared to the standard variety Sappo (100% = 5.14 t/ha);
- the scores for sedimentation value and loaf volume were 7.5 and 7, respectively for Vitus Sejet compared to 8 and 8, respectively, for Sappo; some years later the scores for sedimentation value and loaf volume for Vitus Sejet dropped to 6 and 5, respectively.

Table DK.4. Danish home-bred winter wheat varieties with their year of release, breeder, pedigree and some yield and quality characteristics

Variety	Year	Br.[a]	Pedigree	Yield[b]	Winter-[c] hardiness	Lodging[c] resist.	Sediment. volume[c]	Loaf[c] volume
Anja	1986	Paj	Kranich/Caribo	113*A*	7	3	6	5
Kraka	1986	Paj	Kranich/Caribo	116*A*	8	4	6	5
Sevin Sejet	1991	Sej	B-40/Brigand	113*B*	–	3	2	1
Wase Sejet	1991	Sej	NS-732/Unc04-105	107*B*	7	3	2	2
Gro	1993	Paj	Armada/PF8253	103*C*	6	2		
Laki	1993	Sej	Kraka/Norman	109*C*	6	2		
Lakona	1993	Sej	Kraka/Avalon	99*D*	6	1	6	6
Ure	1993	Hum	selection from Vuka	96*D*	7	2	7	5
Lone	1995	Sej	Sleipner/Rendezvous	103*E*			5	2
Terra	1995	Paj	Kraka/TJB-730/3637	104*E*	6	–	6	5
Frimegu Abed	1996	Abed	KD-6019/Apollo					
Stakado Abed	1996	Abed	AD-7020/AD-7021					
Tagena	1996	Abed	A-1269/Gawain					

[a] Names of breeders have been abbreviated as Paj (Pajbjergfonden), Sej (Sejet), Hum (Hummeluhr).
[b] Yields are related to standard varieties A: Solid (100% = 5.93 t/ha); B: Kraka (100% = 8.01 t/ha); C: Kraka (100% = 7.46 t/ha); D: Kraka (100% = 8.91 t/ha); E: Blanding (100% = 7.81 t/ha) (Blanding is a mixture of Pepital, Hereward, Haven and Hussar).
[c] Score 1–9: a high value indicates a high degree of the characteristic in question.
Source: Groen Viden.

The mean scores for sedimentation value and loaf volume of the four spring wheat varieties from foreign breeders were 7 (6–8) and 6 (4–7), respectively, compared to 6 and 5, respectively, for Vitus Sejet in 1993. Thus, the mean scores for these two quality characteristics for the spring-wheat varieties from foreign breeders are higher than those for the imported winter-wheat varieties, which were 5 and 4, respectively (see Section 2.1.). This is in agreement with the general experience that the grain from spring wheats has a better baking quality than that from winter wheat (Knudsen, 1977).

Based on the areas of certified seed, three spring-wheat varieties were grown in 1984, one of which was bred in Denmark itself. This variety, Sevin Sejet, was sown on 7,2% of the area. Its importance grew during the next four years, covering a mean area of 32.8%, but decreased rapidly after 1988. Only foreign spring-wheat varieties, numbering 3–6 annually, were grown during the period 1991–1997.

J.M.

Acknowledgment

The author wishes to express his gratitude to Mr H. Haldrup (Pajbjergfonden, Odder) for his valuable comments on the manuscript of this chapter and for providing additional information on wheat breeding in Denmark.

References

Åkerman A (1931) Über die Backfähigkeit schwedischer Weizensorten und die Möglichkeit ihrer Verbesserung durch Züchtung. Der Züchter 3: 1–8.

Deneken G, Boesen B (different years) Groen Viden.

Knudsen PH (1977) Agriculture in Denmark. The Agricultural Council of Denmark, Copenhagen.

Jörgensen H (1931) On the separation of 'gassing power' (Diastatic activity) from strength in baking tests. Cereal Chemistry 8: 361–374.

Landbrugsstatistik 1996 (1997) 61. Årgang; Udgivet af Denmarks Statistik.

NN (different years) Officiel Sortsliste.

Rasmussen J (different years) Sorter af Korn.

Salovaara HA, Fjell KM (1995) Wheat usage in northern Europe. In Faridi H, Faubion JM (eds), Wheat end uses around the world. St Paul (MN): AACC, pp 149–166.

Germany

1. Introduction

With an annual production of around 20 Mt of wheat, Germany (G) is the second largest wheat producer in Europe after France. In addition to a vast area of wheat, considerable areas are sown with barley, rye and oats, bringing the total area under cereals to 6.6 million hectares, representing more than one-third of the total arable area in Germany. Barley and wheat are grown on comparable areas (around 2 Mha), whereas the area sown with rye and oats is in decline. The production of rye, which was three times greater than wheat production at the beginning of the 20th century, had decreased to less than one-third of the wheat production by 1995. The introduction of high-yielding hybrid rye varieties in 1990 did not stop this decline. The production of rye is concentrated in the northern part of Germany, on the relatively poor podsols; whereas the production of wheat, barley and oats is spread over the whole country, with the highest concentration of wheat on the rich brown soils of Bavaria. From 1945 to 1989, the ratio of the production of wheat in the Federal Republic of Germany (FRG) compared to the German Democratic Republic (GDR) was approximately three to one.

At the beginning of the 20th century, Germany was a large importer of wheat, second in Europe after the UK. After World War II, an increase in domestic production was pursued through raising the yield levels (without expanding the area under wheat, Table G.1). Very soon, Germany became

Table G.1. Wheat production (1000 mt) and area (1000 ha) under wheat, rye, barley and oats in Germany between 1934 and 1997

Year	Production	Area			
	wheat	wheat	rye	barley	oats
1934–1938	4 058	1 768	2 878	1 248	2 145
1948–1952	3 912	1 492	2 683	1 845	1 678
1961–1965	5 964	1 821	1 958	1 588	1 077
1969–1971	8 471	2 108	1 547	2 102	1 077
1979–1981	11 249	2 340	1 203	2 970	854
1989–1991	15 454	2 479	924	2 596	471
1995–1997	18 851	2 634	840	2 200	307

Source: FAO statistics.

self-sufficient and started to export wheat to other European countries. The net export of bread wheat in 1995 amounted to 3.6 Mt. Approximately one-third of the total wheat production in 1995 was used for milling, one-third was used as cattle feed on the farm and the remaining one-third was exported or used for industrial purposes, such as the production of starch and mixed feeds (DLG Mitteilungen, 7/1997).

Most wheat in Germany is autumn sown. The proportion of spring wheat varies between 10% and 20%, depending on the weather at sowing in autumn and the extent of winter killing. Spelt wheat is still grown on 10 000 to 15 000 ha in the south.

In the early 1990s, the area under durum wheat increased as a result of a temporary EU subvention. In 1995 the area was reduced to 7000 ha.

2. Wheat breeding before 1945

2.1. General remarks

At the beginning of the 20th century, scientific breeding in Germany occupied first place in Europe, both in terms of the number of governmental breeding institutes as the amount of practical work achieved by seed firms. In combination with several renowned handbooks on breeding and scientific journals published in Germany, this created a world reputation for German breeding in those years (Vavilov, 1934).

A large number of seed firms were involved in practical breeding, several of which form the basis of today's German breeding companies. Among the governmental institutions, two are of special interest with regard to wheat breeding for bread-making quality: the Breeding Institute of the University of Halle, where Roemer and Pelshenke began to study wheat quality in the 1920s, and the Breeding Institute of Weihenstephan in Bavaria, where special attention has been given to improving the quality of bread wheat since work started there in 1902. Apart from these two breeding institutes, the 'Reichsanstalt für Getreideverarbeitung' in Berlin has played an important role in studying milling and baking quality of cereals. At the beginning of the 20th century, this was the foremost institute for cereal milling and baking research in the world (Bolling, 1989).

The philosophy behind breeding wheat for quality in Europe and in Germany has been analysed by Isenbeck (1938). Beginning in the late 1870s, the need for highly productive varieties was induced by a large supply of cheap wheat from North America on the world market. Along with improvements in crop husbandry, the introduction of Squarehead (hence the German name 'Dickkopf') types from Great Britain in winter-wheat breeding and

Bordeaux types from France in spring-wheat breeding resulted in a strong increase in productivity, combined with a clear decrease in quality. At this point, developments in different parts of Germany took their own course.

In northern and middle Germany, small- and middle-sized mills were gradually replaced by big mills along the rivers and canals, thus increasing the distance between producers and millers. When the quality of the wheat started to deteriorate, there was no discussion between miller and farmer about their problems. Millers went their own way, tackling the problem by importing high quality wheat from abroad. By mixing imported and home-grown wheat they were able to produce appropriate flours for different purposes (Scharnagel, 1936). As a consequence, most of the wheat varieties grown in middle and northern Germany in the 1930s were derived from Squarehead and Bordeaux types.

In southern Germany things developed along different lines. In Bavaria, for example, small- and middle-sized mills continued to exist. The import of foreign wheat was a less obvious solution due to higher freight costs. As a consequence, the influence of low quality wheats and the expansion of Squarehead types was restricted. These developments may be illustrated by data in Table G.2, in which the increase in kernel yield between 1878 and 1934 is listed for different parts of the country, from north to south. The difference in price paid for imported wheat (9.25 Reichsmark/100 kg) and for German export wheat (5.55 Reichsmark/100 kg) in 1933 (Pelshenke, 1934) illustrates the fact that economic motives definitely played a role in the breeding objectives for wheat.

Table G.2. Kernel yields (kg/ha) of winter wheat in different parts of Germany between 1878 and 1934

Region	1878–1979	1909–1913	1929–1934
Prussia	1380	2310	2230
Saxony	1750	2700	2420
Baden	1150	1810	1780
Bavaria	1320	1630	1890
Germany (average)	1350	2130	2150

Source: Isenbeck (1938).

2.2. Methods used to measure quality

Before the 1930s, the quality of wheat varieties in Germany was generally defined by baking trials during the final stage of breeding. After the introduction of indirect tests, the most important parameters used to describe the quality of the gluten were the 'Quellzahl' and the 'Testzahl'. The Quellzahl was calculated from the 'Kleberquellprüfung' (= gluten-swell-test) as described by Berliner & Koopman (in Pelshenke, 1934). It measured the swelling capacity and dilution of gluten in a solution of lactic acid. The Testzahl was calculated from the 'Schrot-gär-methode' (= wholemeal-yeast method) as described by Pelshenke (1933). This method has been widely used by breeders in many countries as the Pelshenke test.

Wheat material in the final stages of the breeding programme would be submitted to farinograph, extensograph and/or alveograph analyses. A detailed description of these methods is given in Part I of this book.

Lein (1954) pointed out that the gluten quality was estimated twice, both in the Testzahl and in the Quellzahl, and that the importance of the amount of gluten was underestimated. Hoeser (1954) also suggested that breeders should pay more attention to gluten content and proposed to introduce five classes of gluten content for breeding lines under investigation. German literature more often refers to the amount of gluten ('Klebermenge') than to the amount of protein.

The gluten content (K), the Quellzahl (Q) and the Testzahl (P) have been combined in a 'Gütezahl' (G): G = 25K + 100Q + 50P (Pelshenke, 1938; Rosenstiel & Rundfelt, 1963), resulting in three categories of wheat:

- A quality: Gütezahl > 4500
- B quality: Gütezahl 3000–4500
- C quality: Gütezahl < 3000

The Gütezahl was used by the Bundessortenamt for the classification of wheat varieties between 1930 and the early 1960s and has been the subject of quite some discussion (Pelshenke, 1935; Bolling, 1989). The Gütezahl allowed the separation of varieties of low quality but did not differentiate between varieties of medium and good quality (Seibel, 1996). Moreover, it was possible that varieties were classified as A quality with this parameter, yet they would be classified as C quality by a baking test. Pelshenke (1959) suggested that, although the number of A-quality varieties had increased by the use of the Gütezahl, the quality of the gluten had remained the same (see Section 2.5).

2.3. Wheat breeding in northern and middle Germany

Referring to breeding work done before World War I in this part of Germany, Isenbeck (1938) claimed that most breeders had shown no interest in the quality of new varieties. They were convinced that the high productivity of the Squarehead types was incompatible with improved quality. This conclusion, which was totally wrong according to Isenbeck, had a great influence on general opinion and a strongly restraining influence on further breeding for better quality. It was not until after World War I, and especially the 1930s, that the

Table G.3. Yield and gluten quality in the offspring of five spring-wheat crosses

Crosses	Relative kernel yield[a] in:	
	Group A[b]	Group B[c]
Heines Kolben/Marquis	98.2	99.4
Heines Kolben/Garnet	106.7	108.9
Garnet/Peragis	103.3	100.2
Garnet/v.R.S.Dickkopf	123.6	109.2
Garnet/Heine III	98.2	101.9

[a] Yield related to Peragis (100).
[b] Group A: lines with Quellzahl > 15 and Testzahl > 50.
[c] Group B: lines with Quellzahl < 15 and Testzahl < 50.
Source: Vettel & Pelshenke (1934).

quality of home-grown wheat became more important and the possibilities for improvement started to be investigated. One of the breeders who was largely involved in these investigations was Franz Vettel who worked at the seed firm Heine in Hadmersleben. Vettel worked on issues of quality together with Pelshenke, who was working at the University of Halle. Later Pelshenke moved to Berlin and finally became director of the Research Institute at Detmold.

Vettel made crosses between five local spring-wheat varieties (Heines Kolben (a selection from the French Saumur), Peragis, von Rümkers Sommer Dickkopf and Heine III) and two American quality varieties (Marquis and Garnet) with the aim of selecting lines with a mixing value for millers ('Aufmischwert') that was comparable with that of the North American Manitoba wheat. The average quality (1932–1934) of the parent varieties used was summarized (Vettel & Pelshenke, 1934):

	Quellzahl	Testzahl	Gluten content
Garnet	17	137	40.3
Marquis	16	104	42.2
Heines Kolben	11	33	23.5
Peragis	8	27	26.2
v.R.S.Dickkopf	4	23	27.0
Heine III	9	24	22.2

The crosses, made in 1928–1929, were designed so that one of the parents was a quality parent. A rigourous selection on plant type in the offspring of the crosses resulted in more than 90% of the material being discarded, due to the 'exotic' character of Garnet and Marquis. The remaining lines were multiplied and investigated. Vettel expected that the best lines would be ready for release in 3 years. To analyse the data, collected over the years 1931–1934, the lines were divided in two groups: Group A, with relatively a high level, and Group B with relatively a low level of gluten quality, measured as Quellzahl and Testzahl. The relative kernel yields of the two groups, summarized in Table G.3, were compared. The data of a similar analysis, carried out for the amount of gluten, are not presented in this context.

The results indicated the possibility of selecting lines combining either gluten quality and kernel yield or amount of gluten and kernel yield. According to Vettel and Pelshenke, the winter-wheat breeding material of Heine-Hadmersleben justified the same conclusions.

Several spring-wheat varieties that were released in the following years were derived from these crosses made by Vettel. The cross between Garnet and Peragis resulted in the variety (Heines) Koga, an A quality variety released in 1938. Both Koga and Koga II figured in the pedigree of several A quality spring-wheat varieties released in the following years. The combination of Kolben and Peragis could be traced in several varieties, of which the most important was Heines Peko, a high yielding variety with B-C quality that was released in 1947 and was widely grown during the 1950s.

For the analysis of quality, Vettel & Pelshenke (1934) stated that both Quellzahl and Testzahl for the same sample gave more reliable results than the use of one of these methods alone. A further step forward in the test procedure was that the Quellprüfung could be performed with a smaller amount of flour. Thus it was possible to examine the yield of single plants for both gluten quality and amount of gluten. From earlier studies, the characteristic for good quality was presumed to be recessive, which made it possible to start selection as early as in the seeds of the single F_2-plants.

A group of varieties bred by Rudolf Carsten in Schleswig-Holstein (Northern Germany) is considered to be one of the sources of German quality wheats (Lein, 1958). His winter-wheat variety Carstens (Dickkopf) V, released in 1921, became very popular in the 1920s and was grown until the 1950s. There is no concensus about the exact pedigree of Carsten V, but it includes a Squarehead, the very winter-hardy Carstens III and Criewener 104, which was a selection from a German land variety. The winter-wheat variety Carsten V was crossed with the spring-wheat variety Heines Kolben in 1922; some years later, a line from that offspring was crossed with the Canadian variety Garnet and resulted in the variety Carstens Sommer Weizen. According to a leaflet 'Rudolf Carsten 1880–1950' written by an unknown author, two-thirds of the German wheat area in the 1920s was covered with Carstens varieties. It 1928 several crosses were made between Carstens varieties and Russian and

Canadian wheat material. One result was the variety Carstens VI, a variety with a high level of winter-hardiness and a good baking quality.

The names of most of the breeders may be recognized in the names of their varieties, released up to the 1950s. Among the large number of private breeders that were active in northern and mid-Germany in the early years of breeding, Rimpau and Strube deserve mention; both were based near Halberstadt (Sachsen). Wilhelm Rimpau, owner of a small seed firm, was considered by Vavilov (1935) as the father of scientific breeding in Germany. He started making crosses in the 1880s and his variety Rimpaus Bastard was extensively used in further breeding. Rimpau's seed firm has continued under the name von Lochow and later merged with Petkus. Strube is still operating under the same name. Material from these two breeders may be traced in the pedigree of many German varieties.

2.4. Wheat breeding in southern Germany

Plant breeding in southern Germany, started later than in the north. As pointed out earlier, the distance between farmers, millers and plant breeders was smaller in Bavaria than in some other parts of Germany. Moreover, the organization of breeding, testing and commercialization of varieties was in hands of the state, and thus producers and processors had more common interests. One of the results was the continuous attention paid to quality features by breeders. Hence squarehead types were not used as a basis for breeding work, but mainly land varieties appreciated by millers and bakers. Intensification of wheat production in this part of Germany took longer but was reached without the loss of quality characteristics and proved that high yield and good quality could be combined.

The Bayerische Landessaatzuchtanstalt in Weihenstephan has always played an important role in wheat breeding in southern Germany. It was established in 1902 with the objective of combining scientific research with practical breeding. Following encouragement by farmers to breed wheats with good baking quality, an experimental mill and bakery were established there in 1926.

2.4.1. Winter wheat

Scharnagel, who was the first director of the Institute at Weihenstephan, investigated a collection of 42 winter-wheat (and 15 spring-wheat) varieties, which included landvarieties and squarehead types. These were grown, harvested and stored under uniform conditions for three successive years. Seed samples were submitted to uniform milling and baking procedures, resulting in a large amount of data on yield, thousand kernel weight, ash, protein and gluten content, gluten quality and loaf characteristics (Scharnagel, 1929).

Table G.4. Relative kernel yield and loaf volume of some winter-wheat land varieties analysed at Weihenstephan in 1926–1928

Variety	Relative kernel yield	Relative loaf volume
Holzapfels Arras	105	110
Langs Trubilo	105	94
Traubling fr.St.15	103	101
Hauters Pfälzer	103	93
Engelens S 2	102	108
Strubes Gen.von Stocken	101	100
Weihenstephans Diva	99	95
Kraffts Siegerländer	94	110
Criewener 104	92	97
Zapfs Landweizen	86	106

Source: Scharnagel (1929).

From these data important differences in loaf volume between the varieties were detected. Eighteen varieties had combinations of loaf volume and kernel yield that were above average, leading to the conclusion that qualitative and quantitative yield need not exclude one another. Scharnagel was convinced that systematic combination breeding would make it possible to combine useful gluten characteristics in high yielding varieties. Some land varieties that Scharnagel considered to be promising for this purpose have been listed in Table G.4 in order of decreasing kernel yield.

Hoeser (1954) summarized the results in winter-wheat breeding in Weihenstephan since the 1920s. With regard to the origin of material with improved quality, two sources were considered to be important: on the one hand selections from land varieties, such as Zapfs Land(weizen), and on the other hand special combinations of parent varieties, such as Arras from France (see Chapter France, section 4.) and Traublinger Landsorte, which through transgression have resulted in increased quality levels.

An important variety, selected from the cross Trubilo × Arras, was (Langs) Tassilo, released in 1930. It became a major variety and was grown in southern Germany, especially during the 1930s and 1940s. It may also be traced in the pedigree of several Italian varieties. Tassilo had a good gluten quality and according to standards of that time its bread-making quality was good (A). The kernel yield of Tassilo was lower than that of Carsten V, which had a poor quality (C). This classification was based on Gütezahl and a baking test.

Table G.5. Winter-wheat varieties selected from populations of crosses made at Weihenstephan

Variety	Year	Pedigree	Quality group	Loaf volume
Lang-Dörflers Walthari	1941	Arras/Traublinger//Ackermans Jubel	A	495
Schweigers Taca	1942	Tassilo/Carsten V	A	523
Lichtis Braun von Guérard	1946	Tassilo/Darwin	A	–
Erbachshofer NOS Zapfs	1947	Tassilo/Carsten V	B	459
Neuzucht von Reininghaus	1948	Tassilo/Zapfs Land	A	531
Mauerner unbegrannter	1953	Tassilo/Carsten V //Tassilo	A	463
Stauderers Tarzan	1953	Tassilo/Zapfs Land//Tassilo/Carsten V	A	464
Cartsen V (standard)	1921	Carstens III/Dickkopf//Dickkopf/ Criewener 104[a] Dickkopf/ Criewener 104 // B16 de Russie[b]	C	391

[a] According to Zeven & Zeven (1976).
[b] According to Larose et al. (1956).
Source: Hoeser (1954).

Table G.6. Winter-wheat varieties released by private breeders and derived from quality lines from Weihenstephan

Variety	Year	Pedigree	Quality group
Strengs Marchall	1944	Frankenkaiser / Siegfried // Tassilo	A
Toerring II	1949	Hauter/Toerring Alt//Tassilo/Darwin	A
Engelens Festa	1950	Engelens F 4 / Strubes 56 // Tassilo	A(B)
Breustedts Werla	1950	Heine III//Rimpaus Fr.Bastard/Hauter II///Derenb.Silber	A
Breustedts Goten	1951	Heine III//Rimpaus Fr.Bastard/Hauter Il///Ridit/Panzer	AB
Engelens Attila	1953	Engelens F 4 / Strubes 56 // Tassilo	B
Dippes Triumph	1953	Tassilo / Carsten V //Derenburger Silber	B

Source: Hoeser (1954).

The backcross of Tassilo on Arras resulted in the variety Hauter II, released in 1938. Both Tassilo and Hauter have been used as a parent in many crosses and have had an enormous influence on winter-wheat material of southern Germany, as may be seen in Tables G.5 and G.6.

The release of the varieties Walthari and Taca in the early 1940s marked a major step forward in the efforts to combine improved yielding capacity with good baking quality.

Hoeser pointed out that more progress in breeding work was made after methods had been found to differentiate for quality in an early stage of selection. For selection work in Weihenstephan on the amount of gluten, the Quellzahl and the Testzahl were used. Older lines, starting in F_5, F_6 were tested in a baking trial. Later the baking test was expanded with additions of malt or malt and bromate.

Through close cooperation between the Institute at Weihenstephan and private breeding firms in southern Germany, private breeders could make use of the material for further testing and use as parents for their crosses. From the pedigree of the varieties released by the private firms (Table G.6), it is clear that the material from Weihenstephan was frequently used.

2.4.2. Spring wheat

In a review of spring-wheat breeding at Weihenstephan since the 1920s, Hoeser (1954) stated that breeding for quality in spring wheat was relatively easier than in winter wheat, since more venerable gene material of spring wheat was available. As a consequence, work was mainly confined to improving the yielding capacity and disease resistance of existing material. Several of the existing spring-wheat varieties in use had been derived from foreign wheats:

- Lohmans Weender Galizischer Kolben, of Galician origin, grown since 1892
- Strubes roter Schlanstedter, derived from the French variety Bordeaux, grown since 1907
- Janetzkis Früher, of Austrian origin, grown since 1914
- Svalöfs Extra Kolben, derived from Sweden, since 1924
- Heines Kolben, derived from the French variety Saumur, grown since 1920

An important result of the breeding efforts at Weihenstephan was the release of Lichtis Früh (Table G.7), an early ripening spring-wheat variety, selected from a cross involving the Canadian variety Garnet. Lichtis Früh became the most important spring-wheat variety grown in southern Germany after its

Table G.7. Spring-wheat varieties selected from populations of crosses made at Weihenstephan

Variety	Year of release	Pedigree	Quality
Lichtis Früh	1938	Garnet//Janetzkis Früh/Strube line	A
v.Rümkers Erli	1949	Erbachshofer Janetzki / Japhet// Lichtis Früh	A
Lichti II	1952	Lichtis Früh // Hohenheimer Franken	A
Wahrberger Teutonen	1953	Lichtis Früh // Breustedts Teutonen	A

Source: Hoeser (1954).

release in 1938; it was grown until the 1960s. The variety Erli, released in 1949, was grown widely until 1964.

2.5. The 'Kleberweizen-Aktion'

A campaign to stimulate the growth of wheat varieties with improved quality was initiated in 1935 (Schmidt, 1936, 1938; Hoeser, 1954). Varieties that after three years trial had proved to have a certain level of quality, could be classified as 'Kleberweizen'[1]. For certification, farmers had to submit samples for analysis of gluten content, Testzahl, Quellzahl and Gütezahl. Batches with an official certificate were paid an extra allowance of two Reichsmark per 100 kg, which was 10% of the average wheat price during 1936–1938. This extra allowance was an important step forward in the popularity of qualitatively improved wheat varieties and an incentive for breeders to develop this type of variety. Hoeser (1954) calculated that the number of A quality varieties increased from 30% to 70% between 1935 and 1945, while the number of B varieties decreased from 60% to 20%. Between 1945 and 1952, this situation remained more or less stable.

The winter-wheat varieties most widely grown between 1935 and 1950 under the Kleberweizen Aktion were Tassilo, Firlbeck, Hauter, Walthari and Taca, all classified as A quality varieties. The most important spring-wheat varieties were Janetzkis Früh, Lichtis Früh, Wahrburger Spring wheat and N.O.S. Nordgau, again all A quality varieties. The Kleberweizen Aktion stopped in 1950.

Peshenke (1954) argued that the distinction between medium (B) and poor (C) quality wheats had become blurred as a result of the Kleberweizen Aktion. He stated that the average quality level of German wheat varieties was at an alarmingly low level in the early 1950s. Comparing the quality level of three generations of wheat varieties, he listed the the following figures:

	Gluten content	Gluten quality	
		Quellzahl	Testzahl
Land varieties	23,8	14	–
1935–1939	22,0	6	29
1951–1953	18,9	6	31

[1] 'Kleberweizen' litterally means 'gluten wheat'. The term 'Kleberweizen' should not be confused with the term Klebender or Klebriger Weizen, used in the 1970s to indicate varieties that produced a sticky dough (see Sections 3.2.1 and 3.2.2).

Pelshenke stated that the decrease in gluten content was mainly the result of increased yields of modern varieties and was also related to conditions during World War II and the post war period, when farmers were forced to concentrate on producing large quantities of wheat at low levels of fertilizer. This, he said, had resulted in less attention for quality by both farmers and breeders. In order to meet the demands of the milling industry, he proposed raising the gluten content to 23–25%. He suggested that it might be easier to achieve this through the application of nitrogen, and especially late in the growing season, rather than through breeding. In his opinion it was more important for breeders to concentrate on the quality of the gluten. The Swiss variety Probus, as well as several Swedish varieties, demonstrated in his eyes, the possibility of combining high yields with quality.

3. Wheat breeding after 1945

Breeding during and shortly after World War II was related to efforts to increase production in order to become self sufficient in wheat. This meant amongst other things the selection of varieties that could be grown on relatively poor soils and thus take the place of rye. After a vast production increase had been achieved in the course of the 1950s, the quality of the wheat was again a point of attention. From 1945 onwards, breeding work in Germany will be described for the GDR and the FRG seperately, as these two parts of Germany have followed a different course in breeding.

3.1. Wheat breeding in the GDR

The work of Vettel was continued at the breeding station Kloster-Hadmersleben, the former establishment of the seed firm Heine-Hadmersleben. In close cooperation with the University of Halle, Vettel continued his efforts to combine reliability of yield and harvest, yield potential and quality in the same wheat variety (Vettel, 1956).

3.1.1. Winter wheat

The number of winter-wheat varieties with improved quality grown in the GDR in the early 1950s was very low. The only A quality variety recommended was Carsten VI, a variety that was not very popular due to its low productivity. Some of the most important high yielding varieties were Derenburger Silber, Hadmerslebener IV (= Heine IV) and Bastard II, of either B or C quality (based on Gütezahl).

Apart from these three varieties, other crossing parents used by Vettel were Rimpaus Braun, Schreibers Sturm, Admont V40 and Graniatka Sachod-

Table G.8. Winter-wheat varieties in the GDR in 1966 with their year of release, pedigree, area sown in 1966 and quality group

Variety	Year	Pedigree	Area (%)	Quality group
Fanal	1961	Blé 206/Kronen//Derenburger Silber	42	C
Qualitas	1957	Carsten VI / Rimpaus Braun	33	A
Basta	1961	Rimpaus früher Bastard / Standard	8	
Hochland	1955	Svalöf 987//Ceres/Traublinger///Schlötenitzer	6	
Trumpf	1961	Heine IV (= Hadmerslebener IV)	6	B/C
Eros	1958	Halle 813.36 / Heine IV	3	
Muck	1962	Marquis/Sv.Kronen//Rimpaus Braun	2	

Sources: Broekhuizen (1969) and Lupton (1992).

nia (from Poland) for their productivity; and Carsten VI, Tassilo, Marquis (Canada), Blé 206 (France) and Kronen (Sweden) for their quality.

Based on the findings from his breeding material, Vettel was rather optimistic in 1956 about the possibility of creating varieties with quality that could reach the yield levels of the best yielding varieties of those days. Indeed, the variety (Hadmerslebener) Qualitas was released a year later. This A quality variety, selected from a cross between Carsten VI and Rimpaus Braun, was able to compete with the most productive wheat varieties of the day and was widely grown until the early 1970s.

From the pedigree of the recommended varieties grown in the GDR in 1966 (Table G.8), the influence of Vettel's work may be clearly recognized. Among these varieties, Qualitas was the only one with A quality; Fanal was a very high yielding variety.

A complex cross, including Derenburger Silber, Koga, Marquis, Kronen and Hadmerslebener IV resulted in the variety Poros. Subsequently a combination of this variety with Carsten VIII and the Australian variety Record resulted in the variety Alcedo, which was released in 1974. The bread-making and milling quality of Alcedo were very good. It was widely grown until the 1990s, and in 1981 it covered almost half of the winter-wheat area in the GDR.

Considering the home-bred varieties on the List of Varieties of 1984 (Table G.9), the increasing influence of foreign material in the parents is obvious. According to the descriptions of these varieties (a classification was not used in the GDR) they all had good bread-making characteristics. Starting in the 1970s, efforts were made to increase the protein content of existing material. At the Institute for Genetics and Crop Research in Gatersleben, a large number of entries from very different geographical backgrounds were tested for

Table G.9. Winter-wheat varieties released in the DDR between 1974 and 1984

Variety	Year	Pedigree
Alcedo	1974	Record/Poros//Carsten VIII
Compal	1981	Hadm.36044.71//Atlas/Hadm.3071.66/3/Alcedo
Taras	1982	Kavkas/Tadorna//Gaines/Hadm. 338.58
Arkos	1983	Cato/Suwon//Alcedo
Miras	1984	Mir-808//Bezostaya-1/Erythrospernum-1526

Source: GDR List of Varieties 1984.

Table G.10. Relative yield and baking quality of a number of lines selected from a population of a cross between two sister lines from Heines Kolben / Peragis

	Relative yield	Testzahl
Hadm. Stamm 4148	106.6	35
Hadm. Stamm 4149	102.0	38
4149×4148 (24 lines)	104.7	37
range	97.5–113.6	25–52
Lines with transgressive yield and/or quality		
21 764	103.3	49
21 782	109.3	44
21 611	107.6	44
21 704	106.0	42
21 622	109.1	35
21 591	108.8	34
21 643	104.2	34

Source: Vettel (1956).

several years on different sites (Porsche et al., 1979). In addition to Kavkaz, Atlas and Scoutland, commonly used as sources of high protein, several new sources were discovered. The American variety Atlas 66 was widely used as a source for increasing protein content, both for bread making and for animal-feeding value (Richter, 1984).

3.1.2. Spring wheat

As for winter wheat, a superiority of high yielding varieties over varieties with improved quality existed among the spring-wheat varieties grown in the GDR during the early 1950s. The highest yielding and most popular variety was Peko (released in 1947), with B-C quality. Koga, released in 1938, selected

Table G.11. Recommended spring-wheat varieties in the DDR (1966–1968) with their year of release, pedigree and proportion of area cropped with them in 1966 (the pedigree of Garant and Peragis II has been added for information)

Variety	Year	Pedigree	Area (%)
Remo(=Peko)	1944	Heines Kolben/Peragis	10
Capega	1951	Carsten V/Garant	13
Herma	1962	Capega/Garant	42
Derwisch	1962	Stamm 310/Peragis II	18
Carola	1964	Capega/Garant	17
Garant		*Peragis/Garnet*	
Peragis II		*Garnet/Peragis*	

Sources: Broekhuizen (1969) and Lupton (1992).

in a cross between Garnet and Heines Kolben had the best quality (A). A descendant from Koga, Koga II (Koga//Heines Kolben/Räckes Weiszspelz), was released in 1953 and was a very popular variety in the 1960s. It belonged to the category 'Kleberweizen'.

Vettel used the following parents for his crosses:

- A quality: Erli, Rimpaus Langensteiner, Weihenstephaner KJH, Peragis Sommer Weizen II and Hadmerslebener Koga;
- A/B quality: Adlungs Allemannen and Breustedts Teutonen, and;
- B quality: Janetzkis Jabo.

Moreover, several lines derived from crosses involving Kolben, Peragis, Räckes Weiszpelz, Marquis, Garnet and Red Fife were frequently used as a second parent.

Vettel (1956) analysed the offspring of 21 crosses by classifying the material into Group A (Testzahl > 50), Group B (Testzahl = 31–50) and Group C (Testzahl < 30). From comparisons of the relative yields of these groups, it appeared that, for the average of 1954 and 1955, 43.5% of the lines with kernel yields comparable to Peko were of A quality and 56.5% of B quality. From these results Vettel concluded that the combination of yield and quality had been almost reached in the variety Hadmersleben.

Vettel was encouraged by his findings in the offspring of a cross between two sister lines originating from Heines Kolben/Peragis. Several lines showed transgression in both yield and Testzahl. Data for the two sister lines HS 4148 (average 1939–1954) and HS 4149 (average 1940–1954), of some transgressive lines (average 1953–1954) and the average and range of 24 lines of the cross between these sister lines, are presented in Table G.10.

From the pedigree of some of the spring-wheat varieties that were recommended in 1966–1968 (Table G.11), the influence of the work at Halle-Harmersleben is obvious.

Seibel (1996) stated that the potential for bread-making quality in the wheat varieties grown in the GDR at the time of the 'Wende' was remarkably high. That potential was not optimally expressed due to lack of production inputs such as fertilizers. It formed a very good basis, however, for bringing the quality of the German wheat harvests in the following years to levels that are among the highest in Europe.

3.2. Wheat breeding in the FRG

The 1950s in the FRG were characterized by an emphasis on production increases, so varieties with adequate disease resistance and yield reliability were given priority over varieties for quality. It was not until the late 1950s, with the European Community and the end of the availability of cheap American wheat in view, that renewed interest in wheat quality developed (Bolling, 1989).

A vast amount of knowledge has been generated by the Wheat Research Group (Arbeitsgemeinschaft für Getreideforschung) at Detmold, where Pelshenke continued his work after 1945, followed later by, for example, Bolling and Seibel. Since 1954, workshops on matters related to cereal quality have been organized here on a regular basis. The work carried out at Detmold supports private breeders. In view of the large number of breeders and the continuous process of merging and changing of names, the results of their work will be discussed here as a whole, based on several reviews.

3.2.1. The classification of wheat varieties

After years of discussion about the efficiency of a classification based on the Gütezahl, in the early 1960s the value of baking trials was re-evaluated. The correlation between the components of the Gütezahl (gluten content, Quellzahl and Pelshenke value) and the baking test were calculated. The correlation between the Quellzahl and loaf volume appeared to be very low (r = 0.13). Gluten content and Pelshenke value, as well as Zeleny sedimentation value, showed a better correlation with loaf volumes.

From 1963 onwards, a systematic classification of new wheat varieties was introduced and published in the variety list. Based on the loaf volume in three years of standard baking tests (carried out in Berlin, Detmold or Weihenstephan), varieties were classified in three categories A, B or C.

Within a few years time, the standard baking test was replaced by a rapid-mix test (RMT), because the correlation between loaf volume and some essential quality parameters was higher in a rapid-mix test than in the standard baking test. Bolling (1989) calculated the correlation with protein content (r = 0.85) and with sedimentation value (r = 0.81).

The RMT was used for official testing of varieties and for testing of commercial flours. Intensive kneeding of the dough turned out to be a very positive factor in the differentiation of wheat quality. Kneading also revealed that certain varieties produced sticky doughs, for which an additional category was introduced in 1974: the T quality. A variety with this classification could not be processed mechanically unless it was mixed with a non-sticky variety. Such varieties were also indicated by the term 'Klebender Weizen or Klebriger Weizen' (see Part I, Chapter 7), which should not be confused with the term Kleberweizen, used between 1935 and 1950 for varieties with good gluten and bread-making characteristics.

The baking test was further improved by the introduction of the Hagberg falling number. An improvement of the milling capacity of German wheat varieties was pursued in the 1970s, with the objective to reach similar extraction rates as in the French varieties, of which increasing quantities were milled in Germany.

The nomenclature of wheat classes has changed at regular intervals since the 1960s. After a further subdivision of A and B wheats in AI, AII, BI and BII, combined with a Q (for special mixing value) and a T (for dough stickiness) value in 1971, a new classification in A9-A6, B5-B3, C2–1 was adopted in 1982. This classification system, in which the parameter for bread volume played an important role, became the subject of increasing criticism, because it often resulted in batches of quality varieties being refused by the milling industry. In the most recent classification system, introduced in 1995, the number of categories has been reduced to five, including a new category for biscuit wheat:

- E – Elite (improver) wheat (including the previous A8/A9 varieties)
- A – quality wheat (including the previous A6/A7 varieties)
- B – bread wheat
- K – Keks (biscuit) wheat
- C – other wheat

Classification is based on bread volume in a rapid-mix test, elasticity and stickiness of dough, Hagberg falling number, crude protein content, sedimentation value, water absorption and the extraction rate of the flour. It is the first German classification system in which a parameter for milling quality has been included. A detailed description of the criteria for the five categories may be found in the German variety list and in Steinberger et al. (1995).

Table G.12. Some important winter-wheat varieties released between 1950 and 1975 with their year of release, pedigree and quality

Variety	Year of release	Pedigree	Quality group[a]
Heine VII	1950	Hybride A courte paille/Svalöfs Kronen	C
Schernauer	1954	Walthari//Tassilo/Strubes Kreuzung 56	BI
Merlin	1956	Heine VII/Derenburger Silber	C
Jubilar	1961	Schernauer//Taca/Derenburger Silber	BI
Diplomat	1966	Merlin/Format	AII/Q+
Caribo	1968	Carsten VIII/Cappelle	BI
Kranich	1969	Heine 2167/Heine VII//Merlin	BII/T-
Benno	1973	Carsten VIII/Weihenstephan 565.52	BII/T-
Kormoran	1973	Cappelle/Heine 2606//Heine 646	AII/Q-
Disponent	1975	Benno/Florian	AII/Q

[a] Ranging from the best bread-making quality (AI) to biscuit quality (C); Q+ = high mixing value; T- = production of sticky dough.
Source: Hoeser et al. (1979).

3.2.2. Winter wheat

The release of the variety Heine VII in 1950 set new standards for yielding capacity. This variety with short and stiff straw was widely grown throughout Germany, as well as some neighbouring countries. Nevertheless, the market had to wait until the 1960s for the combination of better quality and improved kernel yield, when the varieties Jubilar (1962) and Caribo (1968) were released. An even better quality level in combination with good yielding capacity was achieved in the variety Diplomat, released in 1966. These three varieties were the dominant winter-wheats in the 1970s.

With the introduction of the varieties Kranich and Benno, a new category of wheats was created in Germany, i.e. wheats that produce sticky doughs when processed with modern dough-mixing techniques. A further increase in kernel yield was achieved in the varieties Kormoran and Disponent, both with a high quality level (AII). Table G.12 summarizes this information on all these varieties.

At Weihenstephan, Hoeser continued his efforts to combine increased yielding capacity and stability with optimal quality characteristics by simultaneously selecting on straw quality, winterhardiness, disease resistance and yield, on the one hand, and on sprouting resistance, gluten characteristics and baking quality, on the other hand.

The work of Hoeser et al. (1979) illustrates the results of breeding for bread-making quality since the 1930s. The kernel yield and baking quality of

Table G.13. Bread-making quality (old and new classification), kernel yield (with and without growth inhibitor CCC), protein content and loaf volume of ten important wheat varieties released in southern Germany between 1921 and 1980; results of harvest 1977–1978 at two locations

Variety	Marketed between	Quality		Relative yield		Protein content	Loaf[c] volume
		old	new	-CCC[a]	+CCC[b]		
Carsten V	1921–1960	C	BII	129	100	13.0	88
Tassilo	1930–1960	A	BI/C	100	100	15.9	104
Walthari	1941–1966	A	BI	128	104	14.9	105
Taca	1942–1966	B	BI	108	112	13.7	98
Heine VII	1950–1962	C	BII/C	150	124	13.4	88
Schernauer	1954–1978		BI	136	120	12.7	99
Jubilar	1962–1980		BI	148	127	13.4	100
Caribo	1968–1990		BI	178	147	13.3	100
Disponent	1975–1988		AII	188	153	13.8	114
(Kron)juwel	1980–1990		BI	202	160	13.3	101

[a] 100% (Tassilo) = 3270 kg/ha.
[b] 100% (Tassilo) = 4800 kg/ha
[c]rapid-mix test; 100% (Caribo) = 569 ml.
Source: Hoeser et al. (1979).

10 important winter-wheat varieties, grown in different periods between 1920 and 1975, were compared. With the exception of Carsten V and Heine VII, originating from northern Germany, the varieties were rather closely related and may be considered to reflect the breeding program in southern Germany during this period. The quality of the material was investigated by modern laboratory methods, including a sedimentation test and a rapid-mix baking test. The results are summarized in Table G.13.

The authors stated that within a period of fifty years a simultaneous increase in yield potential and improvement in baking quality had been achieved. The data proved once more that a combination of quality and quantity in one variety was realistic.

These data show that steady increase in kernel yield went hand in hand with decreased protein content in the kernel; this was mainly due to an increased harvest index and a 'dilution'of the protein in the kernel. The same trend was described by Lupton (1982) (see Chapter UK, section 3.2.1). The fact that Tassilo, considered as an A quality wheat in 1930, was classified as BI/C 60 years later illustrates the change in perception that has taken place over the years.

Table G.14. Main winter-wheat varieties with good bread-making characteristics released in the FRG after 1980

Variety	Year	Pedigree	Quality group	Yield[a]
Rektor	1980	Monopol/Kormoran	E	4
Sperber	1982	Robert/Merlin//Kormoran	A	5
Herzog	1986	617-667AD/Kormoran//Kronjuwel	A	6
Borenos	1987	Alcedo/Kenya Civet//Dornburg4056/3/Alcedo	E	6
Zentos	1989	Hadmerslebener 5792.71/Alcedo//Compal	E	5
Bussard	1990	Kranich/Maris Huntsman/Monopol	E	5

[a] Kernel yield is classed with a value ranging from 1 to 9: 1 = very low and 9 = very high; 5 = the average of the group of varieties under consideration.

The highest loaf volume was found in the cultivar Disponent (10 ml more than Tassilo), whereas the protein content of Disponent was 2% lower than that of Tassilo. With the introduction of Disponent, a further increase in yielding capacity was achieved without sacrificing the quality level. The most important varieties with E or A quality released since 1980 have been listed in Table G.14.

The availability and cropping area of winter-wheat varieties with improved bread-making quality in Germany have evolved very effectively. To illustrate the proportions it has assumed, some data from variety lists have been compared for the years 1971, 1981 and 1994. The following tendencies may be derived from these data:

- the number of winter-wheat varieties on the list increased from 37 in 1971 to 68 in 1997, although the lifetime of the varieties has become shorter (N.B. as the number of varieties in Group C was very low, these data were not included in this comparison);
- the number of varieties in quality groups A and E (i.e. with high loaf volumes) increased considerably in 1997 compared to previous years; in 1998 as much as two-thirds of the area under winter wheat was planted with E and A quality varieties;
- the number of varieties in quality group C (i.e. with low loaf volume) was low in each of the years;
- it is obvious that the score for kernel yield of varieties with A (and E) quality is lower than for varieties with B quality; the data in Table G.15 do not indicate that this discrepancy has been reduced between 1971 and 1997.

A closer look at the official list of 1994, comprising 43 varieties with A quality and 21 varieties with B quality, makes clear the relation between

Table G.15. Winter-wheat varieties according to the German (FRG) list of varieties for 1971, 1981 and 1997

Year	Quality groups	n	Area (%)[a]	Yield score[b]	Lifetime (years)	Number of breeders
1971	A	20	47	4.9	7	
	B	15	52	5.4	8	
	C	2	<1	3.0	19	German: 21
1971 total		37				foreign: 1
1981	A	18	43	4.9	5	
	B	28	52	5.6	4	
	C	1	5	6.0	6	German: 19
1981 total		47				foreign: 1
1997	E	10	26	3.8		
	A	26	36	5.7		
	B	26	32	6.2		
	C	9	6	7.6		German: 21
1997 total		71				foreign: 5

[a] Based on the area submitted for seed multiplication.
[b] Kernel yield is classed with a value ranging from 1 to 9: 1 = very low and 9 = very high; 5 = the average of the group of varieties under consideration.

score for kernel yield and year of release of the variety. Considering all 64 varieties, a clear trend of increasing yield score in progressive years can be found. The trend occurs both in A and in B quality varieties. Thus it is evident that the kernel yield of newly released varieties is increasing; a certain arrears in yield for qualitatively better varieties (A) compared with varieties of moderate quality (B) is obvious. The average score for kernel yield is 4.74 for A varieties (n = 43) and 6.67 for the B varieties (n = 21). Analysis of the yield score within the subgroups, ranging from A9 (highest quality) to B3, shows the following results:

A9: 3.33 (n = 6)	B5: 6.50 (n = 8)
A8: 4.33 (n = 6)	B4: 6.55 (n = 11)
A7: 4.67 (n = 12)	B3: 8.00 (n = 2)
A6: 5.37 (n = 19)	

The large difference in yield, between A9 (3.33) and B3 (8.00) varieties, illustrates the persistant discrepancy between quality and quantity in wheat. On the official list in 1997 half of the winter-wheat varieties were E or A class varieties, accounting for almost two-thirds of that year's seed production. Considering the difference in yield score between E/A and B/C quality varieties, one would expect that this situation can only be supported by a premium on the price for quality wheat. Nevertheless, since the 1980s this has hardly been the case, except in years with adverse weather (Seibel, 1996). As, moreover, roughly one-third of total German wheat production is used as cattle feed on the farm, the high proportion of A and E varieties is even more striking.

3.2.3. Spring wheat

At Weihenstephan a large number of crosses have been made, based on the quality varieties Erli and Thatcher (Hoeser, 1958). Erli may be found in the pedigree of Opal, Janus, Arkas and Schirokko, all German spring-wheat varieties with A quality grown between 1960 and 1980. The most important variety grown in that period was Kolibri, released in 1966 by von Lochow-Petkus. Kolibri dominated German spring-wheat hectarage during the 1970s and was popular in Poland, Switzerland, France and the UK as well.

Spring-wheat varieties with A quality, e.g. Ralle, Turbo, Star and Hanno, have come and to some extent gone since. The most recent spring-wheat variety, covering the majority of the area under spring wheat in the early 1990s is Nandu, which was released in 1988 by the same breeder as Kolibri. In 1994 Nandu was classified as an A8 variety but in 1995 it was degraded to Class B under the new classification system.

From comparison of some figures for spring wheat on the official lists of 1971, 1981 and 1994, it appears that the situation is somewhat different from that of winter wheat (see Table G.15). The figures for spring wheat have been listed in Table G.16 according to quality group, number, yield, lifetime and breeder. Apart from two B quality varieties admitted in 1971, all the spring-wheat varieties have A quality. Most of the varieties are home-bred and the number more or less consistant for the three years being compared. There is a trend of earlier replacement as the lifetime of the varieties decreased from 7 years in 1971 to 4 and 5 years in 1981 and 1994, respectively.

3.2.4. Investigations on glutenin and gliadin composition

Systematic analyses of HMW glutenin composition of German wheat varieties have not been carried out on the same scale as in France, Spain, the UK and Eastern European countries.

Odenbach & Mahgoub (1987) studied the HMW glutenin compostion of 107 German-grown wheat varieties and found a clear relation between the

Table G.16. Spring-wheat varieties according to the German (BRD) list of varieties for 1971, 1981 and 1994

	Quality group	n	Yield score	Lifetime (years)	Number of breeders
1971	AI	3	6.8	7	German: 11
	AII	10	3.4		foreign: 1
	BI	2	5.5		
1981	AI	1	7	4	German: 9
	AII	20	5.3		foreign: 1
1994	A9	2	5.5	5	German:8
	A8	6	5.3		foreign: 2
	A7	8	6.4		
	A6	2	6.5		

occurance of specific subunits and the quality group (ranging from A9 to C1) of the varieties. Rogers et al. (1989) investigated the glutenin and gliadin composition of 57 German grown wheat varieties in Cambridge. Their results indicated that some 30% of the variation in bread-making quality, measured as SDS sedimentation volume, was accounted for by variation in the *Glu-1* quality score. This proportion increased to 54% when the score was adjusted for presence of a 1BL/1RS translocation, which occured more frequently in this set of German varieties than in other sets studied by the same authors. The effect of allelic variation at the *Gli-1* and *Gli-2* loci was very low, except from variation caused by the 1BL/1RS translocation.

The HMW glutenin susbunits of the most important varieties and their *Glu-1* scores are presented in Table G.17. Unfortunately, no data are available on the glutenin composition of the older German varieties and no comment can be given on the origin of the different subunits.

Rogers et al. reported finding the subunit pair 3+12 (*Glu-D1*) in the variety Jubilar, which contradicts the results of Odenbach and Mahgoub who reported that it contains the subunit pair 2+12 (*Glu-D1*). Jubilar may have inherited the subunits 3+12 (*Glu-D1*) from the French land variety Arras. The same subunits occur in some older Italian and French varieties, such as Rieti and Vilmorin.

The average quality score for the 57 varieties investigated by Rogers et al. was 5.7, which is comparable to the average of UK-grown varieties (5.2)

Table G.17. HMW glutenin subunit composition and quality classification of German-grown varieties

Variety	Year of release	Class[a]	HMW subunits			*Glu-1* score
			1A	1B	1D	
Merlin	1956	C	N	6+8	2+12	4
Jubilar	1962	BI	N	7	3+12	4
Diplomat	1966	AII	1	7+9	5+10	9
Caribo	1968	BI	N	7	2+12	4
Kranich	1969	BII/T-	N	6+8	2+12	4
Benno	1973	BII/T-	1	6+8	2+12	6
Kormoran	1973	AII	N	7+9	5+10	7
Alcedo	1974		N	7+9	5+10	7[b]
Disponent	1975	AII	1	7+9	5+10	9
Monopol	1975	A9(E)	1	7+9	5+10	9
Kronjuwel	1980	BI	N	6+8	2+12	4
Rektor	1980	A9(E)	N	7+9	5+10	7
Kanzler	1980	A6(B)	N	7+8	2+12	6
Sperber	1982	A7(A)	N	7+9	5+10	7
Arkos	1983		N	7+9	5+10	7
Herzog	1986	A6(A)	N	7+9	2+12	5
Borenos	1987	A8(E)	N	7+9	2+12	5[b]
Zentos	1988	A8(E)	N	7+9	5+10	7[b]
Bussard	1990	A9(E)	1	7+9	5+10	9[c]

[a] The 1995 classification is given between brackets.
[b] After Sasek et al. (1995).
[c] After Zeller (1995).
Sources: Odenbach & Mahgoub (1987) and Rogers et al. (1989).

but much lower than that of Finnish and Australian varieties (8.2 and 8.0, respectively) according to Sonntag et al. (1986).

Variation in HMW glutenin composition of the German-grown varieties investigated was rather low, especially on the *Glu-D1* locus. This was supported by the findings of Moonen et al. (1984), who analysed the glutenin composition of 35 German winter-wheat and 19 spring-wheat varieties. They calculated that 40% of the winter wheat and 90% of the spring-wheat varieties possessed the subunit-pair 5+10 (*Glu-D1*). Fortunately these subunits are related to good bread-making characteristics. Cooke (1995) calculated from a collection of 64 German wheat varieties that almost 60% of the varieties possessed the subunits 5+10 (*Glu-D1*). This proportion is relatively high compared to surrounding countries. Indeed, within Europe a similar proportion was only found in Spain and Russia (see Part I, Table 7.1). The effect of variation in gliadin composition at the *Gli-1* and *Gli-2* loci in the set of varieties investigated by Rogers et al. (1989) was small and, if present, related to a 1BL/1RS rye translocation.

At the Institute for Flour and Protein Research in Garching and the Technical University in München, investigations on glutenin and gliadin composition of wheat have been undertaken since the late 1980s. Wieser et al. (1990) used reversed-phase high performance liquid chromatography (RP-HPLC) to isolate the different gluten components and classify them in three groups:

- a high molecular weight group with HMW glutenin subunits;
- a medium molecular weight group, including ω gliadins;
- a low molecular weight group, including α and γ gliadins and LMW subunits of glutenin.

Characteristic sequences of amino-acids within these groups were responsible for differences in gluten strength. Similarly, the results of a comparative study of wheat and rye cultivars (Kipp et al., 1995) indicated that the amino-acid composition of single subunits could explain the major part of the difference in bread-making characteristics of rye and wheat.

4. Sources of baking quality in German wheat varieties

In their analysis of factors contributing to baking quality in German wheat varieties, Wienhues & Giessen (1957) found that there was nearly no overlap between the genealogy of winter-wheat varieties from northern and southern Germany up to the 1950s. Wheat breeding in the south reflected the clear influence of the breeding institute at Weihenstephan, whereas in the north the work of several private breeders was more pronounced, albeit with some influence from the University of Halle. Wienhues & Giessen stated that, generally speaking, northern varieties were linked to Swedish material, whereas southern varieties were more closely linked to material from France. Moreover, they considered Rimpaus früher Bastard (Squarehead/early American red land variety) and Strubes Kreuzung 56 (Noé/German landvariety) as the two most important varieties on which early wheat breeding in northern Germany was based.

To illustrate the colourful background of some early German wheat varieties, the pedigree of Derenburger Silber (1941), Felix (1943) and Heine VII (1950), all three of poor bread-making quality, is presented with the country of origin of their parents.

Derenburger Silber	Pansar III / Peragis ***Sweden / N. America***
Felix	Tassilo / Carsten // Carsten / Marquillo ***S.Germ. × France / N.Germany (Russia?) // idem / N.America***
Heine VII	Hybride à courte Paille / Svalofs Kronen ***France / Sweden***

A further broadening in the choice of the parents in the following years, contributed to the improvement of the bread-making quality, as in the varieties Rabe (1961) and Diplomat (1966):

Rabe	Heine IV / Svalöf 0871 // Thatcher / Svalöfs Kronen ***Sw. × Germ. / Sweden // N.America / Sweden***
Diplomat	Merlin /4/ Firlbeck I / Marquillo // Taca /3/ Heine VII ***S × N.Germany/4/Germany / N.America // S × N.Germany/3/France × Sweden***

An extensive analysis of the factors contributing to baking quality in German wheat varieties was made by Lein (1958), from the breeding firm Heine-Peragis. He distinguished the following groups of varieties:

- the southern German varieties, based on Tassilo and Hauter;
- the Carsten varieties, partly derived from Eastern European wheats;
- the offspring from Janetzki's Sommer Weizen, possibly in combination with offspring from Garnet;
- the offspring of Thatcher, i.e. mainly Austrian varieties, and some German varieties such as Hanno and Rabe;
- the offspring of wheats collected during the German Hindukush expedition and other German expeditons to the foothills of the Himalayas, in the course of the 1930s.

With regard to the sources of quality, Lein pointed out some specific varieties:

- the French variety Arras, frequently used in crosses made at Weihenstephan in the 1920s, resulting in the winter-wheat cultivars Tassilo, Hauter II, Taca and Werla. Lein suggested that the quality of Arras might be derived from the English variety Yeoman and, through Yeoman, from Red Fife. He pointed out that Yeoman was used in wheat breeding in France, e.g. in the cross that led to Providence, a variety with good baking quality. He suggested that Arras might have also been selected from a cross with Yeoman.
- the North American varieties Marquis and Garnet (derived from Red Fife), frequently used in Hadmersleben and Halle in the 1930s, resulting in the spring-wheat varieties Heines Koga and Peragis Garant. Moreover,

there is an (indirect) influence of Garnet, through the combination with Janetzki, on the varieties Erli, Lichti and Norko;

- the variety Janetzki's frühe Sommerweizen, as a parent of NOS Nordgau and Lichtis früh. Janetzkis frühe Sommerweizen originated from Galicia (southeast Poland) or from the Pannonic Area (Austrian-Hungarian border) and may be related to Red Fife, which was taken along by emigrants from that region on their way to North America.
- Russian and Eastern European land varieties are presumed to be the source of the baking quality of the winter-wheat varieties Carsten VI and Carsten VIII.

During the German Hindukush Expedition of 1935/1936, many wheat samples were collected. Among the material collected in North India and the foothills of the Himalayas some sources with good or very good baking quality were found (Lein, 1949). Crosses made in Germany with such sources led to the lines Heine 2167 and Heine 2174, figuring in the pedigree of the winter wheat Kranich (Heine 2167/Heine VII//Merlin) and the spring wheats Kolibri and Kleiber (both Heine 2174/Peko//KogaII). Lein pointed out that Garnet and Marquis are both derived from sources from North India, particularly from crosses with Gehun and Calcutta.

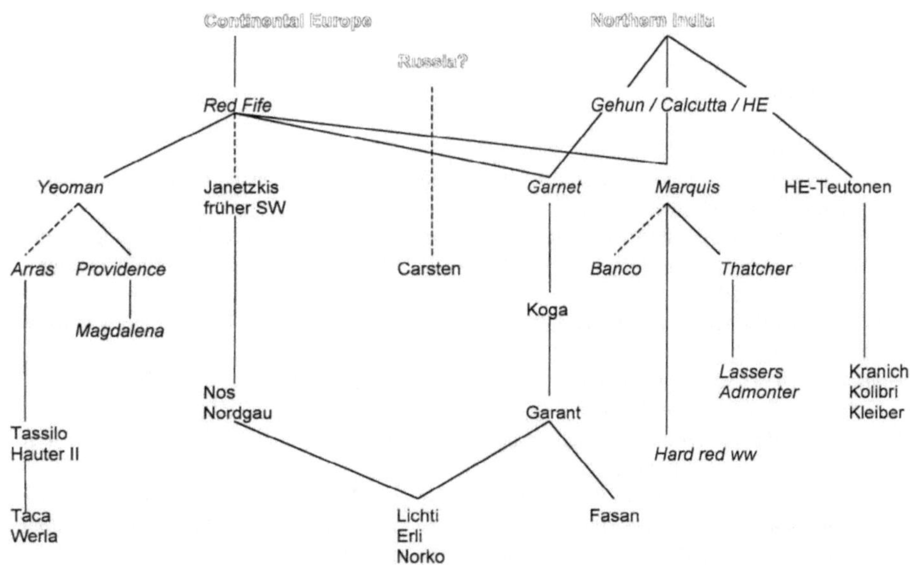

Figure G. 1. Sources of baking quality in German wheat varieties after Lein (1958).
Note: the lines represent derivations of varieties, i.e. not direct descendencies; dotted lines represent presumed derivations; names of non-German varieties have been written in italics.

By stressing the influence of foreign sources on baking quality, Lein did not value the importance of German land varieties in this respect. Land varieties from southern Germany have been used widely in crosses made at Weihenstephan. Hoeser (1954) stressed that in these crosses good baking quality arose through transgression, as both crossing parents were surpassed in baking quality by lines selected from the cross (see Section 2.4.1).

A schematic presentation of the sources surveyed by Lein (1958) is given in Figure G.1. Although Lein admitted that his 'pedigree of quality' – for the sake of clarity – was not fully in agreement with reality, it clearly showed that the characteristic of baking quality in all wheat growing countries goes back to only a few basic forms. Moreover he pointed out that wheat breeders had used only a fraction of the available genotypic variability of the world's wheat collection.

J.M. & D.A.D.

Acknowledgments

The authors wish to thank Dr G. Fischbeck (Institut für Pflanzenbau und Pflanzenzüchtung, Weihenstefan) and Dr W. Seibel (Arbeitsgemeinschaft für Getreideforschung, Detmold) for their help in reviewing this chapter.

References

Anonymus (1970) Rudolf Carsten 1880–1950.

Beschreibende Sortenliste für Getreide 1971–1996. Herausgegeben vom Bundessortenamt. Hannover: Alfred Strohte Verlag.

Bolling H (1989) 25 Jahre Qualitätsuntersuchungen im Rahmen der Wertprüfung in der Bundesrepublik Deutschland. In Bericht über die 14. Getreide-Tagung. Detmold: Granum-Verlag, pp 5–19.

Cooke RJ (1995) Allelic variability at the *Glu-1* loci in wheat varieties. Plant varieties & seeds 8: 97–106.

Hoeser K (1954) 30 Jahre Qualitätsweizenzüchtung in Weihenstephan. In Die Qualitätszüchtung von Brotgetreide. Bericht über die Tagung in Detmold, pp 55–60.

Hoeser K (1958) Stand und Aussichten der Weizenqualitätszüchtung in Weihenstephan. Ber der 2. Getreidetagung Detmold: 17–24.

Hoeser K, Wenisch K, Oppitz K (1979) Ergebnisse vergleichender Untersuchungen neuerer Winterweizensorten mit Sorten aus den Anfängen der Qualitätszüchtung. In Bericht über die 9. Getreidetagung. Detmold: Granum Verlag, pp 20–25.

Isenbeck K (1938) Die Entwicklung des Qualitätsgedankens in der landwirtschaftlichen Pflanzenzüchtung. Kühn-Archiv 50, Festschrift Halle (Saale).

Isenbeck K, von Rosenstiel K (1950) Die Züchtung des Weizens. Berlin und Hamburg: Verlag Paul Parey, 529 pp.

Kipp B, Belitz HD, Seilmeier W, Wieser H (1996) Comparative studies of high molecular subunits of rye and wheat I. Isolation and biochemical characterisation and effects on gluten extensibility. J Cer Sc 23: 227–234.

Lein A (1949) Asiatische Weizensortimente. Kühn Archiv 62: 217–310.

Lein A (1954) Erfahrungen bei der Züchtung neuer Winter- und Sommerweizensorten. In Die Qualitätszuchtung von Brotgetreide. Bericht über die Tagung in Detmold, pp 61–68.

Lein A (1958) Züchterische und pflanzenbauliche Fragen bei Qualitätsweizen. Berichte auf der 2. Getreidetagung in Detmold, pp 199–207.

Moonen JHE, Scheepstra A, Graveland A (1984) Genetische Aspekte des Glutenins in Zusammenhang mit der Backfähigkeit deutscher Weizensorten. Die Mühle u. Mischfuttertechnik 10: 124–126.

Odenbach W, Mahgoub E (1987) Beziehungen zwischen der Backqualität und dem Vorkommen der Gene für die hochmolekularen Untereinheiten des Glutenins bei den deutschen Winter- und Sommerweizensorten. Getreide, Mehl und Brot 41 (7): 195–198.

Odenbach W, Rath C, Waga J, Mahgoub E (1991) Wheat storage protein patterns, a tool for the wheat breeder; problems and aspects. In Proc Eucarpia Cereal section meeting 1991, Schwerin, Germany.

Pelshenke P (1933) A short method for the determination of glutenquality in wheat. Cereal Chemistry (10 (1).

Pelshenke PF (1934) Qualitätszüchtung bei Weizen. Der Züchter 6: 169–172.

Pelshenke PF (1935) Die Bestimmung der Kleberqualität nach der Schrotgärmethode. Z Ges Getreidewesen 22: 46–48.

Pelshenke PF (1938) Die für den Züchter wichtigsten Methoden zur Prüfung der Backfähigkeit. Der Züchter 10: 6–9.

Pelshenke PF (1954) Über die Notwendigkeit und über die Zuchtziele in der Weizenqualitätszüchtung vom Standpunkt der Müllerei und Bäckerei. In Die Qualitätszüchtung von Brotgetreide, Bericht über die Tagung in Detmold, pp 55–60.

Pelshenke PF (1959) Qualitätsfragen bei Getreide und Brot. Arbeiten der DLG 59. Frankfurt am Main: DLG-Verlag.

Porsche W, Lehmann C, Richter K (1979) Ergebnisse von Untersuchungen an proteinreichen Herkunften des Weizens *(Triticum aesivum) L.*. Tagungsbericht der Akademie der Landwirtschaftswissenschaften der DDR 175: 81–86.

Richter K, Herdam H (1979) Erste Ergebnisse von kreuzungen mit Atlas 66. In Aktuelle Probleme der Getreideforschung. Tagungsbericht der Akademie der Landwirtschaftswissenschaften der DDR 119: 137–144.

Richter K (1984) Züchtung auf erhöhten Rohproteingehalt bei Winterweizen. Tagungsbericht Akademie der Landwissenschaften der DDR 224 (2): 521–526.

Rogers WJ, Payne PI, Harinder K (1989) The HMW glutenin subunit and gliadin composition of German grown wheat varieties and their relationship with bread-making quality. Plant Breeding 103: 89–100.

Rosenstiel von K, Rundfeldt H (1963) Zur Frage der Bestimmung der Backfähigkeit bei Weizen. Der Züchter 6 (special issue): 28–38.

Sasek A, Cerny J, Bradova J, Parizek P (1995) Electrophoretic spectra of hordeum barley varieties of domestic and foreign provenance. Gen & Pl br 31: 35–52.

Scharnagel Th (1929) Untersuchungen über die Beziehungen der Weizensorte zur Backfähigkeit. In Forschungen auf dem Gebiete des Pflanzenbaus und der Pflanzenzüchtung. Festschrift zum siebzigsten Geburtstag von Kurt von Rümker. Berlin: Paul Parey, pp 127–145

Scharnagel Th (1936) Probleme der Deutschen Weizenzüchtung. In Sirks MJ (ed), Congrès International de Sélectionneurs de Plantes. Leiden: E.J. Brill, pp 34–36 (vol 2).

Schmidt EA (1936) Ein Jahr Kleberweizen. Mühle 73: 1119–1121.

Schmidt EA (1938) Dar dritte Kleberweizenjahr. Mühle 75: 265–270.

Seibel W (1996) Het stimuleren van de teelt van bakwaardige tarwe in Duitsland. In Themadag Granen 1996. Den Haag: Produktschap GZP.

Seilmeier W, Belitz HD, Wieser H (1991) Separation and quantitative determination of high molecular weight subunits of glutenin from different wheat varieties and genetic variants of the variety Sicco. Z Lebensm Unters u Forsch 192: 124–129.

Sonntag T, Salovaara H, Payne PI (1986) The High Molecular Weight glutenin subunit composition of wheat varieties bred in Finland. J Agric Sc Finland 58: 151–156.

Steinberger J, Meyer D, Zimmermann G, Zeller FJ (1995) Neue Qualitätsgruppen bei Weichweizensorten. Getreide, Mehl und Brot 49 (6): 324–329.

Stuke E (1962) Untersuchungen über die Züchtung von Weizen mit guter Backfähigkeit: (I) Die züchterische Bedeutung von einzelnen Komponenten der Backfähigkeit (II) Uber die Genetik einzelner Komponenten und die Bearbeitung des Zuchtmaterials. Zeitschrift für Pflanzenzüchtung 47 (I): 217–253 and (II): 297–329.

Svenson J (1950) Untersuchungen über die Kleberqualität von Winter- und Sommerweizenstämme der Erntejahre 1947–1949. Der Züchter 20: 356–357.

Vettel F, Pelshenke P (1934) Können wir ertragreiche Qualitätsweizen züchten? Der Züchter 6: 193–199.

Vavilov N (1935) The origin, variation, immunity and breeding of cultivated plants (transl. Starr Chester K). Chronica Botanica 13 (1949): 364 pp.

Vettel F (1956) Stand der Qualitätszüchtung bei Weizen. Sitzungsberichte der Deutsche Akademie der Landwirtschaftswissenschaften zu Berlin 5 (11): 1–32.

Wieser H, Seilmeier W, Belitz HD (1990) Characterization of high molecular weight subunits of glutenin seperated by reverse phase high performance liquid chromatography. J Cer Sc 12: 223–227.

Wienhues F, Giessen JE (1957) Die Abstammung Europäischer Weizensorten. Z Pflanzenzüchtung 37: 217–230.

Zeller FJ (1995) Nützung biochemisch- und molekular-genetischer Methoden zur Identifizierung der Kleberproteine des Weizens. Bericht über die 45. Tagung für Getreidechemie, Juni 1994, Detmold.

The Netherlands and Belgium

1. Introduction

The Netherlands and Belgium together produced 2.9 Mt wheat in 1997, representing a minor fraction of the total European wheat harvest. Wheat production in both countries is very intensive, resulting in the highest average yields in Europe and in the world. On the other hand, the use of high yielding varieties in combination with a humid climate is reflected in the moderate quality of the wheat produced. Wheats from the most productive areas in Belgium and the Netherlands are among the 'mealiest' produced in Europe. To obtain flour that is suitable for the production of bread, flour manufacturers traditionally mix in high quality wheats of foreign origin.

A temporary increase in the area under wheat during the mid–1930s in both countries was caused by stimulating measures by their government (see Section 2). The decrease in total cereal area during the second half of the 20th century (Table NLB.1) has been due to the sharp decrease in the production of oats and rye, which were the main cereal crops during the first half of the century. The use of these cereals as animal feed has been replaced mainly by silage maize. With a sharp fall in the area under barley in the 1970s, wheat became the most important cereal crop; its relative importance in the Netherlands increased from 26% to 70% in the 60 years between 1934–1938 and 1995–1997, respectively.

The proportion of winter and spring wheat varies according to climatic conditions, but farmers prefer to sow winter wheat, in order to obtain maximum yields. Since the 1970s the proportion of spring wheat has been less than 15% of the total wheat area in most years. It may reach up to 50% in years after very severe winters, when frost damage in autumn-sown wheat is so high that re-sowing with spring wheat is necessary. The exceptional situation that the area under winter and spring wheat was equal in a number of consecutive years (1961–1965) was caused by a coincidence of circumstances: heavy stripe rust epidemics that affected most of the winter-wheat varieties; combined with the introduction of some very productive spring-wheat varieties; and the spreading of spring wheat to lighter soils. The most intensive wheat production areas are on the clay regions of lower and central Belgium and in the polders of the southwestern, central and northern provinces of the Netherlands.

Bread forms an important part of the Dutch diet, with two bread-based meals a day. Nevertheless, bread consumption per person decreased from

Table NLB. 1. Area under cereals, winter and spring wheat (1000 ha) and wheat production (1000 t) in the Netherlands and Belgium between 1924 and 1997

Country	Area under wheat	Prod. of wheat	Area under cereals	Area under: winter wheat	Area under: spring wheat	Proportion wheat/ cereals	Proportion wheat/ arable crops
THE NETHERLANDS							
1924–1925	51	139					
1934–1938	142	450	540	121	21	26	15
1948–1952	89	324	490	73	13	18	9
1961–1965	138	606	496	65	67	27	16
1969–1971	146	675	373	99	48	39	21
1979–1981	138	867	225	124	14	62	19
1989–1991	134	1022	192	127	8	70	17
1995–1997	138	1194	198	128	10	70	17
BELGIUM					Year(s)	Area under: ww	Area under: sw
1924–1925	143	374					
1934–1938	176	474	636		1929	150	5
1948–1952	163	525	506		1955	171	20
1961–1965	215	826	509		1965	123	103
1969–1971[a]	209	848	510		1971	152	42
1979–1981[a]	189	949	426		1981	156	11
1989–1991[a]	219	1418	367		1991–1992	197	4
1995–1997[a]	217	1194	333		1993–1994	190	7

[a] including Luxemburg.
Sources: IIA and FAO statistics, Centraal Bureau Statistiek, Broekhuizen (1965) and De Coster (1994).

124.8 to 62.4 kg per year between 1930 and 1973 (Nicolas, 1976). It has remained rather stable thereafter and was 59.2 kg in 1996. A similar development took place in Belgium, where bread consumption decreased from 107 kg per person per year before World War II to 89 kg in 1959 and to 50 kg in 1996.

2. Wheat breeding before 1945

The great crisis in agriculture at the end of the 19th century, initiated by a severe competition from cheap products from North America, Russia and

India, affected the wheat market to such extent that the price of wheat in the Netherlands in 1894 fell to 42% of the price in 1870–1880 (Dorst, 1927). The area under wheat dropped drastically between 1887 and 1897 and the net import of wheat tripled in the same period. Although the crisis was detrimental to several farmers, especially in regions where wheat was a major crop, it also triggered enterprising activities and governmental investment in agricultural education, research and extension.

2.1. The Netherlands

In the Netherlands two groups of varieties were widely grown in 1870: 'Zeeuwse tarwe' and 'Gelderse tarwe'. Zeeuwse tarwe (named after the southwestern province of Zeeland) was appreciated by bakers for its bread-making quality. Within the group of Gelderse tarwe, Gelderse Ris (also called Gelderse rode after its red grain colour) was the most outstanding and also most appreciated for its quality. According to Zeven (1990) the origin of this land variety is likely to be Polish or Russian. The preservation and gradual improvement of these land varieties was provided by a continuous market exchange of seed between various regions. At the start of the agricultural crisis in 1877, farmers showed a growing interest in high yielding varieties from England. This resulted in the introduction of Red Hallett and White Hallett, which proved to be ill adapted to local growing conditions, and later Essex and Squarehead, which were more successful. Essex took the place of Zeeuwse tarwe in the western coastal areas and Squarehead replaced Gelderse tarwe along the northern coast. Moreover, both varieties were used as parents in crosses, the first results of which appeared on the market at the turn of the century.

In the late 1880s, Broekema at the Agricultural School in Wageningen deliberately crossed Squarehead (Rode Dikkop) with Zeeuwse tarwe in order to improve the quality (including the colour of the seed) of Squarehead whilst maintaining its high yielding level (Broekema, 1899). Notably, he did this at a time that the re-discovery of the laws of Mendel had not taken place. The first results of Broekema's efforts were the cultivars Spijk (Squarehead/Zeeuwse) and Duivendaal (Zeeuwse/Squarehead). A backcross of Spijk with Squarehead resulted in the variety Wilhelmina, released in 1901. Subsequent backcrosses on Wilhelmina with Essex resulted in the varieties Juliana and Emma, released in 1921 and 1924. In 1923 Broekema was appointed as a professor of plant breeding and became director of the Institute of Plant Breeding (IVP) in Wageningen.

The most successful of the varieties bred by Broekema was Wilhelmina. It was grown on a large scale in the Netherlands until the 1940s and to some extent in neighbouring countries. Wilhelmina was used as a crossing parent by several European breeders and through the Italian variety Ardito it even appears in the pedigree of Bezostaya-1. An important feature of Wilhelmina was its straw quality, making it possible to apply optimal levels of fertilizer and to facilitate the harvest of the crop. On the other hand, its winter-hardiness was insufficient in cold winters and needed to be improved.

Initially, Wilhelmina was criticized for its poor bread-making quality. However, this feature was not considered to be of primary importance as it could easily be compensated for by mixing in foreign quality wheats, e.g. from Russia and Romania (Dorst, 1927). A comparison of Wilhelmina with the French variety Bordeaux showed that 100 g of Wilhelmina flour resulted in a loaf volume of 444 ml, compared to a loaf volume of 500 ml from the same amount of Bordeaux flour.

Bruyning (1905) investigating the quality of the Dutch wheat harvest, compared the old land varieties to new varieties such as Wilhelmina. He determined the protein content in the seed and the flour, the gluten content of the flour and carried out baking trials. One of his conclusions was that the average loaf volume of these samples was 25% less than the loaf volume obtained by bakeries, which was the result of the mixing with high quality wheats from abroad. (The net import of wheat into the Netherlands between 1901 and 1903 was 272 000 t)

Other wheat cultivars released in the first decades of the 20th century were Witte Dikkop (= White Squarehead), obtained by Mansholt from the same parents as those of Duivendaal, i.e. Zeeuwse tarwe × Squarehead. The varieties Imperiaal II and Miljoen III, both released by the Plant Breeding Institute in Wageningen, were appreciated for their grain quality.

2.2. Belgium

Before 1925, most of the wheats grown in Belgium were of French, Dutch, German or Swedish origin. The first Belgian selections appeared between 1925 and 1930 under the names Trésor 18, Hybride du Centenaire and Prince Leopold (Larose, 1954). Two breeding stations, located in Gembloux and Heverlee, played an important role in Belgian wheat breeding.

The Plant Breeding Institute in Heverlee was established in 1915 as a department of the Free University of Leuven. Important winter wheat varieties released from Heverlee in the years 1938–1939 were: Alba (Trésor/Jacob Cats), Rufus (Teverson/Kroon) and Zanda (Trésor/Staal). The variety Alba was widely grown in Belgium and the Netherlands, covering 49% of the Dutch wheat area in 1950. It was noted for its good bread-making quality).

A major incentive for breeding spring wheat was the extremely cold and long winter of 1941–1942, which made the sowing of alternative wheats[1] unadvisable, thus threatening the supply of bread wheat in Belgium (Dumon & Laeremans, 1956).

A programme was initiated in which high-yielding winter wheats were crossed with spring-wheat varieties that were suitable for late sowing. A cross between the Belgian winter wheat variety Jubilé and the Swedish spring wheat variety Fylgia resulted in two successful spring-wheat varieties: Phoebus and Fylby, registered in 1954 and 1956 respectively. It is likely that the characteristic of late sowing was introduced through the Russian spring wheat variety Ladoga, figuring in the pedigree of Fylgia (= Extra Kolben//Jacinth/Ladoga). The release of Phoebus and Fylby, yielding some 10% more than the older varieties Jufy and Koga II, induced Dumon & Laeremans (1956) to state *'It is likely that farmers in western Europe will grow more and more spring wheat in the future. The yield of these new spring wheat varieties is comparable to the highest yielding winter wheat varieties. It now comes to the improvement of their baking quality, disease resistance and straw characteristics.'* A remarkable increase in the area under spring wheat did follow but fell back again in the 1960s, when improved winter-wheat varieties appeared on the market.

The 'Station pour l'Amélioration des Plantes de Grande Culture' in Gembloux was established in 1913. After some years of selection in various populations, a hybridization programme was initiated in 1924 with special emphasis on productivity. Some ten years later, in 1935, Larose started to lay more emphasis on quality characteristics (Castille & Vandam, 1981).

The first Gembloux varieties appeared on the market in 1937 under the name (Hybride de) Jubilé and Professeur Delos. Jubilé, selected from a cross between Vilmorin 23 and Pansar III, was a high-yielding variety that was widely grown in Belgium (and the UK, under the name Jubilegem). The quality level of Jubilé was lower than that of Professeur Delos. A cross between the French variety Benoist 40 and Professeur Delos resulted in the varieties Blédor and Ministre, of which Blédor was classified as an improver wheat by Larose et al. (1950). From the comparison of Pelshenke values of Blédor with those of its parents, Larose et al. concluded that transgression for this characteristic was possible. The varieties Blédor and Ministre acquired an important place on the Belgian market. Ministre was also grown in the Netherlands, where it covered 22% of the winter-wheat area in 1952.

The quality of these wheat varieties, classified by Larose et al. (1950), is presented in Table NLB.2. Five quality groups were determined on the basis

[1] 'Blés alternatifs' is the French term used for winter-wheat varieties requiring relatively little vernalization which may be sown until the 15th of March.

Table NLB. 2. Classification of Belgian winter-wheat varieties, according to their Pelshenke value

Quality class[a]	P value	Variety	Breeder
Froment de force	> 55	Blédor	Gembloux
Froment très bon	38–55	Professeur Delos	Gembloux
Froment bon	30–38	Alba	Heverlee
		Rufus	Heverlee
		Ministre	Gembloux
Froment passable	25–30	Zanda	Heverlee
Froment médiocre	< 25	Jubilé	Gembloux
		Inst.de Gembloux	Gembloux

[a] The five quality groups may be translated as: improver, very good, good, satisfactory and poor, respectively.
Source: Larose et al. (1950).

of Pelshenke values. It appeared that 11 of the 28 wheat varieties grown in Belgium at the time, including the three most common winter-wheat varieties (Blédor, Ministre and Alba), had Pelshenke values of $\geqslant$30.

Larose et al. (1950) studied the genetic background of the characteristic bread-making quality by analysing Pelshenke values in the F_1 of crosses between poor quality varieties from Gembloux with high quality American varieties. Compared to the parents, the P values in the F_1 were intermediate in most cases, with a tendancy to dominance. Other crosses revealed transgression of the F_1, both in a positive and in a negative way. Larose et al. (1950) studied, moreover, the performance of the most important winter- and spring-wheat varieties in different years and on different locations in Belgium (Table NLB.3). They found large differences in Pelshenke values between years and between locations.

2.3. *The Technical Wheat Committee in the Netherlands*

The world economic crisis of 1929–1930 and the dramatic drop of wheat prices on the grain market urged Dutch and Belgian governments to take measures in order to protect the interests of their national agriculture. In Belgium, a minimum price for home-grown wheat was guaranteed by the government. In the Netherlands, a law was made in 1931 which compelled millers to use at least 25% of home-grown wheat for their bread-making flour. This percentage was increased to 40% in 1933 and finally settled at 35%.

Table NLB. 3. Grain yield and P value of the most important winter- and spring-wheat varieties grown in Belgium in the 1940s: mean values of 11 and 6 trials in different regions in Belgium

Variety	Grain yield		P value
	t/ha	rel.	
Winter wheat (11 trials)			
Blédor	5.830	100	102
Alba	5.280	90.6	57
Ministre	6.210	106.5	32
Spring wheat (6 trials)			
Blé d'Avril	4.136	100	108
Fylgia	3.698	89.4	78
Précoce de Gembloux	3.477	84.1	35

Source: Larose, Legros & Noulard (1950).

This type of measure induced renewed interest in the production of wheat (the area under wheat in the Netherlands reached a record of 150 000 ha in 1934–1936) and in the bread-making quality of home-grown wheat.

In the Netherlands, the Technical Wheat Committee (TTC), with representatives from agricultural research and the milling industry, was established to investigate ways of improving the quality level of the Dutch wheat harvest. One of the members of this committee was Feekes, who became one of the driving forces of cereal research in the Netherlands. Initially, the TTC's efforts were concentrated on milling and bread-making quality, and on the commercial value (grading) of the wheat harvest. The influence of fertilizer level, harvesting method and occurrence of sprouting were investigated in the late 1930s.

Results of the baking trials carried out by the TTC revealed that varieties known for their superior bread-making value, such as Svalöfs Extra Kolben and the French variety Providence, proved to be able to produce a superior flour under Dutch circumstances (see Section 2.3.2). However, the productivity and disease resistance of these varieties were far too low. Indeed, no variety with a productivity comparable to Dutch varieties and a better quality was found.

The committee stated that the most effective tool for improving the quality of Dutch wheats was superior gluten quality in varieties with moderate gluten content. After World War II the task of the TTC was continued by the Cocobro and later the NGC (Netherlands Grain Centre).

Table NLB. 4. Neumann scores for baking quality of one spring-wheat variety and five winter-wheat varieties grown in the Netherlands in 1932 and 1933

Variety	1932	1933	[a]
Van Hoek (sw)	143	153	193
Siegerländer	143	144	180
Imperiaal II	135	143	167
Juliana	130	147	175
Wilhelmina	114	155	165
Carsten V	107	113	143

[a] Treated with chemical flour improver (probably potassium bromate).
Source: de Vries & Feekes (1935).

2.3.1. Winter wheat

The TTC investigated the quality of the wheat harvest of 1932 and 1933, including the most common Dutch-grown varieties (Table NLB.4). Baking trials, carried out with and without chemical flour improvers, were evaluated by a 'Neumann score'. This score included the loaf volume and a value for the crumb structure of the bread (Verslagen TTC, 1934). The introduction of chemical improvers offered new possibilities to use wheats with inferior quality characteristics for bread-making purposes (de Vries, 1933).

The quality score of the spring-wheat variety (Mansholts) Van Hoek appeared to be superior to all the winter-wheat varieties. The German variety Siegerländer, grown mainly in the southern province Limburg, was somewhat superior to Imperiaal II, Juliana and Wilhelmina. At that time, Wilhelmina and Juliana together covered some two-thirds of the winter-wheat acreage in the Netherlands. The quality of Carsten V, mainly grown in the northern provinces, clearly lagged behind the other varieties.

The classification of the winter-wheat varieties presented in Table NLB.5 shows that the land variety Gelderse Ris was the only wheat in the Netherlands classified in Quality Group 1–2. Before the introduction of the Wheat Law in 1931, millers were often willing to pay a higher price for Gelderse Ris wheat because of its excellent quality (Visser, cited by de Vries, 1933). Unfortunately this variety has remained practically unused for further breeding. Only one variety (Elisabeth) has been found to be derived from Gelderse Ris. Most of the other home-bred varieties in Table NLB.5 were classified in Group 3 and 4. The Danish variety Trifolium was grown on a large scale on the peat soils in the northern provinces of the Netherlands.

Table NLB. 5. Comparative trial of winter-wheat varieties bred in the Netherlands and surrounding countries

Variety	Country of origin	Quality group[a]	Variety	Country of origin	Quality group
Gelderse Ris	Netherlands	1–2	Miljoen II	Netherlands	3
Providence	France	1–2	Imperiaal IIa	Netherlands	3
Siegerländer	Germany	1–2	Prins Hendrik	Netherlands	3
Yeoman	UK	1–2	Trifolium	Denmark	3
Vilmorin 27	France	2	Staring	Netherlands	3-4
Emma	Netherlands	2–3	Kruisingsangel	Netherlands	4
Wilobo	Netherlands	2–3	Carsten V	Germany	4
Jubileum	Belgium	2–3	Witte Dikkop	Netherlands	4

[a] The quality groups were determined as: good (1), rather good (2), transition group (3) and unsatisfactory (4).
Source: Feekes & van Dobben (1940, 1941).

2.3.2. Spring wheat

The most important spring-wheat variety grown in the Netherlands during the 1930s was Mansholts van Hoek, selected from a cross between the French varieties Japhet and Gironde and released in 1925. A comparative trial with the Swedish variety Svalöfs Extra Kolben and the German variety Heines Kolben was carried out in the northern region of the province of Groningen. It appeared that the grain of Svalöfs Extra Kolben was much grittier at milling than the seed of the other two varieties. A baking trial (de Vries, 1934) gave the following loaf volumes:

Svalöfs Extra Kolben: 647 (100%)
Heines Kolben: 615 (95%)
Mansholts van Hoek: 590 (91%)

The results proved that it was possible to obtain a product with improved bread-making quality under Dutch growing conditions, which was considered to be a break-through by the members of the TTC.

The baking quality of the most important spring-wheat varieties grown in the Netherlands in the 1930s was investigated by Feekes & van Dobben (1940), along with that of foreign varieties (Table NLB.6). From comparison of the figures in Table NLB.5 and NLB.6, it is apparent that the average quality of the spring-wheat varieties was significantly higher than that of the winter-wheat varieties. Feekes & van Dobben (1940) concluded that, although the foreign varieties as such were not well-adapted to growing conditions in the Netherlands, enough high quality material was available to

Table NLB. 6. Comparative trial of Dutch and foreign spring-wheat varieties

Variety	Country of origin	Quality group
Garnet	USA	1
Svalöfs Extra Kolben II	Sweden	1
Florence/Aurore	France	1
Mansholts van Hoek	Netherlands	1–2
Fylgia	Sweden	1–3
Mansholts Witte	Netherlands	2–3
Blanka	Sweden	2–3
Heines Kolben	Germany	3

Source: Feekes & van Dobben (1940).

Table NLB. 7. Bread-making characteristics of the most important winter- and spring-wheat varieties grown in the Netherlands in 1947–1949

Variety	Released in:		Loaf volume		Protein	W of	BMQ[a]	Area sown
	country	year	ml	rel.	content	Chopin	score	in the NL
Winter wheat								
Alba	Belgium	1938	531	100	9.3	72	5	21–45%
Juliana	Netherlands	1921	496	93	10.5	53	5	10–36%
Staring	Netherlands	1941	499	94	9.6	44	5	19–20%
Carsten V	Germany	1930	426	80	9.9	37	2	11–13%
Spring wheat								
Koga	Germany	1949	526	100	11.0	130	5	0–1%
Blanka	Sweden	1933	512	97	10.3	64	5	57–64%

[a] Bread-making score according to the Descriptive List of Varieties, ranging from 1 to 10 (poorest to best).
Source: Broekhuizen & de Miranda (1951).

breeders, offering possibilities to improve local wheat varieties. In this respect, Broekema (1939) stressed how important it is that the milling and baking industries provide a good description of the required quality characteristics. Detailed information on the quality characteristics of the most important winter- and spring-wheat varieties grown in the Netherlands in 1947–1949 was reported by Broekhuizen & de Miranda (Table NLB.7)

The loaf volume of the Belgian winter-wheat variety Alba was remarkably high and comparable to the German variety Koga. The fact that this high loaf volume was obtained at a lower protein content than that of Koga could indicate that the protein quality of Alba was superior; the W of Chopin did not confirm this assumption, however. Feekes & van Dobben stressed that, although variation in quality existed, the average quality of the Dutch wheat harvest was as a whole not high enough to meet the demands of the baking industry. This was illustrated by the fact that none of the tested varieties received a score higher than 5 on the Descriptive List of Varieties.

3. Wheat breeding after 1945

The productivity and quality of wheat varieties grown in the Netherlands in the years 1957–1962 was compared with those grown during the 1930s by Broekhuizen (1963). Grain yields increased 40%, from 3.2 to 4.5 t/ha, for winter wheat and 37%, from 2.7 to 3.8 t/ha, for spring wheat. On the other hand, loaf volumes dropped 18%, from 510 to 420 ml, whereas the production of a representative flour for Dutch bread making required a loaf volume of 630 ml.

A commission set up by the Ministry of Agriculture concluded that the quality of Dutch wheat production could only be improved by the breeding of improved varieties. In order to stimulate breeders, a regulation was introduced in 1960, that granted a premium to breeders who succeeded in breeding a winter-wheat variety with a baking quality that was 15% higher than the average of the varieties Dippe's Triumph, Felix, Flamingo and Leda, or a spring-wheat variety with a baking quality that was 10% above the average of Carpo, Fasan, Koga II and Opal. Moreover the variety had to be sufficiently resistant to sprouting in the ear. The quality was to be tested over a period of four years. The level of the premium was linked to the amount of certified seed sold, with a maximum of NLG 350 000; granting of the premium was not restricted to Dutch breeders.

A study group of representatives from breeding firms, agricultural organisations, the milling industry and institutes was formed in 1966 to evaluate the regulation. Their conclusion was that the requirements were rather high and could not be realized at short notice. However, it was still seen as an important step to bring the bread-making quality of the home-grown wheat to a level comparable to French varieties. The study group therefore introduced a regulation to award a lower premium for an improvement of the baking quality by 10% for winter wheat or 5% for spring wheat above the current level (Broekhuizen, 1967).

The only variety that has been qualified for this premium is the winter-wheat variety Okapi, obtained from extensive line selection in the variety Caribo (Wouda, 1998). Okapi was released in 1977 by a cooperative of Mansholt, Geertsema and van der Have (MGH) breeders. It was grown on a large scale in the Netherlands until 1989, covering a maximum of 41% of the Dutch area under wheat in 1981. Okapi was also registered in Germany.

Although no other varieties have been awarded the premium, the regulation has been effective in so far that breeders were stimulated to pay attention to quality aspects. This has had a long lasting impact on wheat breeding in the Netherlands.

3.1. The Foundation for Agricultural Plant Breeding (SVP) in Wageningen

Research activities of the Foundation for Plant Breeding (SVP), mainly financed by the Dutch government and partly by farmers, breeders and industry, were aimed at supporting the work of the private breeding firms on a non-competitive basis. By selecting and crossing useful genetic resources, half-products and population material were made available to private breeders for further use in their programmes. Moreover the SVP tested selection methods and criteria.

Varieties with good bread-making characteristics from Scandinavia, Germany, France, Russia and North America were crossed with productive varieties adapted to local growing conditions. The offspring was assessed for yield and baking quality and the most promising lines were eventually used for further crosses. Studies on variation in baking quality within populations from single crosses showed that a rather high percentage of the lines in the first selection year exceeded the mid-parent value of baking quality (Mesdag, 1966). To illustrate this, the results of three crosses with the quality variety Varma are presented in Table NLB.8. Although all three crossing parents Riesebel, Leda and Heine VII belong to the same poor quality group, Pelshenke values in the offspring of the former two were considerably higher than in the offspring of the latter.

Table NLB. 8. Frequency of P values in the offspring of three single cross populations of winter wheat

Cross	Number of lines tested	% of lines better than mid-parent value	% of lines better than best parent
Riesebel/Varma	28	58	54
Leda/Varma	36	53	40
Heine VII/ Varma	25	16	8

Source: Mesdag (1966).

In spite of differences between the populations, they all produced a sufficient number of lines for further breeding, which was considered as a positive result. Similar results were obtained in spring-wheat populations in which the quality was assessed by Zeleny sedimentation values. For selection, SVP prefered the specific sedimentation value (sedimentation value : protein content). By eliminating the influence of the protein content, the specific sedimentation value was considered as a measure for the quality of the protein.

Further results of material obtained from the cross Heine VII/Varma were presented by Mesdag (1975). Ten lines, remaining after two selection cycles on plant characteristics (ear type, straw stiffness, etc.) and quality (sedimentation value, protein content), were submitted to a micro-baking test and compared to the parents. The average loaf volume of the ten lines was 118% (range: 97–122%) relative to the average loaf volume of the parents, Heine VII (100%) and Varma (129%). Lines with the best combination of baking quality and agricultural characterisics were crossed with comparable lines from other crosses and/or with productive varieties and offered to private breeders for further use.

The progress made in combining agricultural and quality characteristics was reported by Mesdag (1979, 1984). Yield and quality of 22 winter-wheat lines, grown in a yield trial, were compared to those of the standard varieties Arminda, Caribo, Nautica and Norda. The average yield of the tested lines was 96% (range: 84–107%) of the average of the standard varieties (100% = 7.2 t/ha). The loaf volumes (in a micro-baking test) of the tested lines varied between 83 and 162% of the average of two standard varieties.[2]

In order to eliminate the influence of protein content, specific loaf volume was calculated as the quotient of loaf volume and protein content. Four lines had a specific loaf volume that was 25% above that of the standard varieties and nine lines had a specific loaf volume that was between 16 and 25% above that of the standard varieties. There was no significant correlation between specific loaf volumes and yield figures, indicating that selection on specific loaf volume could be done at each grain yield level.

A further increase in yield and loaf volume was achieved in a number of lines tested in 1982. Compared to the standard varieties Arminda, Nautica and Okapi, the yield of one of the tested lines was better and four lines had kernel yields between 91% and 97%. The loaf volume of the lines varied from 108% to 120% of the average of the standard varieties.

A programme similar to that described for winter wheat was carried out for spring wheat. In some respects the work on spring wheat was easier, because the yield levels of the foreign parent varieties were higher. Moreover, the spring-wheat programme could be speeded up by growing an extra generation in the southern hemisphere during the European winter. As a matter of fact,

[2] Usually SVP used four standard varieties as a quality reference, but in 1977 the samples of two standard varieties were not usable due to increased α-amylase.

this method was already applied by Koopman in the 1920s by growing a 'winter'-generation of several crops in the East Indies (Koopman, 1963).

3.2. Private breeding firms in the Netherlands

Yielding capacity and disease resistance were always a priority in wheat breeding in the Netherlands. However, the prospect of a premium for a wheat variety with improved quality, enforced by a regulation in 1960, motivated wheat breeders to pay more attention to selection on quality. Literature on their activities in this respect is scarce; the results of their efforts may be judged by the varieties that were released in subsequent years (Table NLB.9). A cooperative breeding effort of three firms, Mansholt, Geertsema and van der Have (MGH), resulted in the release of the successful variety Okapi, which was awarded the premium for improved bread-making quality (see Section 3).

Some information on the work carried out by Cebeco was reported by Mastenbroek (1963) on the occasion of the 25th anniversary of their breeding station. In order to introduce new sources of quality into the crossing programme, foreign varieties were tested for their bread-making quality and growing performance under Dutch conditions. At the same time, material made available by SVP was integrated in the breeding programme. The Pelshenke dough ball test was applied as a selection tool for quality. The first selection was based on a minimum P value of 70–80. After another selection cycle in the field, the P value was determined again and the material submitted to a micro-baking test, carried out at the Institute for Cereals, Flour and Bread (IGMB-TNO) in Wageningen. The best lines were used in further backcrosses, such as the backcross Ring//Opal/Selkirk, made in 1962, which resulted in the spring-wheat variety Sicco (released in 1974). The kernel of Sicco was hard and its bread-making quality was considered to be good (score 8 on the Descriptive List of Varieties). Sicco was grown on a limited scale in the Netherlands, only covering a maximum of 2% of the area under spring wheat. It was more popular in the UK, however, where in 1977 it covered 62% of the area under spring wheat. Another spring-wheat variety with good bread-making characteristics was Minaret, released by the breeding firm Wiersum & Zelder in 1983. It accounted for most of the Dutch and Belgian spring-wheat area in 1989.

The winter-wheat variety Obelisk was grown on a large scale in 1989 and 1990 in the Netherlands, and to some extent in Germany. The winter-wheat variety Accent, released by Van der Have in 1990, was based on a selection made by SVP. Of the varieties listed in Table NLB.14, it is the only variety with subunit 2* (*Glu-A1*), indicating that it may be an interesting source of bread-making quality. Two winter-wheat varieties dominated wheat

Table NLB. 9. Wheat varieties released by Dutch breeders since 1960 with the name of their breeder, pedigree and maximum area sown in the Netherlands

Variety[a]	Year	Br[b]	Pedigree	Maximum area[c]
Winter wheat				
Cleo	1961	Ceb	Heine7/Minister	10% in 1963–64
Flevina	1964	Ceb	Hope/Timstein//3*Heine7	7% in 1965
Manella	1964	Ceb	Alba/Heine7	65% in 1970
Okapi	1967	MGH	selection from Caribo	
Lely	1970	Ceb	Cebeco-30/Flevina	22% in 1977
Clement	1973	Ceb	Cleo//Heine7/Riesebel-57/41	55% in 1975 (+F)
Arminda	1976	Hav	Carsten-854/Ibis	44% in 1985 (+F)
Nautica	1976	Ceb	Mildress/Manella	10% in 1979–80
Citadel	1982	W&Z	complex cross of 24 lines	7% in 1983
Obelisk (B)	1985	W&Z	complex cross of 36 lines	61% in 1989–90 (+G)
Pagode (B)	1986	W&Z	complex cross of 36 lines	12% in 1989
Avir	1988	Ceb	Alcedo/Maris Huntsman	19% in 1987
Accent (B)	1990	Hav	Starke/Vogel-219/Caribo	<1% in 1991[d]
Fletum	1991	Sem	complex	<1% in 1992[d]
Ritmo (B)	1991	Ceb	Hobbit/Line-1320//Wizzard/3/Marksman/Virtue	46% in 1993[d]
Estica	1991	Ceb	Arminda/Virtue	3% in 1993[d]
Eiffel	1992	Ceb	Arminda/Cebeco-323	3% in 1993[d]
Bercy	1994	Ceb	KHP3/Marksman	9% in 1998[d]
Arnaut (B+)	1995	Sem	Urban-3/Arminda//MG-1470/WST-936	<1% in 1997[d]
Spring wheat				
Orca	1961	Geer	Minister/Peko	63% in 1969
Toro	1971	Geer	Minister/Peko//Carpo	42% in 1975
Kaspar	1971	Ceb	Minister/Peko//Heine13037	29% in 1970
Melchior	1974	Ceb	Minister/Peko//Heine13037	51% in 1979
Bastion	1974	Zel	H-12//H-35/Mara	(B, F)
Sicco (B)	1974	Ceb	Ring//Opal/Selkirk	(UK)
Stratos	1980	Hav	Kr-5871/H-401/Orca	46% in 1983
Minaret	1983	W&Z	Bastion/Mironovskaya-808	94% in 1989 (+B)
Vedette	1989	Hav	HPG 522-66/Maris Dove	3% in 1991[d]
Baldus	1990	Ceb	Sicco/4/Sicco/3/N66/MGH653//Kolibri	33% in 1996[d]

[a] Varieties with (B) have been described as varieties suitable for bread-making purposes; (B+) indicates that the variety has improved bread-making quality.
[b] Names of breeders abbreviated from Geertsema (Geer), Mansholt, Geertsema and van der Have (MGH), Cebeco (Ceb), van der Have (Hav), Zelder (Zel), Wiersum & Zelder (W&Z), Semundo (Sem).
[c] Maximum area in the Netherlands; the addition (+F, +G, +B or +UK) indicates that the variety was also grown on a large scale in France, Germany, Belgium or the UK, respectively.
[d] Areas are based on the production of certified seed.
Sources: Lupton (1992) and Dutch List of Recommended Varieties.

production in the Netherlands in the 1990s: Ritmo (from Cebeco) and Vivant (from PBI, UK). Ritmo was classified as a variety with good bread-making characteristics. Apart from the Netherlands, it was also grown extensively in Belgium, France, Germany and Denmark. The variety Arnaut, released in 1995, was the first winter-wheat variety described as possessing 'improved quality'.

3.3. Selection for grain hardness and loafvolume

Most of the wheat varieties on the Dutch recommended list in the 1970s and the early 1980s, produced soft grains with a moderate quality level. Breeders introduced parents with hard grains in their crossing programmes and selected for this character. This resulted in a rapid changeover in the supply of varieties, which is apparant from the figures in Table NLB.10. In the years 1987–1989 the number of varieties producing soft vs. hard grains was more or less equal. In the years 1997–1999 only one variety produced soft grains vs. ten varieties producing hard grains. A similar though less pronounced development took place in the supply of spring wheats. As a consequence, a shortage of biscuit wheat in the mid 1990s necessitated the milling industry to pay a premium for biscuit wheat.

Table NLB. 10. Kernel hardness of Dutch wheat varieties on the Descriptive List of Varieties between 1977 and 1999

Years	Average number of varieties	Kernel hardness	
		soft	hard
Winter wheat			
1977–1979	11	8	3
1987–1989	13	7	6
1997–1999	11	1	10
Spring wheat			
1977–1979	8	4	4
1987–1989	5	3	2
1997–1999	5	1	4

Source: Descriptive List of Varieties.

The selection on loafvolume during the 1970s and 1980s in the Netherlands was based on bread baked in a tin. This baking test is less selective for flourstrength than a test carried out on bread baked on a baking floor, such as the German baking test. Thus, varieties such as Obelisk were introduced, that had a good loaf volume score, but were not appreciated by the milling industry. Starting in the late 1980s breeders decided to pay more attention to the quality of the dough (Roothaan, 1999).

3.4. Wheat breeding in Gembloux and Heverlee

Two winter-wheat varieties, that largely contributed to Belgian wheat production in the 1950s and 1960s were Professeur Marchal and Stella, released from Gembloux in 1957. Both were selected from the multiple cross Rimpau's Bastard II/Prof Delos//Prof Delos/Hybride de Jonquois, and both varieties were the result of intensive selection on winter-hardiness. On the other hand, they lacked sufficient straw stiffness, as well as milling and baking quality. Professeur Marchal was used as a crossing parent, especially by breeders in the UK, resulting in varieties such as Maris Huntsman, Norman and Mardler. All these varieties caused problems in the baking industry due to a sticky dough (see Part I, Section 7.5).

Table NLB. 11. Mean figures of five Belgian winter-wheat varieties, from 14 trials during five years (for agronomical value) and from 22 trials during four years (for quality characteristics)

Variety	Grain[a] yield	Straw[b] stiffn.	Straw length	Protein content	Sedim. value	P value	W Chopin	Loaf volume ml	rel.
Cama	110	8.5	95	10.7	28	98	151	1347	98.3
Mina	109	8.4	100	10.7	24	124	134	1316	96.1
Marco	111	8.3	95	10.8	15	36	71	1351	98.6
P.Marchal	109	6.1	109	10.6	19	33	90	1370	100.0
Cappelle	100	8.2	102	11.5	18	46	125	1454	106.1

[a] Related to Cappelle 100% = 4.97 t/ha.
[b] Straw stiffness on a 0–10 scale; 10 is no lodging.
Source: Noulard & Vandam (1967).

Breeding activities at Gembloux in the mid-1950s were directed towards the improvement of these varieties. Crosses were made to improve their straw quality and their milling and baking characteristics. This resulted in the varieties Cama, Mina and Marco, released in 1967. Noulard & Vandam compared the agronomic value and bread-making characteristics of these new varieties to Professeur Marchal and the French variety Cappelle (Table NLB.11). On the basis of three indirect measures for bread-making quality (sedimentation, Pelshenke and Chopin values) the quality of Cama and Mina had been improved over Professeur Marchal. Strangely enough, this improvement was not reflected in their loaf volumes. On the official Belgian List of Varieties, the technological value of Cama, Mina and Marco was assessed as +, + and

Table NLB. 12. Development of Chopin W values of Gembloux wheat varieties released between 1935 and 1977

Year	Variety	W[a] value	Year	Variety	W value
1935	Prince Leopold	35	1954	Panter	52
	Institut de Gembloux	46		Ministre	57
	Précoce de Gembloux	58		Alfy I	66
1937	Hybride du Jubilé	47		Jufy II	43
	Alter de Gembloux	92	1957	Stella	98
	Professeur Delos	171		Professeur Marchal	99
1949	Prima	51	1967	Cama	176
	Directeur Journée	59		Mina	159
	Blédor	121		Marco	82
			1977	Tina	220
				Zémon	203

[a] The W value of the first three varieties in this list was measured with an extensimeter, the other values (from 1937 onwards) were measured with an alveograph.
Source: Castille & Vandam (1981).

±, respectively, compared to – for Professeur Marchal. Two other winter-wheat varieties with positive quality scores were Leda and Norda, from the Plant Breeding Institute in Heverlee. Cama and Norda covered 43% and 23%, respectively, of the Belgian area under wheat in 1970. The area sown with Cama even increased to 53% in 1975. Leda and Stella were also grown in the Netherlands.

The technological value of a variety accounted for 15% of the points given on the Belgian List of Varieties and was calculated from two years' values of hectolitre weight, W of Chopin and baking trials (Vandam & Castille, 1969).

Castille & Vandam (1981) made a survey of the change in quality (measured as the W of Chopin) of Gembloux varieties released between 1935 and 1977 (Table NLB.12). With the exception of some varieties, an average increase over 40 years is apparent.

From the areas devoted to seed multiplication of wheat in Belgium in the 1990s (Vanorlé, 1996), it appears that the turnover of varieties throughout the years is very quick. Only one of the five main varieties in 1991–1992 was still important in 1995–1996. Nine out of the 10 main winter-wheat varieties in multiplication between 1991 and 1996 were of foreign origin (France, UK, the Netherlands, Germany and Sweden). Among these ten varieties, the only home-bred variety was Capitaine, classified in quality group I-M (Insufficient-Medium quality). Three of the 10 varieties were classified as good quality (B) varieties: Camp Remy, Florin and Soissons, all three of French origin. The Dutch variety Ritmo, classified as a variety of medium quality (M), was the largest variety in multiplication in 1995–1996.

Belgian farmers have a choice among a large number of varieties. Productivity is usually considered the most important criterium. Castille & Vandam (1981) noted that this situation is not likely to change unless a financial reward for quality is guaranteed.

4. Investigations on glutenins and gliadins

4.1. Glutenins

A substantial amount of research on wheat proteins was initiated by Graveland and Moonen at the Institute for Flour, Cereals and Bread in Wageningen (a subdivision of TNO in Zeist). Graveland et al. (1982) described the functional properties of different protein fractions isolated by SDS solution and reported that SDS-insoluble glutenins constituted the most significant protein fraction providing visco-elasticity to the dough. Differences in loaf volume between the good bread-making variety Sicco and the poor bread-making variety Tundra could be explained largely by the difference in amount and molecular composition of the glutenins. Moonen and Graveland used a numbering system for glutenin subunits that was different from the one described by Payne et al. at the PBI in Cambridge, and which was commonly used by most researchers in this field. The finding of Moonen et al. (1983) that subunit 2∗ (*Glu-A1*) and subunits 3+10 (*Glu-D1*) have a positive effect on bread-making quality is consistent with the findings of other investigators, if the conversion according to Table NLB.13. is applied.

Table NLB. 13. Alleles coding for glutenin subunits numbered according to Moonen and Payne

	Chromosome											
	1A		1B						1D			
Moonen et al. (1983)	1	2*	4	5	6	7	8	9	2	3	10	11
Payne et al. (1981)	1	2*	6	7	14	15	8	9	2	5	10	12

Source: Moonen et al. (1983).

Studies of the biochemical properties of the HMW glutenin subunits by use of SDS-PAGE (Moonen et al., 1984) showed that the amino-acid composition of the different subunits was very similar. A closer look at the separate amino-acids revealed that the number and location of the cysteines in the HMW subunits plays a substantial role in the development of the dough, through the development of interchain disulphide bonds. Due to the location of the cysteine residues, subunit 10 appeared to be a better cross-linker than subunit 11 (= subunit 12 according to numbering by Payne) to form an extended disulphide bonded network of glutenin during ripening and dessiccation of the wheat kernels. This could explain the higher level of bread-making quality in wheat material with subunits 5+10 than material with subunits 2+12.

A major contribution to glutenin and gliadin research was made by investigators at CPRO-DLO (formerly SVP) and ATO-DLO in Wageningen, in particular by the group of Kolster. Kolster, Krechting & van Gelder (1988) developed a technique to seperate subunits with minor differences in mobility. By using two-dimensional electrophoresis, consisting of iso-electric focussing (IEF) followed by SDS-PAGE, they demonstrated that HMW subunits with a certain molecular weight could be subdivided on the basis of amino-acid composition. Thus, the difference in bread-making quality between two Dutch varieties with the 'same' HMW subunit composition could be explained.

Kolster & Van Gelder (1990) described a method for studying the quantitative variation of HMW glutenin subunits. The method involved optimization of the SDS-PAGE and the staining of the proteins in the gel, followed by a quantification by laser scanning densitometry (LSD), to determine the amount of dye absorbed by a particular band.

A study of 226 lines of winter wheat grown in the Netherlands (Kolster et al., 1991) revealed that epistatic effects of allelic variation at the HMW glutenin loci contributed to the variation in loaf volume of the lines. The authors stated that allelic variation at *Glu-A1* and *Glu-B1* was only effective in the presence of subunits 5+10 on *Glu-D1*. As a consequence, it was recommended to introduce these subunits into Dutch wheat-breeding programmes in order to improve bread-making quality.

The composition of the most important winter- and spring-wheat varieties released in the Netherlands since the 1970s is listed in Table NLB.14. It is clear that the occurrence of subunits 5+10 is very rare. More than half the varieties have the HMW glutenin composition (N, 6+8, 2+12) or (N, 7, 2+12), which is generally related to poor bread-making characteristics. Subunit 2* (*Glu-A1*) is present in only one variety: Accent. The highest *Glu-1* score is found in the new variety Arnaut, released in 1995.

Weegels et al. (1997) use the term 'gluten macropolymer' (GMP) to indicate the fraction of the glutenin that is not extractable in SDS solution. Changes in GMP structure and quantity during dough resting and mixing were investigated. These changes suggested that specific reactions involving glutenin subunits occur during mixing (depolymerization of GMP) and dur-

Table NLB. 14. HMW glutenin composition and *Glu-1* score of home-bred wheat varieties released in the Netherlands between 1964 and 1995

Variety	HMW glutenin subunits			*Glu-1*
	1A	1B	1D	score
Winter wheat				
Manella	1	7	2+12	6
Lely	1	6+8	2+12	6
Clement	N	6+8	2+12	4
Nautica	1	6+8	2+12	6
Arminda	N	7	2+12	4
Citadel	N	6+8	2+12	4
Pagode	N	7	2+12	4
Obelisk	N	7	2+12	4
	N	20	2+12	4
Accent	2*	6+8	2+12	6
Ritmo	1	6+8	2+12	4
Arnaut	1	7+9	5+10	9
Spring wheat				
Sicco	1	7	5+10	8
Orca	N	7	2+12	4
Toro	1	7	2+12	6
Kaspar	N	7	2+12	4
Melchior	N	7	3+12	4
Stratos	N	6+8	2+12	4
Minaret	1	7	5+10	8
Adonis	1	7	2+12	6[a]
Bastion	1	7+9	2+12	7[a]

[a] Data obtained by converting results of Moonen et al. (1982) to numbering of Payne et al. (1981).
Source: Moonen et al. (1982), Kolster & van Gelder (1991), Rai (1998).

ing resting (re-polymerization of the extractable glutenin polymers). The GMP is one of the parameters used for the classification of wheat varieties in the Netherlands.

Verbruggen et al. (1998) described a procedure to isolate LMW and HMW glutenin subunits simultaneously, which may then be used for further fractionation for rheological and bread-making experiments.

4.2. Gliadins

Doekes (1969) analysed the gliadin composition of 80 varieties and selected lines of winter and spring wheat and classified them in five morphological groups, each with a similar gliadin pattern.

Janssen et al. (1990) compared the rheological behaviour of the gluten from Obelisk (poor bread-making quality) and Katepwa (good bread-making quality). They concluded that the ratio of gliadins to glutenins was a prime factor determining the rheological characteristics of gluten and that only at a high gliadin to glutenin ratio, the origin of the fractions was important.

Van Lonkhuyzen et al. (1991) analysed 32 wheat samples of the same HMW glutenin subunit composition but different gliadin composition and bread-making quality. They found that 82% of the variation in bread-making quality could be explained by four gliadin peaks, as identified by RP-HPLC.

J.M. & D.A.D.

Acknowledgments

The authors wish to thank Ir D. de Coster (Clovis, Matton, Tiegem, Belgium) and Ir M.E. Roothaan (Advanta Van der Have, Rilland, The Netherlands) for their valuable comments on the manuscript of this chapter. We are grateful to Dr R.K. Rai (Cebeco Zaden, Lelystad) and to Dr Ir C.A.M. Marcelis (Produktschap GZP, The Hague) for providing additional data.

References

Broekema L (1899) Duivendaal- en Spijktarwe. Maandblad Vereniging van Oud leerlingen der Rijks Landbouw School, No. 127.

Broekema C (1939) Opening Kwekersdag 1939. Landbouwkundig Tijdschrift 51: 793–808.

Broekhuizen S (1951) De Technische Tarwe Commissie (1931–1941). Cocobro Jaarboekje 1: 29–42.

Broekhuizen S, de Miranda H (1951). De bakwaarde van de in Nederland verbouwde tarwerassen. Landbouwkundig Tijdschrift 63: 776–788.

Broekhuizen S (1963) Het streven naar verbetering der bakkwaliteit van de in Nederland verbouwde tarwe. Jaarboekje NGC 8: 65–73.

Broekhuizen S (1967) Tiende Jaarverslag van de Stichting Nederlands Graan-Centrum: Werkgroep Bakkwaliteit. Jaarboekje 1966–1967 NGC 10: 13–16.

Broekhuizen S (1969) Atlas of the cereal-growing areas in Europe. Agro-ecological Atlas of Cereal Growing in Europe, vol II, 157 pp. Wageningen: Pudoc; Amsterdam, London, New York: Elsevier Publishing Company.

Bruyning FF (1905) La valeur boulangère du froment. Archives du Musée Teyler, Serie II, 9. Les Heritiers, Loosjes, Haarlem, 217 pp.

Castille JP, Vandam J (1981) Bakwaarde van Belgische tarwerassen. Landbouwtijdschrift 34: 1431–1441.

Castille JP, Vandam J (1985) L'Influence du facteur Variété sur la qualité, in: Qualité et Teneur en Protéines des Blés. Note Technique du Centre de Recherches Agronomiques de l'Etat à Gembloux, No 6/40, juin 1985: 5–16.

Coster D de (1994) Stand und Perspektiven der Sortenentwicklung in Belgien. Bericht über die 45.Arbeitstagung 1994 der Arbeitsgemeinschaft der Saatzuchtleiter im Rahmen der Vereinigung österreichischer Pflanzenzüchter, BAL Gumpenstein, pp 73–80.

Doekes GJ (1969) Wheat grain proteins; Analysis of varieties by starch-gel electrophoresis (PhD thesis). Rotterdam: Bronder Offset B.V.

Dorst JC (1927) La culture du froment aux Pays-Bas pendant les cinquante dernières années. Actes de la 1ere Conférence Internationale du Blé, Rome, avril 1927, Imprimerie de l'Institut International de l'Agriculture: 417–422.

Dumon AG, Laeremans R (1956) Het kweken van zomertarwerassen. Agricultura 1956, Band IV, 2e reeks, No. 1: 63–80.

Feekes W, Van Dobben WH (1940) Verslagen van de Technische Tarwe Commissie XIV Bakonderzoek bij de Rassenproeven 1933-'38. Drukkerij Gebroeders Hoitsema NV, Groningen, pp 339–429.

Feekes W (1978) Adaptionsregionen Europas, Getreide-Ekotypen, Modifikationen, Sortenbenützung. Bericht über die Arbeitstagung 1978 der Arbeitsgemeinschaft der Saatzuchtleiter in Rahmen der Vereinigung Oesterreichicher Pflanzenzüchter, Gumpenstein, pp 3–12.

Graveland A, Bosveld P, Lichtendonk WJ, Moonen JHE, Scheepstra A (1982) Extraction and fractionation of wheat flour proteins. J Sci Food Agric 33: 1117–1128.

Janssen AM, Vereijken JM, Beintema JJ, Witholt B, van Vliet T (1990) Rheological studies of gluten. In Bushuk W, Tkachuk R (eds), Gluten Proteins 1990. St Paul (MN) American Association of Cereal Chemists, pp 180–186.

Kolster P, Krechting CF, van Gelder WMJ (1988) Variation in high molecular weight glutenin subunits of *Triticum aestivum* and *T. turgidum* spp. *dicoccoides*. Euphytica supplement December 1988: 141–146.

Kolster P, van Gelder WMJ (1990) A method for the quantification of individual high molecular weight glutenin subunits using SDS-PAGE and laser scanning densitometry. In Bushuk W, Tkachuk R (eds), Gluten Proteins 1990. St Paul (MN) American Association of Cereal Chemists, pp 349–359.

Kolster P, van Gelder WMJ (1992) Variation in type and relative amounts of the high molecular weight glutenin subunits in Dutch wheat varieties. In Kolster P (ed), High molecular weight glutenin subunits of wheat: qualitative and quantitative variation in relation to bread-making quality. Thesis Wageningen, 135 pp.

Kolster P, van Eeuwijk F.A., van Gelder WMJ (1991) Additive and epistatic effects of allelic variation at the high molecular weight glutenin subunit loci in determining the bread-making quality of breeding lines of wheat. Euphytica 55: 277–285.

Kolster P, Krechting KF, van Gelder WMJ (1993) Variation in type and relative amounts of the high molecular weight glutenin subunits in Dutch wheat varieties. J Sci Food Agric 61: 167–174.

Koopman C (1963) Geschiedenis en ontwikkeling van het veredelingsbedrijf Cebeco. In 25 jaar veredelingsbedrijf Cebeco, pp 8–22.

Larose E, Legros R, Noulard L (1950) L'amélioration des froments en Belgique en vue du perfectionnement de leur valeur d'utilisation. VIIIe Congrès International des Industries Agricoles, Bruxelles: 1–28.

Larose E (1954) Veertig jaar activiteit van het Rijksproefstation voor Plantenveredeling te Gembloux. Landbouwtijdschrift 7: 1–16.

Larose E, Moës A, Bonnier Ch, Noulard L (1956) Classification et identification des races de froment et d'epautre en Belgique. Bulletin de l'Institut Agronomique et des Stations de Recherches de Gembloux 13: 7–113.

Larose E (1966) L'importance à attribuer à la sélection et l'amélioration des plantes cultiveés dans les progrès réalisés par l'agriculture Belge au cours des cinquante dernières anneés, les perspectives pouvant être envisagées pour l'avenir. Annales de Gembloux 72 (3): 269–310.

Lonkhuyzen van HJ, Hamer RJ, Schreuder C (1991) Influence of specific gliadins on the bread-making quality of wheat. Cereal Chemistry 69 (2): 174–177.

Mastenbroek C (1963) Tarwe. In 25 jaar veredelingsbedrijf Cebeco, pp 49–65.

Mesdag J (1966) Ervaringen met de veredeling op bakkwaliteit bij tarwe. In Lange W, Lamberts H, Wit F (eds), Ontwikkelingen in de plantenveredeling (special publication Stichting Voor Plantenveredeling), Wageningen, pp: 194–201.

Mesdag J (1975) Quality improvement of European wheats. Proc 2nd Int Winter wheat Conf, Zagreb 1975: 341–350.

Mesdag J (1979) Combinatie van bakkwaliteit en opbrengst als veredelingsdoel bij tarwe. Zaadbelangen 33: 254–256.

Mesdag J (1985) Developments in breeding winter wheat for bread-making quality in some north-western European countries. Neth J Agr Sc 33 (3): 215–234.

Moonen JHE, Scheepstra A, Graveland A (1982) Use of the SDS-sedimentation test and SDS-polyacrylamidgel electrophoresis for screening breeder's samples of wheat for bread-making quality. Euphytica 31: 677–690.

Moonen JHE, Scheepstra A, Graveland A (1983) The positive effects of the high molecular weight subunits 3+10 and 2∗ of glutenin on the bread-making quality of wheat cultivars. Euphytica 32: 735–742.

Moonen JHE, Marseille JP, Scheepstra A, Graveland A (1984) Genetic aspects and biochemical properties of the high molecular weight subunits of glutenin in relation to bread-making quality. In Graveland A, Moonen JHE (eds), Gluten proteins; Proc of the 2nd Int Workshop on Gluten Proteins. TNO, Wageningen, 218 pp.

Moorsel JF van (1934) De Tarwewet 1931. N. Samson N.V., Alphen aan de Rijn, 156 pp.

Nicolas E (1976) Over Graan, Meel en Brood. Zeist: CIVO-TNO publicatie.

Noulard L, Vandam JB (1967) Drie nieuwe variëteiten van wintertarwe. Landbouwtijdschrift 6: 851–855.

Payne PI, Corfield KG, Holt LM, Blackman JA (1981) Correlations between the inheritance of certain high-molecular-weight subunits of glutenin and bread-making quality in progenies of six crosses of bread wheat. J Sc Food Agric 32: 51–60.

Rai RK (1998) personal communication Lelystad, Wageningen.

Rogiers L, Vercruysse-Dewitte D, Coucke P (1991) Characterization and identification of glutenin HMW subunits in Belgian wheat cultivars differing in breadmaking quality. In Bushuk W, Tkachuk R (eds), Gluten Proteins. St Paul (MN) American Association of Cereal Chemists, pp 603–614.

Roothaan ME (1999) personal communication Rilland, Wageningen.

Vandam JB, Castille JP (1969) Amélioration du froment et qualité technologique. Colloque Pain et Santé, Bruxelles, avril 1969: 53–84.

Vanorlé L (1996) Le paiement de la qualité du froment panifiable. Publications de l'Institut Economique Agricole (I.E.A.) du Ministère des Classes Moyenne et de l'Agriculture, Etudes No 75, pp 1–56.

Verbruggen IM, Veraverbeke WS, Vandamme A, Delcour JA (1998) Simultaneous isolation of wheat high molecular weight and low molecular weight glutenin subunits. J Cer Sci 28: 25–32

Verslagen van de Technische Tarwe Commissie (1934–1942). Groningen: Drukkerij Gebroeders Hoitsema.

Vries O de (1933) Kwaliteit en bakeigenschappen van tarwe; overzicht van de resultaten van de T.T.C. Landbouwkundig Tijdschrift 45: 876.

Vries O de (1934) Een goede baktarwe in Nederland. Correspondentieblad voor de Rijkslandbouwvoorlichtingsdienst No.2.

Vries O de, Feekes W (1935) Recherches sur la qualité boulangère de blé dans les Pays-Bas. Praeadvies Congres van de Fédération Internationale des Sélectionneurs. Bruxelles, julliet 1935.

Weegels PL, Hamer RJ, Schofield JD (1997) Depolymerisation and Re-polymerisation of wheat glutenin during dough processing (II) Changes in composition. J Cer Sci 25: 155–163.

Wouda KP (1998) Personal communication.

Zeven AC (1990) Landraces and improved cultivars of bread wheat and other wheat types grown in the Netherlands up to 1944. Wageningen Agricultural University Papers 90–2, 103 pp.

The United Kingdom

1. Introduction

The United Kingdom (UK), comprising England, Scotland, Wales and Northern Ireland, is one of the main wheat producers of Europe. Wheat has been grown in the British Isles since prehistoric times and an extensive survey on the use of wheat throughout the history of Great Britain has been given by Percival (1934). As early as the 11th or 12th centuries, several kinds of bread were made that varied in quality and appearance. The change over from rye, oats and barley to wheat as the chief or only bread corn was complete throughout the country by the 1820s or 1830s, and from then on the use of wheat flour for bread making has been universal. The terms 'strength' and 'weakness' of wheat grains and flour were in common use in the UK at the beginning of the 19th century.

The area under wheat has undergone large fluctuations throughout the centuries. A vast production increase over the second half of the 20th century (Table UK.1), resulting in an increased area and yields that have almost tripled, has been enabled by a combination of improved cultural methods and a very efficient wheat breeding programme. Whereas before World War II the UK imported more than three-quarters of the wheat for her home consumption, this amount had been reduced to less than one-quarter by 1990.

The share of wheat in the total cereal production of the UK increased from one- to two-thirds during 1950–1995. During these years wheat mainly filled the space left by oats, of which the area grown decreased dramatically from 1 500 000 ha in 1945 to 100 000 ha in 1997. The proportion of oats is somewhat higher in Wales and Scotland than in England, due to climatic differences. Oatmeal is used for animal feed, oatcakes, porridge and haggis.

The British climate, with mild winters and cool summers, is favourable for cereal growing and yields are among the highest recorded in any country where wheat is a major crop. The greatest limiting factor for wheat growing in the UK is abundance of rainfall at ripening in certain parts of the country. Although the climate is best suited for the production of high yields of biscuit and feed wheats, it offers opportunities for improving bread-making quality (Jenkins, 1984).

The main wheat growing areas in England are located in the South-East, East Anglia and the Midlands. Highest yields are obtained in Norfolk. Yields in the western part of England, Wales and Northern Ireland are somewhat

Table UK.1. Wheat acreage (1000 ha) and production (1000 t) in the UK since 1852

Year	Area	Production
1852	1 500	2 518
1900	747	1 585
1934–38	754	1 743
1948–1952	881	2 397
1961–1965	870	3 520
1969–1971	980	4 140
1979–1981	1 434	8 116
1989–1991	2 026	14 143
1995–1997	1 954	15 180

Sources: Percival (1934); FAO statistics.

less satisfactory, mainly due to higher rainfall. In Scotland, the best wheat areas lie on the banks of the Forth and the Tay, extending up the coast as far as Aberdeen where relatively high yields may be obtained. Most of the wheat grown in the UK is winter wheat. Very little spring wheat was grown before the 1950s. Nowadays, the proportion of spring wheat varies between roughly 3% and 7% of the area under wheat, depending on weather and on the availability of suitable varieties. Practically all the spring wheat is used for bread making.

The UK was more or less self-supporting in wheat until the Industrial Revolution at the end of the 18th century, when it began to import some supplies from India and Russia. The main sources of wheat to meet the growing demands of the increasing population during the 19th and 20th centuries were North America, Australia and Argentina. These countries supplied 80–90% of the wheat imports in the 1930s (Astor & Rowntree, 1938), and bakeries in the UK depended on these strong wheats for the production of bread. In the 1960s, a typical British loaf would have been made from 60% Canadian spring wheat, 20% poor-quality British wheat and 20% filler wheat, usually from Australia.

Because the UK had been a wheat deficient country for more than a century, British millers were highly experienced in the international trade of wheat. In 1960, when European demand for wheat had seriously diminished, the UK was the most important wheat buyer on the world market, handling about 20% of the world trade. After the UK joined the European Union (EU) in 1974, American and Australian wheats were gradually replaced by wheat from EU countries. In 1992, the main EU countries exporting to the UK were France, Germany and Ireland.

The vast production increase of wheat in the UK since World War II, combined with improved varieties and the addition of gluten to bread flours, have reduced wheat imports to a minimum. The introduction of a wheat classification scheme, devised in 1975 by The Home-Grown Cereals Authority to make home-grown wheat more attractive to farmers and millers, has undoubtedly contributed to these developments. The content of home-grown wheat in bread-making grists rose from approximately 10% to 70% between 1973 and 1983 (Annual Report PBI, 1985). During the 1990s this content decreased in favour of imports of high protein quality wheat from EU countries, lying around 1 Mt annually.

Exceptional weather conditions aside, UK farmers have the ability to produce more wheat in total than the national requirement, which for all purposes amounts to some 10 Mt. The British use about 5 Mt of wheat for flour milling per year, two-thirds of which is used for bread making. A further 1 Mt of soft wheat is used for making biscuit and cake flours, with other uses (e.g. household purposes, such as the typical British puddings and pies) completing the balance. It may be interesting to note that half of the wheat produced in Scotland, approximately 1 Mt, is used for distilling, producing the famous Scotch Whisky. The chief requirement for this industry is high yielding, soft milling varieties with a high carbohydrate content, from which a high yield of alcohol can be obtained.

2. Wheat breeding before 1945

2.1. Early varieties and squareheads

Many of the wheats grown in the UK in the 18th and first half of the 19th century were of good bread-making quality. Their origin and a rating of their quality (Percival, 1934) are listed in Table UK.2.

Due to the fall in price of wheat and the high cost of rent, rates, labour and other inputs, farmers were compelled to make yield per acre their chief aim. In consequence, wheats of the squarehead type and other heavy cropping varieties of inferior quality started to spread widely from about 1870 onwards (Percival, 1934). Although the exact origin of the squarehead wheats is not known, Great Britain is the first country in which their appearance and use has been recorded (Vavilov, 1935).

The earliest breeding activities in the UK are associated with names of Le Couteur, Hallett and Sheriff. Le Couteur started wheat breeding by individual selection. Hallett introduced the classic pedigree method of repeated selection, choosing heads from the best plants and among the heads those

Table UK.2. Quality and origin of some early bread wheats in the UK as described by Percival (1934)

Name	Quality	Origin
Old Welsh White Chaff Red	superior[a]	Wales
Rough Chaff White[b]	fine	Essex, Kent, Sussex
Red Stettin[c]	high	Baltic region
Old Irish Red Chaff	good	Central Ireland
Irish Coney Island	good	Coney Island (Ireland)
Price's Prolific	high	raised about 1886 by Mr Price (Gloucestershire)
Badger	average	selection from Squarehead
Starling II	high	selection from Wilhelmina (Netherlands)[d]
Fox	superior	selection from Blé à duvet (France)

[a] Superior to the average English wheat.
[b] Grown extensively until 1880 in Essex, Kent, Sussex and other southern and south-midland counties under the names Velvet Chaff, Hedge Wheat, Old Hoary and Tunstall.
[c] Cultivated for more than 100 years in Central Ireland, and doubtlessly originally imported from the Baltic Port of Stettin, from which considerable amounts of wheat were imported between 1784 and 1814.
[d] Selected by Percival from Wilhelmina, probably after a natural cross pollination.

with the best grain (Vavilov, 1935). In Scotland, during the second half of the 19th century, Sheriff, who used to examine fields of growing crops in different parts of the country for individual plants superior to the rest, introduced several new types. He also produced a number of hybrids between 1856 and 1870, none of which, however, proved superior to the varieties commonly cultivated (Percival, 1934).

Sheriff was closely involved in the propagation of the squarehead wheats throughout Britain and Europe. These highly productive wheats with stiff straw were characterized by short, dense ears, almost square in cross section. Carried to Denmark by a pupil of Sheriff in 1874, from where they spread to Germany, Sweden, the Netherlands, Belgium and France, squarehead wheats were frequently used in wheat breeding programmes in these countries. This resulted in a large number of very productive varieties, unfortunately producing grains of poor bread-making quality.

2.2. The Plant Breeding Institute in Cambridge

Extensive work on wheat hybridization in England was undertaken by Biffen at Cambridge University, starting in 1898. He continued his work as director of the Cambridge University Plant Breeding Institute from 1912 onwards. This Institute, which remained closely linked to the University until 1952, has played a dominant role in British wheat breeding. It worked side by side with the National Institute for Agricultural Botany (NIAB) in Cambridge, responsible for the testing of new varieties since 1920 and issuing an annual list of classified and recommended varieties.

Biffen was intrigued by the question whether it was possible to combine the strength of American wheats with the yielding capacity of English wheat varieties. That quality and quantity of wheat were inversely related had already been acknowledged by Andrews in 1853, who wrote: 'It should always be borne in mind that those wheats which are prolific are generally of inferior quality, but as a rule the most prolific is the most profitable' (cited from Percival, 1934).

Having studied the inheritance of the physical texture and the nitrogen content of wheat endosperm, Biffen concluded that the 'hard, translucent character of the grain' was genetically dominant over the 'soft, opaque condition'. Moreover he stated (1908) that 'strength is inherited in its entirety'. With this knowledge in mind, Biffen made several crosses with the Canadian spring-wheat variety Red Fife. This variety, originally selected in Ontario (Canada) from a sample of wheat from Dantzig (Gdansk) by a Scotsman called Fife, has been used extensively by breeders all over the world as a source of bread-making quality.

From the offspring of Red Fife and an English wheat, Biffen isolated the variety 'Burgoyne's Fife', with a strength of 90–100% of that of Red Fife. It was appreciated by millers, who were willing to pay the same price as they did for strong American wheats. Although the variety had a good yielding capacity (Wood, 1914), for some reason it was never grown on a large scale.

Biffens first success of wide acclaim was the release of the winter-wheat variety Little Joss in 1910. Little Joss was selected from the progeny of a cross between the old English variety Squareheads Master (a very 'heavy cropper' under English conditions), and a Russian spring-wheat variety with good resistance to yellow rust, Ghirka. Little Joss became very popular, especially in the Black Fen area, owing to its capacity to fill out and ripen its grain evenly though lodged (Bell, 1950).

Little Joss was followed in 1916 by Yeoman, resulting from a cross between White-chaffed Browick (an English squarehead wheat) and Red Fife. In Yeoman, English farmers were provided for the first time with a high yielding winter wheat of good bread-making quality. Moreover, Yeoman was 'definitely superior in yield to Squareheads Master on heavy soils in high state of fertility' (NIAB trials, 1924–1928). Squareheads Master, Yeoman and Little Joss were the most widely grown winter-wheat varieties during the late 1920s and 1930s (Lupton, 1992). Little Joss and Yeoman remained on the NIAB recommended list until 1957.

Biffen used to screen his breeding material (in F_3 and F_4) on gluten quality by applying the chewing test. It was described as follows: 'This test is based on the physical characteristics of the gluten when isolated by chewing some 20 or 30 grains for 10 minutes or so. The starchy contents disappear, thus leaving a pellet of gluten in the mouth. A shrewd idea of its toughness can be formed by stretching the pellet' (Biffen & Engledow, 1926).

A description of the type of wheat bakers were interested in was given by Percival (1934):

> The baker prefers flour of the strong class, because the number of loaves which can be made from a given weight of it is greater than from an equal weight of weak flour. This arises from the fact that strong flour imbibes more water and retains more after baking than weak flour, and the *water sells as bread.* For example a 280 lb. sack of strong Canadian flour will absorb as much as 80 quarts of water, and from the dough 400–440 lbs. of bread (i.e. 100–110 loaves) can be baked. A sack of the same weight of flour made from an average English wheat will not take up more than 62–64 quarts of water and the yield of bread will not exceed about 360 lbs. (i.e. 90 loaves).

Biffen's work at the Plant Breeding Institute was continued by Engledow, who was the breeder of the next major winter wheat to come from Cambridge: Holdfast, selected from the cross of Yeoman with the Canadian variety White Fife and released in 1936. Holdfast set a new standard for bread-making quality for 50 years to come (Annual Report PBI, 1985). It showed improvements in grain yield, straw strength and disease resistance but it had a white grain and was thus liable to sprouting during wet harvests. It remained on the NIAB recommended list until 1985.

During the 1930s, the majority of the wheat varieties grown in the UK were home-bred. A restricted area was sown with the Dutch variety Wilhelmina and the French variety Bersée. The proportion of home-bred varieties slowly decreased during the 1940s and was very low between 1955 and 1975, due to the success of the French variety Cappelle Desprez. The importance of wheat varieties with improved bread-making quality before 1945 was limited, as may be illustrated by the fact that the proportion of the English wheat acreage sown with Yeoman and Holdfast was no more than 12% in 1926 and also in 1945 (Lupton, 1992).

It is remarkable that Biffen and Engledow did not use any of the wheats listed in Table UK.2. as a quality source for their wheat breeding programmes. Nor was this done at a later stage by breeders at the Plant Breeding Institute or private companies. With the knowledge on protein quality acquired since the 1980s (see Section 3.5), renewed interest in these old land varieties might be fruitful.

2.3. Private breeders

The influence of private breeding companies was rather limited during the first half of the 20th century. Gartons Ltd of Warrington released the variety Redman, selected from the cross Yeoman with Squareheads Master. The performance of this variety was comparable to Yeoman, both in bread-making (Earnshaw, 1949) and in yielding capacity. However, the area sown with Redman never exceeded 5% of the total wheat area. One very successful variety was Hybrid 46, released by Marsters Ltd in 1946, resulting from a cross between the short-strawed French variety Benoist 40 and other varieties including Yeoman. Hybrid 46 was a high yielding variety with biscuit-making quality; it was grown for more than 20 years and used as a reference variety until 1968. Other private breeding companies working on wheat during these years were Rothwell Plant Breeders and Miln & Co. Their work will be discussed in Section 3.3.

3. Wheat breeding after 1945

The attempt to stabilise the wheat acreage at a considerably higher level than before World War II and the emphasis placed on high yields, confronted wheat breeders with new challenges. An increasing wheat acreage implied that varieties were needed that performed well on less favourable soil types. On the other hand, lodging resistant varieties were needed to withstand the most extreme high fertility conditions. In both cases, adequate agronomic performance had to be combined with an acceptable level of milling and baking quality.

The exceptional situation existed between 1955 and 1975 that the English wheat crop was dominated by the French variety Cappelle Desprez. Against this background it was remarkable that the proportion of home-bred wheat varieties increased from 3% in 1964 to a record 92% in 1982.

Following the enforcement of the Plant Varieties and Seeds Act in 1964, enabling plant breeders to obtain property rights on new varieties, the share of private breeding companies in British wheat breeding increased, although the Plant Breeding Institute (PBI) in Cambridge remained the most important wheat breeder in the UK. The proportion of the UK wheat area sown with PBI winter-wheat varieties was about 85% in 1986 but by 1996 had fallen to 50%, due to the introduction of varieties from private breeders in the UK and from abroad.

3.1. The classification of wheat varieties

The National Institute for Agricultural Botany (NIAB) in Cambridge issues an annual list of classified and recommended varieties. The classification for quality is based on the result of baking trials for bread and biscuits, endosperm texture (hard or soft), Hagberg falling number, specific weight, Zeleny volume, protein content and thousand grain weight.

On the NIAB classified list, wheat varieties are classified into categories A, B, C and D for yield and for bread- and biscuit-making properties. The recommended list gives a score for bread- and biscuit-making quality ranging from 5–8. Score 8 corresponds to the best available varieties, varieties that score 7 are 'consistently acceptable for all bread-making process', varieties that score 6 are 'suitable for some bread-making process' and varieties that score 5 are 'not favoured for bread making' (NIAB Recommended List, 1996). Yield trials for the recommended list are conducted throughout the UK, coordinated by the NIAB and funded by the Home-Grown Cereals Authority (HGCA).

The HGCA, which is funded from farm levies on cereals and oilseeds, classes wheat varieties in three categories according to their potential for bread making:

I: 'prefered bread-making varieties' includes varieties most suitable for inclusion in bread-making grists;

II: 'acceptable bread-making varieties' includes varieties with more limited but useful potential for inclusion in bread-making grists;

III: 'non bread-making varieties' includes varieties most favoured in biscuit grists, and those predominantly used in animal feed but which may be incorporated in certain grists.

The suitability for bread making in this classification is determined in controlled baking tests in which volume and appearance of loaves are recorded. Certain properties of the flour are taken into account, as well. Varieties marketed for bread making are required to have an acceptable protein level, good specific weight, high Hagberg falling number, hard endosperm texture and a good milling score. Ease of milling is considered a varietal characteristic that cannot be influenced by husbandry or climatic conditions and is a measure of the ability to extract large amounts of flour of good colour.

The National Association of British and Irish Millers (NABIM) produces an annual list of preferred wheat varieties to assist farmers in making their marketing plans and planting decisions. Also, it clearly indicates how millers view varieties for milling.

3.2. *The Plant Breeding Institute in Cambridge*

Following the decision of the Agricultural Research Council (ARC) in 1952 to establish the Plant Breeding Institute in Cambridge, separate from the university, the scale of wheat breeding operations and the intensity of selection work greatly expanded. This may be illustrated by the increase in the number of F_3 ear rows of winter wheat from 6 000 to 45 000 between 1964 and 1983. The mandate of the Plant Breeding Institute (PBI), funded by the ARC and the HGCA, was to work on the improvement of all important crops grown in England. With its headquarters in Trumpington, the PBI remained a governmental institution until 1987. The combination of basic research with breeding activities at the PBI formed the basis of a very efficient and successful wheat breeding programme. Since 1987, basic research work has been continued by the Institute of Plant Science Research (IPSR), under the Agricultural and Food Research Council (AFRC). The applied breeding work has been continued under the name Plant Breeding International.

3.2.1. *Investigations on bread-making quality*

The breeding policy at PBI has always been that half of the work in wheat breeding should be aimed at developing bread-making varieties (Annual Report PBI, 1986). During the 1950s, PBI researchers paid a lot of attention to the inverse association of grain protein content and grain yield. Bell & Bingham (1957) proposed the following strategy for wheat breeding: 'With the high yield capacity that is expected in the UK, it would seem that there must be a definite limit with regard to the upper range of grain protein content. In the breeding of wheats for bread baking it will probably be necessary to work at lower protein levels and to manoeuvre, within the best protein quality, for the particular objective in mind'. In other words, they were aware that their major tool in quality breeding might be the quality of the protein rather than the amount of the protein. This was based on the knowledge that the amount of protein in the grain was primarily dependent on environment and agricultural practice and less on genotype, although it had a 'heritable component'. On the other hand, the stress/strain properties of the dough were found to be more strongly inherited (Bingham, 1962). Despite the efforts put into selection on quality characteristics, breeders also had to face reality, as expressed by Bingham (Annual Report PBI, 1968): 'The miller in England pays only a slight premium for home grown wheats with good baking quality. Thus a new variety should have a high yielding capacity to be accepted by the growers'.

The adoption of new agricultural practices, especially with respect to nitrogen application, has greatly influenced the problem of the protein yield of the grain: higher levels, split application and application just before or during anthesis have resulted in more efficient uptake of nitrogen. On the other hand, possibilities of late N application have been reduced over the past 10–15 years for environmental reasons, resulting in a renewed interest in efficiency of N uptake and translocation

Lupton (1982) presented the results of a comparative trial with eight winter-wheat varieties released between 1908 and 1980. Protein yields, as well as total yield, in the new varieties were 10–15% higher than in the older varieties. The harvest index had increased from 0.36 in the oldest variety (Little Joss) to 0.51 in the newest variety (Norman). The nitrogen harvest index, however, showed little variation between these varieties. The overall result was that the nitrogen percentage in the grain of the modern varieties was lower than in the older ones.

Bingham (1983) stated that although varietal differences in nitrogen production per unit area of land do exist, the effect of growing conditions is at least four times greater and 'the responsibility for this character will remain largely in the hands of the grower'.

Since the 1980s, a vast amount of knowledge has been acquired on the quality of proteins and in this respect the 1980s have been characterized as the 'decade of the HMW subunit'. The underlying research to a large extent has been the concern of Dr Peter I. Payne and his co-workers at the PBI in Cambridge. Through their work and the training of several visiting scientists, knowledge on glutenin and gliadin composition of the wheat grain has increased worldwide.

According to Payne (1987) the protein (gluten) composition of bread wheat is controlled by nine major genes located on chromosomes 1 and 6, grouped as *Glu-1*, *Gli-1* and *Gli-2*, which code for subunits of high and low molecular glutenins and for gliadins. The most straightforward link with bread-making quality was found in the high molecular weight (HMW) glutenins (Payne et al., 1987a), but ongoing investigations in low molecular weight (LMW) glutenins and gliadins (Payne et al., 1990) have shown that their influence is not to be underestimated. Analyses of the genes on *Gli-1* and *Gli-2* showed that especially the low molecular weight (LMW) glutenin subunits coded by the genes on *Gli-A1* influenced dough properties.

The relation between the HMW glutenin subunit composition of wheat genotypes and their bread-making quality, measured as the result of SDS sedimentation, has been translated into a *Glu-1* score by Payne et al. (1987a). A description of this scoring system is given in Part I, Chapter 7. Results of the analyses of British wheat varieties will be discussed in Section 3.5.

Now, at the close of the 20th century, the inverse relationship between grain yield and grain protein content continues to be a constraint for the breeding of high yielding varieties with improved bread-making quality, although the gap between bread-making and fill wheats has greatly narrowed. However, the problem is seen from another perspective, in which 'protein quality' and 'genetic transformation' are important concepts. The identification of proteins that confer good bread-making quality, so that they may be combined by breeding or by genetic transformation (Shewry et al., 1995), is the basis of the breeding of a new generation of wheat varieties with improved quality.

3.2.2. Tools in quality breeding

Bell & Bingham (1957) summarized the objectives of selection for bread and biscuit types at PBI:

Bread	*Biscuit*
Red grain and smooth skin	Red grain and smooth skin
Hard milling	Soft milling
High water absorption	Low water absorption
Suitable extensometer curve	Suitable extensometer curve
(high resistance; medium extensibility)	(low resistance; high extensibility)
Protein content above a minimum	Protein content below a maximum
White flour	White flour

The baking quality of PBI breeding material in the 1950s and 1960s was assessed by microkjeldahl nitrogen analyses and with the Henry Simon absorption meter and extensometer, which measure the water absorption of flour and the physical properties of unyeasted dough in terms of load-extension curves. The hardness of the grain was measured as the tenacity with which the endosperm adheres to the bran after milling with a Brabender stone mill, resulting in what is known as a bran score, ranging from 1 (soft) to 5 (hard). This distinct classification appeared to be largely independent of environmental conditions. The variety Maris Widgeon has to a large extent been the result of a selection for a suitable combination of bran score and extensometer score.

The Pelshenke test, which had been controversial in the UK since its introduction in 1931, was re-evaluated by Pushman & Bingham in 1975. They pointed out that the Pelshenke test was independent of variation in α-amylase activity and protein content of the grain, which gave it an advantage over small-scale baking tests. Pushman & Bingham suggested that the test was particularly suitable for the selection of early generation material in a quality breeding programme: it should be possible to discard 'non-starters' in the F_2 generation by combining a small-scale milling test and the Pelshenke test, irrespective of seasonal conditions.

The intensity of selection on grain-protein content increased after the introduction of near-infrared reflectance (NIR) analysis. Starting in 1980, it became a routine method at PBI for the determination of protein content of wholemeal wheat flours, enabling selections to be made as early as the F_2 generation. In the years that followed, the technique was used to screen simultaneously for protein content, grinding hardness and grain texture.

In the same period, SDS sedimentation tests were introduced in the wheat breeding programme. According to Bingham (1983) 'the SDS sedimentation test was far superior to the Pelshenke and Zeleny tests in ability to predict loaf volume. Compared with the Zeleny it gives a much better indication of protein quality, is less affected by protein content and uses a simply prepared wholemeal, whereas the Zeleny needs a white flour prepared to an exact specification.' Thus the same sample could be used for SDS sedimentation, NIR analysis and α-amylase determination. Compared to other European countries, SDS sedimentation has been used extensively in the UK.

Between 1984 and 1986, electrophoresis by SDS-PAGE was integrated as a routine method for the screening of early generation material (F_2–F_4). Moreover, it was used to assist in chosing parents for the breeding programme and for checking the genetic uniformity of advanced lines (Annual Report PBI, 1986). The results of electrophoretic screening of glutenins of the most important varieties with bread-making quality will be discussed in Section 3.5.

3.2.3. Winter-wheat varieties bred at PBI

Between the mid 1950s and the mid 1970s, the British wheat area was dominated by the French variety Cappelle Desprez; in 1962 it covered a maximum of 84% of the area under wheat in the UK. This high yielding variety was notable for its durable disease resistance but was not classified as a variety with bread-making quality (score 5 on the Recommended List).

The first PBI winter-wheat variety with a quality level comparable to that of Holdfast was Maris Widgeon,[1] released in 1964 and selected in a cross between Cappelle Desprez and the good bread-making variety Holdfast. Maris Widgeon was recorded as having hard grains with a very good milling quality, similar to that of the reference variety Elite Lepeuple. Its bread-making quality was very good: slightly superior to Elite Lepeuple. The bread-making quality of Elite Lepeuple in France was classified as average, comparable to that of Cappelle Desprez.

Maris Widgeon was not grown on a large scale as premiums for quality at that time were to low. It did, however, play a key role in the pedigree of several later varieties. Crosses of Maris Widgeon with dwarf types resulted

[1] The prefix Maris in the name of PBI varieties released in the 1960s and 1970s, was derived from Maris Lane, the address of PBI headquarters in Trumpington, Cambridge.

in some high yielding varieties with good quality, such as Maris Freeman, Wizard, Sentry and Bounty. However, no home-bred winter-wheat variety on the NIAB recommended list has received a bread-making quality score comparable to that of Maris Widgeon ever since. Maris Widgeon is still grown by 'organic' farmers in the UK.

The release of Maris Huntsman in 1972 set new standards in yielding ability and disease resistance. It was grown on a large scale in the UK, covering more than one-third of the winter-wheat area in 1975. It was also grown on a limited scale in France, where it was used as a reference for yielding capacity but was classified as 'non-panifiable'. Although the milling quality of Maris Huntsman was satisfactory, its bread-making quality was very poor, due to a tendency to produce sticky dough. Closer investigation of this characteristic has shown that this is due to its high endogenic α amylase content, which was most probably inherited from the Belgian variety Professeur Marchal. The same problem occured in the varieties Fenman, Kinsman, Mardler and Norman. The variety Hobbit, released in 1976, outyielded Maris Huntsman by 2% and had the same poor bread-making quality. It was the first semi-dwarf variety to be widely grown in the UK, covering 25% of wheat acreages in 1979.

It was not until 1979 that a new home-bred variety with a satisfactory level of bread-making quality was released. This was the variety Bounty, resulting from a complex cross involving the high yielding varieties Cappelle and Hybrid 46, as well as the variety Maris Widgeon, from which it inherited its quality. Bounty was grown for a short period only, as it was soon replaced by the higher yielding and more successful variety Avalon, which in 1980, a year after its release, covered 29% of the winter-wheat area; Avalon was grown until the early 1990s. During the intervening period it competed with high yielding varieties such as Norman and Galahad. Avalon appeared to have about 0,5–1% more protein than would be expected in relation to its yield (Annual Report PBI, 1985), which was considered to be a slight step forward on the way to increasing the protein level of varieties with bread-making quality. Moulin, released in 1981, whose parentage included Maris Widgeon, Hobbit and a Mexican line, was, like Avalon, classified as a Class I variety by the HGCA, but it was not grown on a large scale. Mercia, however, with a quality level comparable to Moulin, was grown widely for several years after its release in 1986, covering 26% of the winter-wheat area in 1988. It was still on the Recommended List of 1996. Mercia, outyielding Avalon by 4%, is an early maturing variety with a quality that is widely accepted by the baking industry for all purposes. The pedigrees of the above mentioned varieties are presented in Table UK.3.

Table UK.3. Pedigree of Maris Huntsman, Hobbit and the most important PBI varieties with more than average bread-making quality released between 1964 and 1997

Variety	Year	Pedigree
M. Widgeon	1964	Holdfast/Cappelle Desprez
M. Huntsman	1972	CI 12633/Cappelle Desprez//Hybrid 46/3/Prof. Marchal
M. Freeman	1974	M. Widgeon/M. Ranger
Hobbit	1976	Prof. Marchal/Marne Desprez//VG 91443/TJB 16
Bounty	1979	CI 12633/Cappelle//Hybrid 46/3/Widgeon/4/Widgeon/5/Durin
Avalon	1980	CI 12633/Cappelle//Hybrid 46/3/Widgeon/4/Widgeon/5/ Durin/Bilbo
Moulin	1981	M. Widgeon/Cimmyt line/Hobbit sib
Brimstone	1985	54–218/Hobbit/Hustler
Mercia	1986	Talent/Virtue/Flanders
Hereward	1991	Norman sib/Disponent
Abbott	1997	Avalon/Brimstone/Torfida sib

Sources: Lupton (1992); NIAB Recommended List of Varieties.

To illustrate the broadening genetic basis of the quality of PBI varieties over the years, some remarks on their glutenin composition are appropriate. Analyses of the HMW glutenin composition of varieties mentioned in this section has shown that the quality of Maris Widgeon and its derivates (Bounty and Avalon) can be ascribed to the presence of subunit 1 on *Glu-1A* locus, whose origin can be traced back to Red Fife. In contrast, the quality of Moulin is not based on HMW subunit 1 but on the subunit pair $17 + 18$ on the *Glu-1B* locus, inherited from the Mexican line Yecora/Ciano 67. The variety Mercia had neither subunit 1 nor $17 + 18$ but a very good bread-making quality, based on the subunit pair $5 + 10$ on the *Glu-1D* locus, which it inherited from the French variety Flanders.

A major step forward in combining high yield potential and bread-making quality was the release, in 1991, of Hereward, a variety combining excellent bread-making and milling characteristics (HGCA Class I) with satisfactory yield potential. The parentage of Hereward includes the German variety Disponent and a sib of the high yielding variety Norman. The HMW glutenin composition of Hereward, comprising the subunits $7 + 9$ (*Glu-1B*) and $3 + 12$ (*Glu-1D*), is somewhat contradictory to its quality level.

Hereward and Mercia have been able to compete with high yielding varieties such as Beaver, Riband and Brigadier (ICI) due to their relatively high yielding capacity, reinforced by a premium for quality wheat paid by the milling industry since 1987. The progress that has been made in combining

Table UK.4. Relative yields without fungicide treatment (YLD) and scores for milling quality (M) and bread-making quality (B) of winter-wheat varieties on the NIAB recommended list of cereals of 1974, 1981, 1986, 1991 and 1996; each group of data is based on at least 3 years trials

	1974			1981			1986			1991			1996		
	YLD	M	B	YLD	M	B	YLD	M	B	YLD	M	B	YLD	M	B
Cappelle	97	5	5												
M. Widgeon	98	9	8												
M. Freeman	107	7	7												
M. Huntsman	120	5	3	100	5	3									
Bounty				100	7	6									
Avalon				105	7	6	99	7	6						
Mercia							103	7	6	94	7	6	88	–	7
Hereward										105	7	7	97	–	8
Beaver										112	4	2			
Spark													96	–	7
Brigadier													101	–	–

a steady increase of yielding capacity and improved bread-making quality is illustrated by the data presented in Table UK.4. Maris Huntsman, a very high yielding variety in the early 1970s, was overtaken 10 years later by Avalon, a variety with much better bread-making characteristics. Another 10 years later, the yield of Hereward surpassed that of Avalon and had slightly better bread-making quality as well. Hereward was still able to compete with the high yielding feed variety Brigadier, in 1996. The relative yield figures in Table UK.4 are without fungicide treatment. After fungicide treatment, Hereward and Spark both yield 114% of the average, compared to 127% for Brigadier. In 1997, Abbott, with a bread-making classification between that of Mercia and Hereward, was added to the Recommended list.

3.3. Private breeders

Two private breeding companies that released several wheat varieties during the 1950s were Miln & Co. in Chester and Marsters Ltd in Kings Lynn (Norfolk). In 1972 the two companies merged into Miln Marsters Ltd, which via Hilleshög AB and Zeneca became part of the ICI group.

Although the variety Hybrid 46 from Marsters Ltd was very successful, the varieties bred by Milns & Co. were more interesting from a point of view of bread-making quality. The varieties Dominator (Jubilegem/Atle), released in 1957 and Milfast (Atle/Holdfast), released in 1958, both had very good milling and bread-making quality. The bread-making quality was apparently inherited from Atle, a spring-wheat variety from Sweden, that had been crossed with the Belgian variety Jubilegem (= Jubilé) and the PBI variety Holdfast.

Dominator was similar in milling quality to Holdfast but slightly inferior in bread-making quality. The milling and bread-making qualities of Milfast were described as slightly superior to Holdfast. Moreover both Dominator and Milfast yielded 103% relative to Bersée (NIAB trials 1953–1956). Both varieties were favourable exceptions to the apparent negative correlation between bread-making quality and yielding capacity (Mesdag, 1985). However, Dominator and Milfast have never been grown on a large scale.

A large number of wheat varieties have been released by Rothwell Plant Breeders (RPB), a branch of the Nickerson Group, since the 1970s. They now operate under the name NRPB, in short Nickerson.

The RPB varieties Aquila and Armada were widely grown between 1977 and 1983 but were not interesting from a quality point of view. Mission and Boxer, released in the 1980s, had improved bread-making quality but were not satisfactory enough in their yielding capacity. The variety Spark, released in 1993, selected from a cross between Moulin (PBI) and the spring wheat Tonic (Nickerson), scored well in bread-making tests and was acceptable to all millers for bread making. RPB/Nickerson has been more successful in the selection for bread-making quality in spring wheat. Varieties of these will be discussed in Section 3.4.

3.4. Spring-wheat breeding

Until 1970 the area under spring wheat in the UK was dominated by Swedish varieties such as Atle, Svenno and the Belgian variety Jufy. The release of the PBI variety Maris Dove formed a short intermezzo in this foreign 'symphony' but soon had to cede its place to Sappo (Sweden) and Sicco (Netherlands).

It was only after the release of Highbury in 1978 that the UK spring-wheat acreage was again dominated by home-bred varieties. Highbury (Sona227//Svenno/Jufy) was a high yielding bearded variety with good bread-making characteristics. Both Highbury (Category A) and Maris Dove (Category B) were accepted by the National Association of British and Irish Millers as potential varieties for use in bread-making grists.

Hotspur (Highbury//Sirius/Ciano67), released in 1985, was the most recent PBI variety classified as Category A for bread-making quality. Later PBI spring-wheat varieties such as Wembley, Sandown, Solitaire, Concerto, Saracen and Troy were all classified in bread-making quality Category B.

Since the 1980s, NRPB/Nickerson has released a number of spring-wheat varieties with good bread-making quality, of which Broom and Tonic have been the most widely grown. Broom was the most important spring-wheat variety grown between 1980 and 1985, and Tonic covered 57% of the spring-wheat acreage in 1988. Both varieties were classified in Category B for bread-making quality and in Category A for grain protein content. Other RPB/Nickerson spring-wheat varieties are Musket, released in 1983, and Cub, Dollar and Rascal, released in the early 1990s.

The two leading spring-wheat varieties in 1996 and 1997 were Chabliz and Shiraz, both released in 1995 by Cambridge Plant Breeders, Twyford. Chabliz (Jerico/Tonic) and Shiraz (Jerico/Axona) are both high yielding varieties with good bread-making characteristics. Shiraz has a potential for bread making but is unusual in having a soft endosperm. It may have special uses.

One could say that the average quality of spring-wheat varieties grown in the UK over the past 50 years has been very high (Figure UK.1). Apart from exceptions, such as the variety Jufy (very popular in the early 1960s), the major part of the spring-wheat harvest was suitable for use in millers' grists.

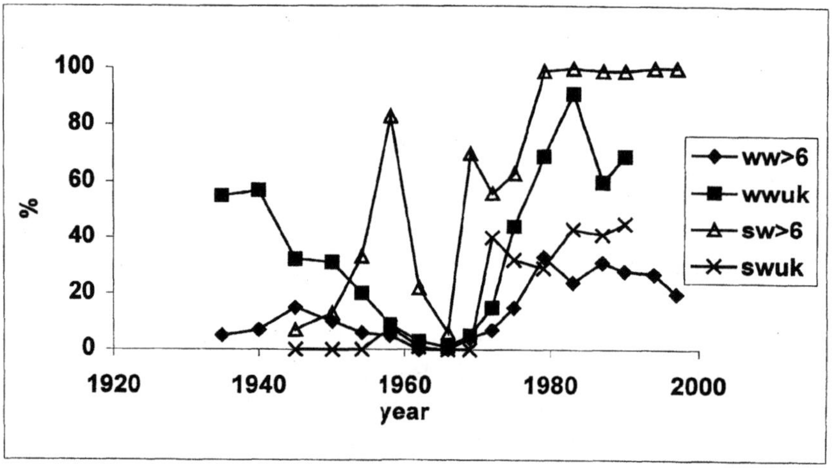

Figure UK. 1. Figure 1: Proportion of winter and spring wheat in the UK that has been home-bred (wwuk and swuk) and proportion of winter and spring-wheat varieties with a quality score higher than 6 (ww > 6 and sw > 6) during the second half of the 20th century.

3.5. Glutenins and gliadins in British wheat varieties

Analyses of the glutenin composition of 84 British-grown varieties (Payne et al., 1987a) revealed that the HMW glutenin subunits occuring most frequently were the null allele of chromosome 1A, subunits 6 + 8 and 7 of chromosome

1B and subunits 2 + 12 of chromosome 1D. All these subunits are commonly associated with poor bread-making quality. Fourty-three percent of the varieties had one of the two glutenin patterns (N, 6 + 8, 2 + 12) or (N, 7, 2 + 12). The subunits 20 and 13 + 16, commonly occuring in foreign wheat varieties, were not present in any of the British varieties, whereas the 2* allele on the *Glu-1A* locus was very rare in the British material.

The results of the study indicated that 47–60% of the variation in bread-making quality was accounted for by variation in HMW subunits of glutenin. After adjusting the scores for the presence of the 1BL/1RS rye translocation, which causes increased stickiness of the dough, the proportion of variation accounted for was raised to 55–67%.

The glutenin composition and the classification of the most important wheat varieties with bread-making quality is listed in Table UK.5. The subunit pair 5 + 10 of the *Glu-1D* locus occurs in all but one of the spring-wheat varieties and also in the winter-wheat varieties Chieftain, Mercia and Spark. Chieftain, however, is classified in bread-making quality Category C; this is probably related to the presence of a 1BL/1RS translocation, causing sticky dough. Mercia, with a *Glu-1* score of 6, has been widely accepted for its bread-making value for more than ten years.

The basis of bread-making quality in UK home-bred varieties was rather limited until the 1980s. A study of the pedigree of the PBI varieties with bread-making quality, from Holdfast up to Bounty and Avalon (Payne et al., 1979), showed that this quality was significantly associated with the glutenin subunit 1, inherited through a long ancestral line from the Canadian variety Red Fife. As mentioned in Section 3.2.2, the genetic basis of the bread-making quality in UK wheat varieties has become more diversified since.

The gliadin composition of wheat proteins in the variety Chinese Spring has been studied by Payne et al. (1987b, 1990). Three storage-protein loci on the short arms of chromosomes 1A, 1B and 1D, defined as *Gli-1A*, *Gli-1B* and *Gli-1D*, code for three protein families: ω-gliadins, γ-gliadins and low molecular weight (LMW) glutenin subunits, respectively. The loci on chromosome 6, defined as *Gli-2*, code for α- and ß-gliadins. Payne et al. concluded that of the three sets of proteins coded at *Gli-1*, the LMW glutenin subunits were primarily responsible for the differences in SDS sedimentation volume and the elasticity and extensibility of the dough. The allelic variation on the *Gli-2* locus had no significant effect on bread-making characteristics. Payne et al. emphasized that their conclusions were based on results from genotypes with weak dough-mixing properties and that the relative effects of ω-gliadins, γ-gliadins and LMW glutenins might be quite different in varieties with strong mixing doughs.

Table UK.5. Classification (NIAB) categories for yield and bread-making quality (bmq) and glutenin composition of winter- and spring-wheat varieties with bread-making quality bred in the UK compared to Cappelle and Maris Huntsman; the varieties have been listed in order of the year of release

Variety	Classification		HMW glutenin subunits			*Glu-1*
	yield	bmq	1A	1B	1D	score
Winter wheat						
Cappelle	D	C	N	7	2 + 12	4
Maris Widgeon	D	A	1	7	2 + 12	6
Maris Huntsman	A	D	N	6 + 8	2 + 12	4
Maris Freeman	C	B	1	7	2 + 12	6
Bounty	B	B	1	7	2 + 12	6
Avalon	A	B	1	7 + 8	2 + 12	6
Chieftain	A	C	N	7 + 8	5 + 10	8
Sentry	C	B	1	7	3 + 12	7
Mission	A	B	N	6 + 8	2 + 12	4
Moulin	A	B	N	17 + 18	2 + 12	6
Brimstone	B	B	N	6 + 8	2 + 12	4
Mercia	B	B	N	6 + 8	5 + 10	6
Rendezvous	B	B	1	6 + 8	3 + 12	6
Pastiche	B	B	1	7 + 8	4 + 12	5
Hereward	A	B	N	7 + 9	3 + 12	5
Spark	A	B	N	7 + 8	5 + 10	8
Abbott	A	B	1	6 + 8	2 + 12	4
Spring wheat						
Maris Dove	B	B	N	14 + 15	5 + 10	8
Highbury	A	A	N	17 + 18	2 + 12	6
Broom	A	B	N	7 + 9	5 + 10	7
Musket	C	B	1	7 + 8	5 + 10	10
Hotspur	D	A	N	6 + 8	5 + 10	6
Solitaire	B	B	2*	17 + 18	5 + 10	10
Tonic	B	B	N	7 + 8	5 + 10	8
Shabliz			1	7 + 9	5 + 10	9
Shiraz			1	14 + 15	5 + 10	10

Sources: Payne et al. (1987a), Payne (1998) and NIAB data

Recent studies of Shewry & Tatham (1997) confirmed that disulphide bonds play a key role in determining the structure and properties of wheat gluten proteins. They found that the extensibility of the dough was mainly influenced by the low sulphur-containing prolamins, i.e. most notably the ω-gliadins and the HMW glutenins. The importance of disulphide bonds for the quality of wheat had been described 30 years before by Belderok (1966), who argued that the quality of wheat is determined by its flour protein content and its disulfide to thiol ratio (SS:SH).

In a study (Cooke, 1995), carried out at NIAB, Cambridge, the HMW glutenin subunit compostion of 697 currently grown wheat varieties from 20 countries all over the world was analysed. The varieties were divided on the basis of their glutenin composition into 63 groups, of which 4 groups accounted for over one-third of the varieties tested. Differences between various countries in distribution of glutenin alleles and in level of gene diversity were notable. For more details, see Part I, Chapter 7.

The knowledge on the importance of different protein groups for gluten strength and bread-making quality has proved to be a very useful tool in wheat breeding programmes, enabling a more efficient choice of parents and selection of offspring. How large the impact will be of experiments with transgenic wheat plants recovered in laboratories, remains to be seen. Shewry et al. (1995) reported that particle bombardment of tissue culture material from the feed cultivar Brigadier resulted in transformants with a manipulated gluten structure.

J.M. & D.A.D.

Acknowledgements

The authors wish to thank Dr F.G.H. Lupton and Dr P.I. Payne (PBI, Cambridge) for their valuable comments on this chapter.

References

Annual Report Plant Breeding Institute Cambridge, 1958–1968.

Astor V, Seebohm Rowntree B (1938) British Agriculture; the principles of future policy. London: Longmans, Green & Co, 469 pp.

Belderok B (1966) Betekenis van thiol- en disulfidegroepen voor de veredeling van tarwe op bakkwaliteit. IGMB Mededeling 168, TNO Wageningen.

Bell GDH, Bingham J (1957) Grain quality; a genetic and plant breeding character. The Agric Review 111: 10–22.

Biffen RH, Engledow FL (1926) Wheat breeding investigations at the PBI, Cambridge. Research Monograph 4, Min Agr Fish, 114 pp.

Bingham J (1962) The contribution of plant breeding to grain quality in wheat. Monograph 16 Soc Chem Ind London.

Bingham J (1983) The yield of cereals. Monograph 1 Royal Agr Soc England, 5–11.

Bingham J, Lupton FGH (1987) Production of new varieties: an integrated research approach to plant breeding. In: Lupton FGH (ed), Wheat Breeding; its scientific basis. London: Chapman & Hall, 566 pp.

Cooke RJ (1995) Allelic variability at the *Glu-1* loci in wheat varieties. Plant Varieties & Seeds 8: 97–106.

Earnshaw F (1949) A description of the recommended varieties of wheat and barley. J Nat Inst Agr Bot 5: 283–361.

Jenkins G (1984) Winter and spring wheat; a review of breeding objectives. Annual Report Plant Breeding Institute. Cambridge.

Lupton FGH (1982) Recent advances in cereal breeding. Neth J Agr Sc 30(1): 11–23.

Lupton FGH (1987) Wheat breeding; its scientific basis. London: Chapman & Hall, 566 pp.

Lupton FGH (1992) Changes in varietal distribution of cereals in central and western Europe. Agro-ecological Atlas of Cereal Growing in Europe, vol IV, 161 pp. Wageningen: Pudoc.

Mesdag J (1985) Developments in breeding wheat for bread-baking quality in some north-western European countries. Neth J Agr Sc 33: 215–234.

NIAB recommended list of varieties 1974–1996.

Payne PI, Corfield KG, Blackman JA (1979) Identification of a HMW subunit of glutenin whose presence correlates with bread-making quality in wheats of related pedigree. Theor Appl Gen 55: 153–159.

Payne PI (1987) Genetics of wheat storage proteins and the effect of allelic variation on bread-making quality. Ann Rev Pl Phys 38: 141–153.

Payne PI, Nightingale MA, Krattiger AF, Holt LM (1987a) The relationship between HMW glutenin subunit composition and the bread-making quality of British grown wheat varieties. J Sc Food Agr 40: 51–65.

Payne PI, Seekings JA, Worland AJ, Jarvis MG, Holt LM (1987b) Allelic variation of glutenin subunits and gliadins and its effect on bread-making quality in wheat: analysis of F_5 progeny from Chinese Spring x Chinese Spring (Hope 1A). J Cer Sc 6: 103–118.

Payne PI, Seekings JA, Kaur H (1990) Contribution of allelic variation of glutenin subunits and gliadins to bread-making quality. In: Bushuk W, Tkachuk R (ed), Gluten Proteins. St Paul, MN: AACC.

Percival J (1934) Wheat in Great Britain. Reading, Berks: Leighton & Shinfield, 125 pp.

Pushman FM, Bingham J (1975) A re-appraisal of the Pelshenke test as a screening technique in the breeding of bread-making wheats for the United Kingdom. J Agr Sc 85: 221–231.

Shewry PR, Tatham AS, Barro F, Barcelo P, Lazeri P (1995) Biotechnology of bread making: unraveling and manipulating the multi protein complex. Biotechnology 13: 1185–1190.

Shewry PR, Tatham AS (1997) Disulphide bonds in wheat gluten proteins. J Cer Sc 25: 207–227.

Vavilov N (1935) The origin, variation, immunity and breeding of cultivated plants (transl. Starr Chester K) Chronica Botanica 13 (1949): 364 pp.

Wood TB (1914) The story of a loaf of bread. Cambridge University Press, 140 pp.

France

1. Introduction

Bread is an essential element of the French diet and the French 'baguette' is famous all over the world. With a production of 30 Mt, France (F) is the fifth largest wheat producer after China, India, USA and the former Soviet Union. Within western Europe, France is the most important wheat producer, followed by Germany and the UK. Whereas France once imported wheat (from the Soviet Union and later from America), in the 1950s it became an exporter and in 1995 it was the third exporter of wheat on the world list after the US and Canada. Moreover, France is the largest exporter of flour in the world. However, limited amounts (250 000 t in 1994) of improver wheat ('blé améliorant') with very high protein content ($\geqslant$ 14%) still have to be imported, traditionally from North America and in recent years from Germany. Serious efforts are made to become self-supporting in this respect as well (Caillez, 1994). In spite of a spectacular increase in wheat production, the area under wheat remained relatively stable since 1815 (Table F.1), with a maximum of 7 million hectares in 1893.

Half of the 1995 wheat production was designated for national use: 4.4 Mt for human consumption (57 M inhabitants) and 9.1 Mt for animal feeds. The remainder was used for industrial processing (data from Ministry of Agriculture, Paris). The daily bread consumption of the French decreased from 552 g in 1880 (Grandeau, 1885) to 217 g in 1995 and it seems likely that it will continue to decrease.

Most of the bread wheat grown in France is winter wheat. The area under spring wheat is about 150 000 ha in normal years, i.e. some 3% of the total wheat area. Winter-wheat varieties requiring relatively little vernalization ('blés alternatifs') may be sown until 15 March. Durum wheat was grown on a modest 230 000 ha in 1995.

The most important growing areas for bread wheat are located in the Parisian Basin (including Beauce and Brie), in Northern France (including Picardie and Champagne-Ardenne) and in the Loire Basin. Apart from these regions, where the highest yields are obtained, wheat is grown on all soil types, including poor soils. Due to this fact, average yields tend to be lower in France than in surrounding countries.

For wheat growing, France can be divided roughly into three climatic zones: the northwestern and central part, where occasional frost occurs; the northeastern part, with regular frost; and the southern part, without frost and

Table F.1. Wheat area (1000 ha) and production 1000 t) in France between 1815 and 1997

Year	Area	Production
1815	4 592	2 960
1850	5 951	6 600
1900	6 713	8 860
1924–1925	5 563	8 328
1934–1938	5 224	8 143
1948–1952	4 264	7 791
1961–1965	4 265	12 495
1969–1971	3 892	14 112
1979–1981	4 473	22 362
1989–1991	5 102	33 171
1995–1997	4 966	33 585

Sources: Anonymus (1923), IIA (1927) and FAO (1951–1998) statistics.

with very warm summers (serious brown rust epidemics, scorch[1] damage) and strong winds. On the French variety list, the yielding capacity of varieties is evaluated seperately for the north and the south of the country.

Apart from a large number of private breeding companies, wheat breeding has been undertaken by the INRA, the National Agricultural Research Institute in Versailles, founded in 1921. A Permanent Technical Committee for the Selection of cultivated plants (CTPS), founded in 1929, decides on the admission of varieties to the National List. Analyses of distinction, uniformity and stability, and of the agronomical and tecnological value of potential varieties are coordinated by the GEVES (a public institution under the wings of the INRA), the Ministry of Agriculture and the GNIS (National Group of Seed and Plant material). The ENSMIC, the National High School for Milling and Cereal Industries plays an important role in the research on technological value. The Technical Institute of Cereals and Forage crops (ITCF) carries out the testing of varieties in the different regions.

[1] The French use the term 'échaudage', i.e. the termination of the life-cycle due to high temperatures and the depletion of the water available in the soil (called 'hay curing' in the USA).

2. Wheat breeding before 1945

During the 19th century, most of the wheat grown in France consisted of land varieties, that had been adapted to regional circumstances. Essential characteristics were earliness for varieties grown in the south and cold resistance for varieties grown in the east. From 1850–1860 onwards, the 'Blés d'Aquitaine' (from the southwest) such as Noé and its descendants Japhet, Gros Bleu and Bordeaux, started to spread over the country. Noé, derived from a selection made by a farmer in a wheatlot from Odessa and later propagated by the Marquis de Noé, can be traced in the pedigree of numerous French (and Italian) wheat varieties. Wienhues & Giessen (1957) stated that Noé was present in the pedigree of 99% of the varieties used in France in 1950. Noé was an early variety with short straw, susceptible to yellow rust. Unfortunately no exact data are available on its bread-making quality, which was classified as 'good' by Garola (1914).

Around 1870–1880 wheats from England, e.g. Victoria, Chiddam and Prince Albert, were introduced into northern France. These varieties were later than the local French wheats, but were appreciated for their stiff straw and large ears (Bustarret, 1948).

2.1. De Vilmorin

The foundations of wheat breeding in France were laid by the de Vilmorin family. The Vilmorin House, in Verrières-le-Buisson, on the outskirts of Paris, may be called the cradle of scientific breeding in France; its history goes back to 1775. After a period of working mainly on beetroot and sugarbeet, wheatbreeding was initiated in 1873 by Henri de Vilmorin. Reports of experimental crosses between different wheat species (*T. polonicum, T. turgidum, T. monococcum*) were published by him as early as 1880.

The publication of Mendel's laws on heredity around 1900 did not cause any drastic changes in de Vilmorin's work, although things were clarified and breeding was undertaken more efficiently (de Vilmorin, 1951).

The combination of bread wheats from Aquitaine and England resulted in a series of high yielding, early maturing and lodging resistant varieties. The first successful variety was Dattel, which caused a sensation amongst French farmers, mainly due to its uniformity. It was followed by Bon fermier and Hâtif inversable, both high yielding varieties with good baking characteristics. Hâtif inversable (released in 1898), an early, lodging resistant variety, was described by Schribaux (1923) as a typical 'blé de conciliation' combining very high yields with excellent bread-making results. Bon fermier and Hâtif inversable were grown on a wide scale in France until the 1920s; they were grown successfully in Italy as well.

The work of Henri de Vilmorin was continued by his son Philippe, who introduced the concept of 'composed crosses'. This resulted in a series of varieties, released in the late 1920s, such as:

- Hybride des Alliés (Massy/Parsel//Japhet/Parsel), a variety with medium baking quality, adapted to relatively poor soils, especially in the western part of France. It had a very low cold requirement and could be sown in northern France even in February or March. Several of the French 'blés alternatifs' have inherited this character from Alliés.
- Vilmorin 23 (Melbor/Grosse tête//Japhet/Parsel), a high yielding variety with poor baking quality.
- Vilmorin 27 (see Figure F.1), an early, high yielding variety with a high level of resistance to yellow and stem rust, very good lodging resistance and a Chopin W value of 110, which was very high at that time.

These mentioned varieties, especially Vilmorin 23 and 27, were grown on a large scale until the early 1950s. Of the 69 000 ha under seed production of wheat varieties in 1949, 14 432 ha were sown with Vilmorin 27 and 9624 ha with Vilmorin 23 (de Vilmorin, 1948).

Vilmorin 27 was the result of the endeavour to improve on the baking quality of the existing material. One of the ancestors of Vilmorin 27 is Blé seigle, a land variety described by Henri de Vilmorin in 1880 as a wheat with remarkable quality. The name of this wheat (translated: rye wheat) refers to its ability to perform on very light soils with a low pH.

Until the end of World War II baking trials were the main criterion for assessing baking quality in France. Official baking trials were carried out by the 'Grand Moulins de Paris' and later at the 'Ecole Française de Meunerie'. After the war the Chopin alveograph test replaced this method. Only in 1965 was the baking trial re-introduced as an official quality criterion (see Section 3.2.2).

The traditional French loaf is a hearth loaf, baked on a floor and not in a tin. The most typical bread is the 'baguette', a long, thin loaf with a crispy crust. Traditionally, the numerous small bakeries in French villages and cities

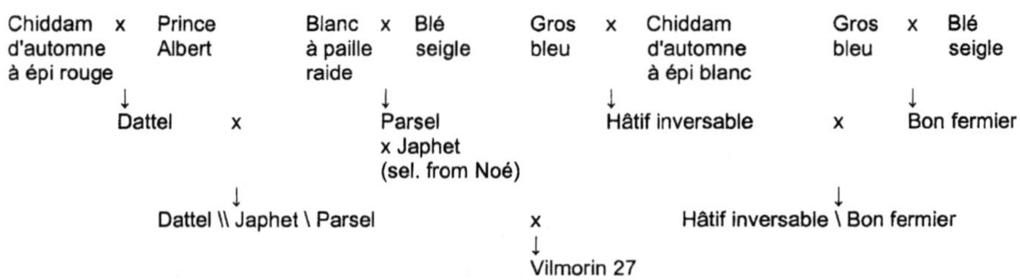

Figure F.1. The pedigree of the winter-wheat variety Vilmorin 27.

produced fresh bread throughout the day. As most French are used to buy fresh bread at least once a day, the shelf-life of the bread was of minor importance. However, due to changes in lifestyle of the French, shelf-life has become an increasingly important characteristic over the past years.

The introduction of the Chopin alveograph was welcomed as a useful testing tool for breeders and as a precise and objective measuring technique in comparison to, for example, the Saunder's chewing test (Schribaux, 1929). Through close cooperation with Marcel Chopin, the De Vilmorin House had disposal over a small-scale extensimeter from 1924 onwards. This made it possible to analyse small quantities of seed in the course of a breeding programme, instead of having to wait for F_5 or F_6 for sufficient seed. A collection of 94 wheat varieties and land varieties grown in France were analysed by Chopin in 1933. The data referring to the varieties mentioned in this paragraph are presented in Table F.2. PLM 1, a selection in the land variety 'Mouton', from the Jura, has been included because of its very high Chopin value. The name PLM refers to the name of a transport company operating in the east of France. At the beginning of the 20th century transport companies played an important role in the testing, selection and diffusion of varieties. PLM 1 can be traced in the pedigree of several INRA and Desprez varieties (see Section 2.2).

The quality of Vilmorin 27, measured as the W of Chopin, was a considerable improvement on older varieties. However, French millers generally complained about the quality of the new high yielding varieties. They considered the poor quality to be a consequence of the use of wheats from Great Britain as parents. However, changing bread-making and milling techniques also demanded another quality standard (Alabouvette, 1934). Yet, yield also

Table F.2. Chopin values (W) of some French winter-wheat varieties, harvested in 1932 and their proportion of total area under winter-wheat in 1932

Variety	W of Chopin	Proportion of total area (%)
Bon fermier	84.0	10.3
Hâtif inversable	58.5	3.5
Alliés	84.8	1.6
Japhet	83.0	4.2
Vilmorin 23	42.9	18.9
Vilmorin 27	105.0	5.3
PLM 1	154.3	0.02

Source: Alabouvette (1934).

Table F.3. Quality parents used by De Vilmorin for the breeding of improver wheat

Variety	Origin	Chopin values	
		G	W
Red Fife	Canada	8.2	180
Marquis	Canada	8.5	180
Garnet	Canada	8.3	280
de Cermelles	France	9.0	170
Florence-Aurore	Australia/France	7.5	220
Pusa 422	India	8.0	180
Polish land variety	Poland	9.6	350
Yaroslaw wheat	Ukraine	9.9	450

Source: de Vilmorin (1948).

needed special attention: the first 'Semaine Nationale du Blé' was organised in 1923 with the aim of France attaining self-sufficiency in wheat, i.e. becoming independant of imports from USA and Canada.

De Vilmorin initiated in 1926 a hybridization programme of strong wheats with high W values and the best French varieties in order to obtain high yielding varieties with improver quality ('blé améliorant'). The quality parents that were used and their country of origin and Chopin values (measured in 1924–1925) are listed in Table F.3.

In addition to Chopin's extensimeter, from 1934 onwards breeders started to use the Pelshenke test for analyses in F_3 or even in F_2, in order to discard very poor quality material. As soon as sufficient seed was available, a Brabender farinograph was used for the analysis of the advanced material.

Until World War I the firm De Vilmorin had a monopoly on French wheat breeding. Indeed, more than 40% of the French wheat area in 1937 was covered with Vilmorin and Vilmorin-Andrieux varieties. The name Vilmorin remained visible in the French wheat breeding world until 1970 when its cereal department was taken over by UCOPAC.

2.2. Other breeders

Starting early in the 20th century, but especially after World War I, the number of private breeders in France increased considerably and the market became overrun with a vast number of new varieties. Regulations for the introduction of new varieties were set in order to prevent varieties from being brought onto the market that lacked sufficient genetic purity or agronomical and technological value. A Permanent Technical Committee for the Selection of cultivated plants (CTPS) was set up and the first official variety list ('Catalogue') appeared in 1932.

Breeding issues upon which French wheat breeders focussed are reflected in the number of points (in paranthesis) that could be earned for a variety: yield (35); milling and baking value (20); earliness (5); resistance to frost (8), lodging (10), stem rust (8), yellow rust (6), leaf rust (4) and loose smut (4); a total of 100 points.

Wienhues & Giessen (1957) pointed out that the majority of French wheat varieties available in the 1950s had a very narrow basis. They stated that this could be due to the fact that the most important breeding stations were in areas where a mild sea climate dominated. Under these favourable circumstances, little attention was given to cold and rust resistance.

The use of foreign genetic sources has become diversified largely under the influence of plant breeding institutes.

2.2.1. Plant breeding institutes

The Institute for Agricultural Research (IRA) was founded in 1921 by the Ministry of Agriculture, with a Central Station in Versailles and regional stations in, amongst others, Clermont-Ferrand, Montpellier and Dijon. The Central Station was directed first by Schribaux, later by Alabouvette (1928) and by Crépin (1937), all of whom had earned a reputation for themselves in wheat breeding.

Schribaux was convinced that quantity and quality could be combined in one variety and introduced the term 'blé de conciliation' (Figure F.2) for this type of wheat (Schribaux, 1923). Moreover, he stated in 1901 that France, growing mainly early and high quality varieties such as Bordeaux and Japhet, contrasted with Germany, Belgium, Sweden and Denmark where late varieties from England dominated the market.

Schribaux used several foreign parents in his crosses. Like de Vilmorin, he used squarehead types ('épi carré') from England, but he also introduced Hungarian (Magyarovar, Székács), Russian (Krelof) and Italian (Rieti) wheats. Varieties derived from Italian genitors proved to be successful, especially in the south of France. Some of the varieties, resulting from crosses made by Schribaux between 1912 and 1918, all classified as medium baking-quality varieties were:

- Institut Agronomique (Hâtif inversable/Italian line)
- Hybride à courte paille (Epi carré/Krelof//Bon fermier)
  Préparateur Etienne (Rieti/Epi carré//Hâtif inversable)
- Bon fermal (Alliés/Bon fermier)
- Inversal (Alliés/Hâtif inversable).

Figure F.2. Poster presented at the 'Semaine du blé' held in 1923 in Paris.
(Translation: The bread presented at this buffet was made of the type of wheat flour that M. Schribaux named 'Blé de Conciliation'. This wheat gives satisfaction to producers (because of the quantity), to millers and bakeries (because of the quality) and to the consumer (because of the bread). This bread was made of pure flour (extraction rate 74%) of wheat that yielded 5200 kg/ha whereas the average yield in France is 1400 kg/ha. This proves that excellent bread can be made from high yielding wheats.

Alabouvette apparently was only referring in part to Schribaux when he stated in 1934 that 'French breeders have been hypnotized by the Vilmorin varieties and have not sufficiently used quality sources from abroad. Hâtif inversable and Alliés have been used wrongly as quality parents'.

The agricultural research institute (IRA) became the 'Institut National de la Recherche Agronomique' (INRA) in 1946. Two important INRA varieties, classified as 'blé améliorant' (Class I) were Renfort, released in 1944, and Magdalena (K-8/Székácz//Providence), a variety with relatively little winter-hardiness, released in 1949. Renfort resulted from a cross between Vilmorin 27 and PLM 1, which was a selection from a very strong land variety (Mouton) from the Jura (see Table F.2 and F.4) The most successful INRA variety was

Table F.4. Early Desprez varieties with good bread-making quality, compared to PLM and Hybride du Joncquois

Variety	Year of release	Pedigree	Class[a]
PLM	1927	selection in land variety Mouton	I
Hybride du Joncquois	ca.1924	Vilmorin 23/Institut Agronomique	V
Flandres Desprez	1940	Vilmorin 27/PLM	II
Chartres Desprez	1943	Japhet/Parsel/Hybride à courte paille	I-II
Cappelle Desprez	1946	Vilmorin 27/Hybride du Joncquois	II
Lille Desprez	1950	HdJoncquois/Vilmorin 27//Vilmorin 27	I
Ardennes	1953	Hybride du Joncquois/PLM	I
Marne	1954	HdJoncqois/Vilmorin 27//HdJoncqois/PLM	I-II
Artois Desprez	1957	Blé des Dômes/Pévèle	II

[a] Classification as used by Jonard (1951), described in Section 3.2.2.

Etoile de Choisy, released in 1950, a very early (14 days earlier than Vilmorin 27) and productive variety with lodging and cold resistance but with medium baking quality. It was the result of a cross between Mon Desir/Ardito and K-3/Mouton à épi roux, backcrossed with Mouton à épi roux. From Mouton it inherited both its cold resistance and some of its quality. Its lodging resistance came from Ardito. K-3 was a line from Schribaux, derived from Noé. Etoile de Choisy was mainly grown in southern France and it remained a leading variety until the 1960s.

2.2.2. Private breeders

La Maison Florimond Desprez in Cappelle, not far from the border between Belgium and northern France, was founded in 1830. Like La Maison De Vilmorin, it was a family enterprise, initially working mainly on sugarbeet.

Breeding work on wheat started in the 1920s, resulting in a vast number of Desprez varieties, which were released from 1937 onwards.

One of its first varieties was Hybride du Joncquois, originating from a cross made in 1924 between Vilmorin 23 and Institut Agronomique. This variety with very poor baking quality itself was frequently used as a parent by Desprez and nevertheless produced several varieties with very good bread-making quality. The Desprez varieties belonging to Classes I and II are listed in Table Fra.4. Apparently the quality of PLM (Class I) and Vilmorin 27 (Class II) could easily compensate for the poor quality of Hybride du Joncquois (Class V).

The release of Cappelle (Desprez) in 1946 was a breakthrough: it was grown on a large scale for 25 years, until the early 1970s. Moreover it was grown on a very large scale in the UK between 1955 and 1975; indeed, in 1962 it covered 84% of the UK wheat area. Cappelle was greatly appreciated for its durable disease resistance.

The bread-making quality of Cappelle was comparable to that of Vilmorin 27, which was generally used as a standard variety for quality, but the yield of Cappelle was considerably higher than that of Vilmorin 27. The relative yield figures of some important winter-wheat varieties (with their year of registration) that have been released before and after Cappelle may illustrate the great leap forward that was achieved with this variety.

Vilmorin 27 (1927):	88%
Cappelle (1946):	100%
Magdalena (1949):	89%
Etoile de Choisy (1950):	95%

Amongst the numerous breeders working on wheat during the first half of the 20th century, only three are mentioned here: Tourneur, Blondeau and Benoist. Some of the most important varieties released by these breeders are listed in Table F.5.

Bustarret stated in 1948 that wheat yields in northern France had increased with 50% between 1920 and 1939, mainly due to the introduction of high yielding varieties and the use of nitrogen. Whereas in 1925 half the French wheat consisted of land varieties and 20–25% of 'Blés d'Aquitaine', in 1949 less than 10% of the wheat consisted of land varieties. Only a few foreign varieties were grown: Goldendrop and Teverson (from England) in the west; some Italian varieties in the south; and some German varieties in the east.

Table F.5. The most important varieties released before 1950 by Benoist, Tourneur and Blondeau

Breeder Variety	Year of release	Pedigree	Class[a]
Benoist			
Hybride 40	1927	selection in Wilson (sel.in Hât.Inv.)	III
Tadépi	1949	Hybride 40/Hybride du Joncquois	III
Tourneur			
Inversable Bordeaux		Hâtif Inversable/Rouge de Bordeaux	III-II
Vague d'épis	1943	Chanteclair/V27	II
Blondeau			
Hybride de Bersée	1936	Alliés/Vilmorin 23 (alternative)	V
Yga	1938	Vilmorin 27/Red Fife	II
Petit Quinquin	1943	Vilmorin 23/Inst.Agr.//Providence	II-III

[a] Classification according to Jonard (1951), as described in Section 3.2.

3. Wheat breeding after 1945

3.1. Winter wheat

3.1.1. Breeding issues

After a period of marketing problems caused by a steady decrease in bread consumption in the years prior to World War II and by the disruption of the war itself, a new incentive was given to wheat breeders by a vast increase in wheat exports in the 1950s. Beeding aims were influenced by these developments. Yield remained the most important issue, including achieving a favourable response to increasing quantities of fertilizer. Apart from resistance to diseases (mainly rusts and footrot), cold resistance was an important issue for northern and central France, especially after 3 million hectares of wheat were destroyed in the winter of 1955–1956. Specific issues for the south were earliness, resistance to stem rust, heat tolerance and resistance to scorching or shrivelling ('échaudage'). With regard to bread-making quality, varieties now had to meet the demands of the European market, which meant that the French standard for wheat quality, expressed as the W of Chopin, was no longer sufficient (Mesdag, 1975).

3.1.2. Quality classification

Between 1945 and 1965, the classification of French wheat varieties was based only on their Chopin W value. Five quality classes for bread-making quality ('force boulangère') were used (Jonard, 1951):

I	$W \geqslant 150$	very good
II	$100 < W < 150$	good
III	$60 < W < 100$	medium
IV	$50 < W < 60$	poor
V	$W \leqslant 50$	bad

After 1952, new varieties were required to have a W value of 100 or more. Breeders strived for wheats with a W analogous to that of Vilmorin 27, which was 120. After French breeders started to use Canadian wheats for crossing purposes, this standard rapidly increased. By 1984 the average W value of French varieties was 160–180 (Roussel, 1984) and in 1996 over 200.

The use of Chopin values as a standard for quality was questioned by several breeders (de Vilmorin, 1958). The introduction of hard endosperm in modern varieties has modified the texture of the protein, leading to increased Chopin W values. This is related to a higher dough stiffness (P), due to the increased hydration necessary to obtain a dough that is suitable for the production of French bread (Rousset, 1998). Thus, the ITCF classification includes a description of the character grain hardness ranging from 'extra soft' to 'hard'.

Baking trials were re-introduced in 1965, partly under the influence of EC regulations. Baking trials were carried out in two variations: under normal and under intensive kneading. Bread-making trials are done by the CNERNA (Centre National de Coordination des Etudes et Recherches sur la Nutrition et l'Alimentation). The bread-making score is calculated from 13 parameters, measured from the dough, the bread and the crumb.

The evolution of the techniques used in French bakeries (intensive kneading, volumetric cutting, controlled fermentation and freezing of dough) has gradually lead to higher demands on protein content and bread-making quality. Until 1995, the following categories, based on Chopin W values and bread-making tests, were used for the classification of wheat varieties.

A = improver wheat, including 'blé de force' and 'blé améliorant'
B1 = good quality, equal to or better than Capitole (ww), or better than Bastion (sw)
B2 = medium to good bread-making quality
C1 = rather poor bread-making quality
C2 = poor quality, almost unsuitable for bread making ('impanifiable')
D1 = unsuitable for bread making but with a high protein level
D2 = unsuitable for bread making (CNM = 'collants/non machinables' standing for sticky dough, e.g. Maris Huntsman and Clement)

Moreover, the ITCF used four categories: BAF (≈ A), BPS (≈ B1), BPC (≈ B2) and BAU (other uses than bread). In addition, the milling industry gives a recommendation for certain varieties: •• and • (bread-making quality), C ('correcteur'), F ('force'), ++ and + (biscuit-making quality). In 1995, a combination of the CTPS and ITCF classification was introduced:

A = 'blé améliorant' = improver wheat

BPS = 'blé panifiable supérieur' = superior bread-making wheat

BPC = 'blé panifiable courant' = normal bread-making wheat

BB = 'blé biscuitier' = biscuit-making wheat

BAU = 'blé pour autres usages' = wheat for other purposes

An additional recommendation given by the milling industry may influence the commercial value of a variety.

3.1.3. Breeding for improved quality

De Vilmorin wrote in 1948 that the bread-making quality of wheat behaves as a polygenetic recessive character. He stated that 'if both parents have a good quality, the offspring will not contain plants with a very low quality and therefore there is no need to analyse the quality in the early generations. If however one of the parents has a very low quality, such as Carstens V, Jarl or Bordeaux, an analysis of the quality of the offspring is necessary from F_2, F_3 onwards in order to select the few good individuals'.

Zeleny or SDS sedimentation and protein content are commonly used on the breeding stations to assess the quality of F_5 material. From F_6 onwards, a more detailed analysis (extraction rate, falling number, Chopin W value, baking trial) is generally carried out by specialized laboratories (Grand Moulins de Paris). At Benoist plant breeding station the 'methode eliminatoire' was applied, meaning that all material between F_3 and F_4 was analysed with a mixograph, in order to eliminate poor quality material at an early stage (Prins et al., 1984).

A comparison of 46 technological parameters used for the evaluation of bread-making quality that was carried out at the INRA (Branlard et al., 1991) showed that Pelshenke, modified Zeleny (using 70% instead of 20% extraction rate flour) and mixograph analyses were all highly hereditable and well correlated with bread-making quality. These three tests, all of which require small amounts of seed, could be combined in a breeding index for improving the efficiency of selection for wheat quality.

Analysis of gliadin and glutenin composition by electrophoresis opened up new possibilities for breeders. In 1977, Rousset, working at the INRA, proposed a breeding scheme, that was based on an assumption made by Sozinov & Poperelya in the Soviet Union, that there is a relation between the gliadin pattern of wheat and the gluten quality.

Simple three- or four-way crosses with 'strong wheat' genitor(s) of which the electrophoretic gliadin diagram is known and varieties with good agronomical characteristics.

↓

Population composed of highly heterozygotic plants

↓

Bulk method for a few years (= breeding for environmental adaptation)

↓

Grain selection based on electrophoretic gliadins diagram: electrophoresis on a half grain, embryo with other half grain is sown. Maintain only seedlings with an electrophoretic diagram resembling that of the strong wheat genitors

Twenty years later, selection on gliadin and glutenin composition of wheat grains has become common practice among wheat breeders. Research on the relation between bread-making quality and the gliadin pattern of wheat varieties has been carried out by Branlard and his co-workers at INRA. Their work will be discussed in Section 3.3.

With regard to the correlation between yield and quality traits, Rousset et al. (1988) stated that the negative correlation between yield and protein content, and, consequently, between yield and Zeleny test performance tends to weaken after one cycle of selection. Accordingly, selection for both yield and technological quality should be easier in a younger population. The results of their experiments showed that a recurrent selection programme with short cycles and index selection might be interesting in multi-trait breeding programmes in wheat. A comparison of different selection indices was described by Branlard et al. (1992).

3.1.4. Results of quality breeding since 1945

A steady increase in quality has been achieved since 1945. The most important varieties released up to 1970 are presented in Table F.6. Champlein and Joss are examples of varieties in which yield increase has been given priority over quality. In Hardi and Capitole, however, a combined yield increase and quality improvement has been achieved (Mesdag, 1975). A further increase in yield while maintaining bread-making quality has been achieved since the 1970s, as may be seen from the data in Table F.7. The English variety Maris Huntsman, which was commonly grown in France, has been listed in Tables F.6 and F.7 as a reference between the two generations of wheat varieties. The yielding capacity of Maris Huntsman, which was extremely high in 1973, has largely been surpassed by later varieties with excellent bread-making characteristics. Courtôt (INRA) was the first semi-dwarf variety with good bread-making quality grown in France.

Table F.6. Yield and quality of the most important varieties released between 1945 and 1970, relative to Maris Huntsman

Variety	Year	Pedigree	Yield (rel.)	Quality W	Quality score[a]
Cappelle	1946	Vilmorin 27/Hybr.du Joncquois	94	120	6
Champlein	1959	Yga/Tadépi	100		4
Moisson	1962	Cappelle//80–3/EdChoisy	97		5, 5
Capitole	1964	Cappelle//80–3/EdChoisy	98	170	7
Joss	1966	Cambier 194/Tadépi//Cappelle	102	75	3, 5
Hardi	1969	Cappelle/Thatcher	102		7
Top	1970	TF354/Cappelle	102		6
M.Huntsman	1973	Cappelle*4/Hybrid46//*2Prof . Marchal[b]	109		3

[a] Quality score, based on Chopin W value, ranges from 9 (= very good) to 2 (= very poor).
[b] This pedigree is different from the pedigree given by Lupton (1992).
Source: Mesdag (1975).

Table F.7. French bread-wheat varieties released after 1970 compared to Hardi and M. Huntsman

Variety	Year	W	Quality classification[a]			Relative yield[b]	
			Gévès	ITCF	milling	North	South
Hardi	1969	7.5	B1	BPS		84	
M.Huntsman	1973	3	D2	BAU		89	
Darius	1974	8	A	BAF	••C	80	
Courtôt	1974	7		BPS			
Camp Rémy	1980	6.5	B1	BPS	••	89	83
Récital	1986	7	B1	BPS	••	95	92
Soissons	1988	7	B2	BPS	••	100	100
Baroudeur	1988	6.5	B1	BPS	••	101	
Renan	1989	8	B1	BPS		88	
Qualital	1991	8	A	BAF	••F		91
Aztec	1994	5	B1	BPS			111
Paindor	1995	6		BPS	•	101	
Rapor	1997	7.5		BPS			106
Isengrain	1997	5.5		BPS		110	

[a]Explanation of symbols is given in Section 3.1.2.
[b] Related to Soissons.
Source: Semences et Progrès.

Camp Rémy (Unisigma), a variety with excellent bread-making quality and a Chopin W value of 150, was proposed for a bonus under the policy to decrease the amount of high protein wheat being imported from Germany. This cultivar has long been used as a reference for quality.

Récital (Benoist) was much appreciated by the milling industry but unfortunately it was very sensitive to mildew and rusts.

Soissons (Desprez), released in 1988, has become a very successful variety due to its high yielding capacity amd relatively good bread-making characteristics; it is much appreciated by the milling industry. In 1997 it was still the most commonly grown winter-wheat variety. Roussel stated in 1984 that of the varieties grown in that year, only Hardi, Castan, Capitole and Camp Rémy (together 19% of the total area) were satisfactory from the miller's point of view. Fidel, the main variety (covering 22% of the wheat area) in 1984, was not appreciated by the milling industry, despite its relatively good bread-making characteristics (W = 117).

Ten years later, in 1994, the situation had improved considerably. More than half of the area under wheat in France was covered with bread wheat varieties with superior bread-making quality, of which Soissons and Sidéral were the most important. Chasseray (1996) noted a tendency to replace BPS and BPC varieties by high yielding BAU varieties: in 1996 51.3% of the area under wheat was sown with BPS varieties (compared to 57% in 1995) and 20.7% with BPC varieties (compared to 22.5% in 1995). This trend is expected to continue, with a further decrease of the area sown with Soissons, until a new alternative appears on the market.

3.2. Spring wheat

Practically all the spring wheat grown in France before 1945 was of foreign origin with varieties such as Fylgia and Extra Kolben from Sweden. Traditionally grown French varieties were Chiddam de Mars and Saumur de Mars (the suffix 'de Mars', meaning 'of March', was generally used for spring varieties).

The importance of spring wheat increased somewhat in the 1950s, filling some of the area left free by the decrease of oats and taking over part of the area occupied by barley. Spring wheat was (and still is) mainly an alternative crop, grown when the sowing of winter wheat had not been possible or when the crop had been destroyed by cold. Varieties that behave as spring wheats in northern France may be sown in autumn in the South.

The main aim pursued at Vilmorin in the 1950s was breeding a variety that was earlier and shorter than Fylgia, and with better quality and stem-rust resistance (or very early maturity in order to escape epidemics). Parents with short straw were used to breed varieties that would be able to stand up to high amounts of fertilizer. American, Canadian and Australian parents were used to improve the quality. Apart from good milling and baking values, unfortunately these parents also introduced susceptibility to yellow rust. Experience had proved that 'by crossing a first class American wheat such as Manitoba with any French wheat, even one of a deplorably low milling and baking quality, one could obtain very good lines with a high W of Chopin' (de Vilmorin, 1958).

Rex Vilmorin, released in 1962, was a very successful variety. It covered 68% of the spring-wheat hectarage in 1968 and remained on the variety list until 1990. Resulting from a cross between Thatcher (USA) and the winter wheat Vilmorin 29, Rex combined productivity, lodging resistance and low sensitivity to stem rust with very good bread-making quality (Mayer, 1962).

One year later, Florence-Aurore was registered as a variety. It was introduced from Tunesia, where it had been selected in a cross made by Schribaux in the 1920s between Florence (from Australia) and Aurore (from Canada), both parents with excellent bread-making characteristics, which they had inherited from Red Fife (Crépin, 1934). Florence-Aurore, which like its parents had excellent quality characteristics, covered 39% of the spring-wheat acreage in 1964 and remained a leading variety until 1996.

Prinqual, released in 1978, is an American variety with improver quality comparable to Florence-Aurore, but with better productivity. Like Florence-Aurore it has been grown on vast areas in the south of France. Indeed, in 1996 it was the third most widely grown spring-wheat variety.

Ventura & Furio, released in 1985 and 1991, and classified as B1 and B2 wheats (i.e. no improver wheats but with good bread-making qualities) respectively, brought a further improvement in yield. Furio covered more than 40% of the spring-wheat area in 1996. Some data on these varieties are presented in Table F.8. The productivity is related to Bastion (NL) and Wim (B), both high yielding spring-wheat varieties with poor quality characteristics.

3.3. Analyses of HMW glutenins

Branlard & Le Blanc (1985) at INRA investigated the HMW glutenin subunit composition of a collection of 195 French wheat varieties. Eighteen bands of different mobilities were observed, one of which had not been described by Payne et al. (1983). This band, with a mobility situated between Band 10 and 12, was named Band 11. It is controlled by a gene on chromosome 1D and associated with Band 2 of the same chromosome. The combination 2+11 was found in 6 varieties.

The most frequent diagrams in this study were: N,7,2+12 (25 cvs); N,7+8,2+12 (17 cvs); N,7+8,5+10 (17 cvs); N,7+9,2+12 (16 cvs), N,7,5+10

Table F.8. Quality and productivity (related to Bastion and Wim, respectively) of spring-wheat varieties

Variety	Year of release	Quality class	W score	Yield (rel.)
Rex[a]	1962	A	8.5	
Florence-Aurore[a]	1963	A	9	
Bastion[a]	1976	B2	6	100
Prinqual[a]	1978	A	9	91
Ventura[a]	1985	B1	7.5	106
Florence-Aurore[b]	1963	A	9	73
Prinqual[b]	1978	A	9	85
Wim[b]	1979	C1	4	100
Ventura[b]	1985	B1	7, 5	98
Furio[b]	1991	B2	6, 5	101

Sources: [a] variety list 1988; [b] Semences et Progrès 1996.

(14 cvs); N,7+8,4+12 (13 cvs) and N,6+8,2+12 (12 cvs). Through comparison with the world collection, diagrams (N,7,2+12), (N,7+8,2+12) and (N,7+9,2+12) were identified as typical for French wheat. The allele coding for glutenin subunits 3+12 is also characteristic for French cultivars and is probably derived from Noé (Table F.9). This allele is exclusively found in French and some British varieties (Cooke, 1995).

The variation in patterns found in the study was much larger in the spring wheats than in the winter wheats, which may be due to spring-wheat breeding having been based on a larger variety of foreign varieties. Of the varieties available in 1994 only one-quarter had one of the three typical 'French patterns', which appear to be gradually fading out. Cappelle Desprez, with glutenin pattern (N,7,2+12) has apparently left its traces in numerous varieties.

The *Glu-D1* subunit 5+10, which is considered to be related to good breadmaking characteristics, is present in several winter and spring-wheat varieties. The oldest French variety in which this band was encountered was Magdalena, an INRA variety with the Hungarian Széckacz as one of its parents.

Of the 123 winter-wheat varieties on the 1994 variety list, 10 varieties had the highest *Glu-1* score (i.e. 10), and 19 varieties had a *Glu-1* score 8 or 9, implying that one-quarter of the winter-wheat varieties listed had a high quality score. Obviously, for spring wheat the average was higher: 3 of the 9 varieties scored 9, 4 varieties scored 8, one scored 7 and one scored 6. The fact that the HMW glutenin composition of wheat varieties is mentioned

Table F.9. The HMW glutenin subunit compostion and *Glu-1* score of some important French bread wheat varieties released since 1900

Variety	Year	Glutenin subunits			*Glu-1*
		Glu-A1	*Glu-B1*	*Glu-D1*	score
Winter wheat					
Noé		1	7+9	3+12	7
Vilmorin 23	1909	N	7	4+12	3
Vilmorin 27	1927	N	7	3+12	4
Hybride de Bersée	1936	1	7+8	4+12	7
Vague d'Epis	1943	1	7	2+12	6
Cappelle Desprez	1946	N	7	2+12	4
Magdalena	1949	N	7+9	5+10	7
Etoile de Choisy	1950	N	7+8	2+12	6
Vilmorin 53	1952	N	20	3+12	4
Champlein	1959	N	7+8	4+12	5
Moisson	1962	N	7+8	2+12	6
Capitole	1964	N	7+9	2+12	5
Joss	1966	N	6+8	2+12	4
Hardi	1969	2*	7	2+12	6
Top	1970	N	7+8	3+12	6
Talent	1973	N	7+9	4+12	4
Darius	1974	N	7	2+12	4
Fidel	1978	N	7+8	5+10	8
Camp Rémy	1980	2*	7	3+12	6
Festival	1981	N	7+9	2+12	5
Scipion	1981	N	7+8	2+12	6
Soissons	1987	2*	7+8	5+10	10
Sidéral	1990	N	7+9	2+12	5
Arfort	1990	2*	17+18	5+10	10
Qualital	1991	2*	7	5+10	8
Amélio	1991	N	7	5+10	6
Bonpain	1993	2*	17+18	5+10	10
Aztec	1994	N	17+18	3+12	6
Paindor	1995	1	7	2+12	6
Rapor	1997	1	7+9	5+10	9
Isengrain	1997	N	7+8	5+10	8
Spring wheat					
Rex	1962	1	7+9	5+10	9
Florence-Aurore	1963	2*	7+9	5+10	9
Prinqual	1978	2*	17+18	2+12	8
Ventura	1985	2*	7+8	2+12	8
Furio	1991	N	7	5+10	6

Sources: Branlard & Le Blanc (1985), Variety Bulletin (1994), Semences et Progrès (1997), Rousset (1998) and Branlard (1998).

on the French variety list, may illustrate the importance given to this quality measure.

Two *Glu-B1* subunits occur on the list which are not included in the scoring system used by Payne et al.; they are subunit 21 in the winter-wheat variety Foison and the subunit pair 14+16 in the spring-wheat variety Axona.

Branlard & Dardevet (1993) investigated the genetic background of the quality of the variety Darius. Darius was classified as a top quality variety, but has a 'poor' HMW glutenin pattern (N,7,2+12) and a *Glu-1* score of 4. Investigation of the progenies of crosses of Darius with Corin, Capitole and Courtot showed that the quality of Darius may be related to its gliadin composition. The null allele of Darius, characterized by the absence of the *Gli-D1* encoded ω-gliadins, was associated significantly with higher dough tenacity P and strength W. It is not clear whether there is an interaction between the *Gli-D1* null allele and HMW glutenin alleles.

The influence of the *Gli-D1* null allele has also been reported by Russian scientists in a bread wheat line derived from the cross Priboj/Bezostaya-2. Branlard & Dardevet suggested that it may be useful to introduce this allele in wheat breeding programmes for quality improvement.

The association between dough strength (Alveograph W value) and the gliadin pattern of a number of French varieties was investigated by Metakovski et al. (1997). A total of 60 different gliadin alleles was identified in the 80 cultivars studied. They were divided in four groups (I to IV): the best ($275<W<530$), good ($235<W<275$), poor ($120<W<140$) and the worst ($70<W<120$) varieties. Although the occurence of certain alleles was significantly higher in group I than in group IV, the results of this study were not in agreement with the results of similar studies in Italian and Russian wheat cultivars.

D.A.D.

Acknowledgments

The author wishes to thank Michel Rousset (INRA, Domaine Brunehaut), Philippe Lonnet (Florimond Desprez, Cappelle) and Gérard Branlard (INRA, Clermont Ferrand) for their help with reviewing this chapter and for providing addtitional information where necessary.

References

Alabouvette L (1934) La qualité des blés cultivés en France et le problème de son amélioration. Annales Agronomiques 4: 207–252.

Anonymus (1923) Le Blé: production, industries de transformation, commerce. Compte-rendu des travaux de la semaine nationale du blé. Paris: Office de Publications Commerciales.

Branlard G (1998) Personal communication, Clermont Ferrand, Wageningen.
Branlard G, Pichon M (1984) La qualité et la sélection du blé tendre. Bulletin Technique d'Information 392/393: 477–482.
Branlard G, Dardevet MJ (1985) Diversity of grain proteins and bread wheat quality. I. Correlation between gliadin bands and flour quality characteristics. J Cer Sc 3: 329–343.
Branlard G, Le Blanc A (1985) Les sous unités gluténines de haut poids moléculaire des blés tendres et des blés durs cultivés en France. Agronomie 5 (6): 467–477.
Branlard G, Rousset M, Loisel W, Autran JC (1991) Comparison of 46 technological parameters used in breeding for bread wheat quality evaluation. J Gen Br 45: 263–280.
Branlard G, Pierre J, Rousset M (1992) Selection indices for quality evaluation in wheat breeding. Theor Appl Gen 84: 57–64.
Branlard G, Dardevet M, Nieto-Taladriz MT, Khelifi D (1993) Allelic diversity of the omega gliadins as revealed by SDS-PAGE; their possible implication in quality variation. In Hoseney (ed), (1994) Gluten proteins; principles of cereal science and technology. St Paul, MN: American Association of Cereal Chemists.
Branlard G, Dardevet M (1994) A null Gli-D1 allele with a positive effect on bread wheat quality. J Cer Sc 20: 235–244.
Bustarret (1948) Etat actuel de l'amélioration du blé en France. In Le Pain; annales des journées scientifiques du Centre National de Coordination des Etudes et Recherches sur la Nutrition et l'Alimentation (Paris).
Caillez B (1994) Le blé améliorant, un marché à reconquérir. Cultivar 364: 46–47.
Chasseray P (1996) Caractérisation des blés de la récolte 1996. Industries des Céréales nov-dec 1996: 9–14.
Crépin C (1934) Valeur boulangère et amélioration des blés. Le Sélectionneur 3: 65–69.
Garola CV (1914) Céréales (Encyclopédie Agricole). Paris: Librairie Baillière et fils.
Grandeau L (1885) La production agricole en France, son présent et son avenir. Paris: Berger-Levrault & Cie.
Jonard P (1951) Les blés tendres cultivés en France. Paris: INRA.
Mayer R (1962) L'amélioration des plantes en France. Annales de l'amélioration des plantes 12 (hors série).
Mesdag J (1975) Quality improvement of European wheats. In Proc 2nd Int Winter wheat Conf, Zagreb: 341–350.
Metakovski EV, Felix I, Branlard G (1997) Association between dough quality (W value) and certain gliadin alleles in French common wheat cultivars. J Cer Sc 26: 371–373.
Prins LAJ, Belderok B, Mesdag J, van der Woude K (1984) Verslag van een studiereis naar Frankrijk. Wageningen: NGC publ.
Roussel P (1984) La qualité des blés tendres. Cultivar 176, 33–35.
Rousset M (1977) Breeding for improvement of utilization value of bread wheat. Cer Res Comm, 5(2): 139–144.
Rousset M, Kervella J, Pichon M, Branlard G, Doussinault G, Picard E (1988) Effectiveness of recurrent selection for bread-making quality breeding in wheat (*Triticum aestivum* L.em.Thell.). In Proc 7th Int Wheat Genetics Symp, Cambridge.
Rousset M (1998) personal communication, Wageningen.
Schribaux E (1923) Recherche des meilleurs blés au double point de vue du rendement et des qualités boulangères. In Le Blé; production, industries de transformation, commerce. Compte rendu des travaux de la semaine nationale du blé à Paris, pp 138–147.
Schribaux E (1929) La valeur boulangère des blés cultivés en France. Bulletin de l'Association des sélectionneurs de plantes de grande culture 2: 102–127.

Vilmorin de M (1948) La création des blés d'automne de bonne qualité boulangère. In Le Pain. Paris: Centre National de la Recherche Scientifique.

Vilmorin de M (1951) Cent ans de sélection généalogique et cinquante ans de génétique appliquée à Verrières. Zeitschrift für Pflanzenzüchtung 29(3): 288–301.

Vilmorin de R (1958) Wheat breeding in France; problems and methods. In Proc 1st Int Wheat Gen Symp, Manitoba, Canada, pp 49–54.

Wienhues F, Giessen JE (1957) Die Abstammung Europäischer Weizensorten. Zeitschrift für Pflanzenzüchtung 37(3): 217–230.

Spain

1. Introduction

Two factors that have had a remarkable influence on the culture of wheat in Spain (SP) are the complex history and the specific geographic situation of the country. Spanish history is characterized by a succession of civilizations, of which the Roman and the Arab have had a large impact on agriculture. Spain, originally inhabited by Iberians, later joined by Kelts, was invaded by the Romans from 200 BC onwards. One of the elements introduced by the Romans was a complex watersystem that changed agriculture there to a large extent. After invasions by Vandals, Sueves, Alans and Goths, Spain was invaded by the Moors in the year 711 AD. Initially these North African 'landlords' occupied most of the country but soon they were restricted mainly to the province of Andalucia, in the south. Through the occupation of the Moors, which lasted up to the 16th century, Spain became a member of the Islamic world, which has left its traces on Spanish culture and agriculture. Moreover, each of these civilizations introduced its own crops and plant material and thus greatly influenced the diversity of botanical species in Spain.

Due to its geographical situation, Spain has an exceptional range of growing conditions. Growing conditions on the high central plateau, which has a relatively cool land climate, are totally different from the southern coastal areas, where the climate is Mediterranean. The northern part has its own specific growing conditions. An important factor that characterizes the Spanish climate is a lack of rain, causing large differences in annual yield and large variations in total production. The irregularity in rainfall, not only from year to year but also within the year, necessitates the breeding of drought tolerant varieties.

2. General remarks on wheat growing in Spain

Up until the 1970s the area under wheat in Spain was stable at around four million hectares, comparable with France and Italy. Between 1970 and 1980, however, a fundamental redistribution of cereal crops led to a considerable decrease in the area under wheat (Table SP.1). Whereas the total area under cereals remained stable, the area under wheat, oats and rye decreased in the favour of barley. This is related to the fact that barley yields in Spain are generally higher than those of wheat, combined with the level of cereal prices.

Table SP. 1. Area (1000 ha) and production (1000 t) of cereal crops in Spain between 1924 and 1997

	Wheat		Barley		Rye		Oats		Total	
	area	prod	area	prod	area	prod	area	prod	area	prod
1924–1925	4 270	3 870								
1934–1938	4 591	4 392	1 895	2 394	593	551	776	670		
1948–1952	4 162	3 625	1 557	1 909	622	482	623	519	7 379	7 352
1961–1965	4 161	4 365	1 420	1 959	442	382	534	447	7 111	8 676
1969–1971	3 727	4 734	2 235	3 922	320	292	481	504	7 406	11 821
1979–1981	2 628	4 510	3 520	6 571	219	239	453	527	7 391	14 709
1989–1991	2 182	5 236	4 361	9 346	204	279	344	474	7 756	19 306
1995–1997	2 046	4 646	3 554	8 091	167	231	380	473	6 720	17 734

Sources: IIA and FAO statistics.

A considerable increase in cereal yields took place during the 1970s, mainly due to improved agricultural practices (e.g. irrigation), the use of fertilizers and the introduction of higher yielding varieties.

2.1. Botanical diversity

The Iberian Peninsula is endowed with a great diversity of botanical varieties of *Triticum*. Vavilov (1935) ascribed this to 'the antiquity of the culture in the region and its nearness to the place of origin of the 28 chromosome species as well as to the exceptional range of growing conditions'. *Triticum aestivum*, *Triticum durum* and *Triticum turgidum* were all cultivated in considerable quantities in Spain during the first decades of the 20th century. *Triticum spelta* was grown in the mountains of Asturia and *Triticum dicoccum* was grown over a sizeable area in the Pyrenees by the Basques; *Triticum monococcum* and *Triticum polonicum* were grown occasionally. The diversity of *Triticum* species gradually decreased so that by 1950 70% of the Spanish wheat acreage was covered with bread wheat, 25% with durum and 5% with spelt and other wheat species. By 1965 the proportion of durum wheat had fallen to less than 10% of the total wheat area, but renewed interest under the influence of EU subventions has resulted in an increase in the area under durum wheat to approximately 30% of the total wheat area in 1997. The redistribution of bread, durum and spelt wheat in the most important wheat growing areas between 1965 and 1994 is presented in Table SP.2.

Wheat culture in Spain was pursued along traditional lines for much longer than in many other European countries. Although varieties from foreign and later from Spanish breeding programmes have been introduced since the

Table SP. 2. Area (1000 ha) of bread, durum and spelt wheat in the most important wheat growing areas of Spain in 1965 and 1994

	Bread wheat		Durum wheat		Spelt wheat
	1965	1994	1965	1994	1965
Castilia (new and old) 1 576	735	45	81	4	
Aragon & Catalufia	718	174	–	124	3
Andaluciá	499	147	200	394	1
Spain, total	3 914	1 322	335	648	6

Sources: Broekhuizen (1968); Carillo (1998).

1950s, land varieties are still grown, especially in Galicia. The reason for the widespread evolution of these land varieties and for the fact that they have survived for a relatively long period may partly be found in the infrastructure of the country, with a large number of isolated places. Moreover, the great diversity of ecological conditions makes the replacement of these land varieties with breeding products a tedious job. Former Spanish wheat collections have been described by several authors, such as Gadea (1954) and Sanchez Monge (1957).

2.2. Wheat growing areas and wheat consumption

Wheat can be grown all over Spain. In higher areas it is often the only crop possible. Wheat, as well as barley, is mainly grown as a winter crop. Sowing is done in autumn and Northern European spring-wheat varieties may be used for this purpose. In the higher areas of the central plateau, sowing in spring may be preferred because the risk of winter killing.

The agricultural regions of Spain are shown in Figure SP.1. Up to the 1970s, wheat was mainly grown in the Duero region (including the province Castilla la vieja and the fertile plains of Leon) and in the central region (including Castilla la nueva). One part of Castilla la vieja was called the 'Tierra de pan'(land of the bread), which once was the largest granary of Spain. Other important wheat growing areas have always been, in the northeast the Ebro valley (including Aragon) and Cataluña, and in the south the rich plains of Andalucia.

Although a large part of the wheat area in Castilla was replaced by barley between 1970 and 1980, it remains the most important wheat growing area of Spain (Table SP.3). In Andalucia, the area under bread wheat has decreased in favour of durum wheat since 1991, under the influence of an EU subvention encouraging the cultivation of durum wheat in this region.

Table SP. 3. Distribution of bread and durum wheat over different regions of Spain in 1996–1997

Region	Proportion (%)		Region	Proportion (%)	
	bread	durum		bread	durum
Castilla-Leon	44	1	Navarra	6	1
Castilla-La Mancha	15	5	Cataluña	5	–
Andalucia	9	64	Extremadura	4	7
Aragon	7	22	other regions	10	0

Source: Carrillo (1998).

Total Spanish wheat consumption is about 4 Mt annually. The consumption per person is higher in Spain than in any other EC country. According to the authorities, this is due to the fact that apart from feeding almost 40 million Spanish inhabitants, the same number of tourists has to be fed each year (Strzybny, 1982).

2.3. Irrigation

Cereals in Spain are cultivated under one of two regimes: with or without irrigation. The breeding of varieties presents different problems according to the regime of cultivation. In non-irrigated areas, drought resistance is of primary importance. Resistance against the 'asurado', a shrivelling of the grain as a result of strong hot winds that blow during the milky stage of maturation, is another important objective for wheat breeders. The area of cereals cultivated under irrigation in Spain varies between 200 000 and 300 000 ha. Generally speaking, wheat under irrigation is used as bread wheat. Most of the varieties grown under irrigation are imported (French, Italian) or home-bred varieties. Traditional varieties are usually not grown under irrigation.

3. Wheat-breeding institutes

In 1927, the Centro del Cerealicultura was created in Madrid by Don Marcelino de Arana y Franco, who began to select and cross wheat. He was the first person to introduce foreign varieties into Spain, such as Manitoba from Canada, Senatore Capelli and Ardito from Italy, and Hybrido L-4 from France. Unfortunately he saw little of the results of his work as he was killed during the Spanish Civil War. The Centro del Cerealicultura is now one of the breeding stations of the Instituto Nacional de Investigaciones Agronomicas (INIA) in Madrid. This institute, funded by the Ministry of Agriculture, has several breeding stations throughout the country. Another important wheat-breeding institute is the experimental station Aula Dei at Zaragoza, in the Ebro valley. Founded in 1948 by the Council for Scientific Research (CSIC), it was directed for many years by Enrique Sanchez Monge, who played an important role in Spanish plant breeding in general and wheat breeding in particular. In recent years the Institute in Zaragoza has focused on barley breeding. The most important wheat-breeding centres are indicated in Figure SP 1.

Figure SP. 1. The agricultural regions of Spain and the locations of the most important wheat-breeding institutes. (Design: Lucas Janssen)

4. Wheat breeding in Spain

4.1. Objectives at the start

Apart from the period from 1927 to 1936, when Marcelino worked on wheat at the Centro del Cerealicultura, no mention is made in the literature of wheat breeding in Spain during the first half of the 20th century. In 1949, Sanchez Monge advocated the need for an inventary of the Spanish wheat collection as a starting point for wheat breeding in Spain. In the same year Gadea, working at INIA, wrote that it would be 'very sad to loose the treasure of such a rich flora, result of centuries of natural selection'. Moreover he advocated the need for an inventory of this 'National heritage' and its conservation in a gene bank, to be used by breeders with the ultimate aim of improving Spanish wheat production. The basis for this inventory had already been laid in the early 19th century by two botanists: Lagasca and Clemente (Molina & Peña, 1952). An extensive collection of diploid, tetraploid and hexaploid species of wheat is nowadays maintained at the plant genetics department (CRF) of INIA in Madrid.

Following the example of American and other European breeding programmes, Sanchez Monge (1949) suggested that through hybridization Spain should be able to produce the first Spanish wheat varieties within 12–15 years. Indeed, several Spanish bred varieties, such as Pané 247, were released in 1965.

The most important issue for Spain in those days was to become self-sufficient in wheat, as large quantitites of wheat were imported from Argentina. To increase yields, a high level of drought resistance and a reduced plant height were required. Although flour characteristics were measured, bread-making quality was not a breeding issue. Quality was only important for wheat grown under irrigation and foreign varieties were generally used for that purpose.

Breeding objectives in the 1970s were the same as at the start: short stature, suppression of inproductive tillers (typical of the old drought resistant varieties), uniform ear height, resistance to drought and yellow rust and enough baking quality for the export of annual surpluses (Sanchez Monge, 1977). Twenty years later, drought resistance and quality were still the main objectives of wheat breeding (Carrillo, 1998), and surpluses for export had fallen to a minimum, such that Spain imported more than 2 Mt of wheat in 1994.

4.2. Spanish land varieties

Traditional, indigenous Spanish wheat varieties were composed of a mixture of land varieties. The most common land varieties of bread wheat grown in Spain around the middle of the century were:

Los Candeales: a group of bearded, short, drought resistant wheats with square heads, and white ears, grains and flour. Included in this group were: candeales corrientes (yellow) and candeales finos (pale). Two commonly grown varieties were *El Candeal Alcala Castilla* and *El Candeal blanco*. Both varieties were from central Spain and produced white flour with an excellent quality, producing a very white, rather compact bread that was much appreciated in Castilla. El candeal was also generally appreciated for its good water absorbing capacity and easy milling.

Los Chamorros (i.e. 'the boldheaded'): a group of awnless varieties from Castilla; this type was more primitive than the Candeales.

Las Jejas: with amongst others *La Jeja blanca* with short straw, compact ears, grown in the north of Spain.

Los Grandales: tall wheats from Galicia, high yielding and with good quality

Los Empedrados (i.e. 'pavement' wheats): composed of a mixture of red and white grained types; these are an extreme example of the heterogeneity of land varieties.

Los Aragones: grown in the area around Aragon and extremely rich in gluten. Most of these land varieties have lost their importance, with the exception of a selection from Los Chamorros.

4.3. Imported and home-bred varieties

The first imported variety grown in Spain was Manitoba, from Canada. It was widely grown between 1930 and 1950 and used as a reference for quality characteristics. Other imported bread wheat varieties came mainly from Italy and France, and later from Mexico. In irrigated areas, the local varieties were first replaced by imported varieties, such as Etoile de Choisy from France and Mara from Italy, and later by home-bred varieties such as Pané 247, a cross between the French hybrid L-4 and the Italian variety Mentana. Pané 247 was used as a parent for further improvement of leaf rust resistance and quality.

The most important Spanish bread wheat varieties grown in 1965 were: Aragon 03 (22%), Candeal (9%), Rojos (7%), Pané 247 (6%), Cabezorro (6%), Negrillo (5%) and Chamorro (3%). Apart from Pané 247, these were all selections from land varieties. Foreign varieties grown in Spain were Estrella (=Etoile de Choisy) (9%) from France, Impeto (6%) and Mara (3%) from Italy, and Ariana (6%) from Tunesia (Broekhuizen, 1968).

Twenty-five years later, in 1990, Spanish-bred wheat varieties were still not grown on a large scale; Pané 247 still covered only 5% of the Spanish bread-wheat area. The most widely grown variety was Anza (34%), followed by Yecora (9%), Cajeme (7%) and Rinconada (6%). The former three are Mexican (CIMMYT) varieties and Rinconada was bred in Spain from CIMMYT material. Two varieties bred in Spain were Marius (Cadet/3/Thatcher/Vilmorin 27//Ariana/L:FU) and Astral (Fortunato/Yga//Florence Aurore/3/G4), together covering 11% of the wheat area (Lupton, 1992).

In 1997 more than half of the Spanish wheat hectarage was under home-bred varieties: Marius (28%), Astral (19%), Pané 247 (8%), Alcotan (6%), Rinconada (5%), Chamorro (2%) and Fiel (2%). The most widely grown foreign cultivar was Soissons (12%).

4.4. Methods to determine baking quality

The earliest reports of systematic analyses of wheat flour characteristics are from the Centro de Cerealicultura in Madrid (Anonymus, 1940). Some results of wheat harvested in 1937 are presented in Table SP.4. From these data it appears that protein level and fermentation index of Manitoba, both irrigated and non-irrigated, were clearly far above those of local Spanish varieties. Candeal, famous for its fine white flour, gave good flour yields with a high gluten content but apparently of another quality than Manitoba. Gluten and protein level of Hybrido L-4 and Empedrado were low. Two terms that were commonly used to describe the bread-making quality of Spanish wheats, were 'specific gluten quality' (defined as the quotient of fermentation index and percentage dry gluten) and 'protein quality' (defined as the quotient of fermetation index and percentage of protein).

Table SP. 4. Quality characteristics of some traditional Spanish wheat varieties grown in the region Valladolid, harvested in 1937

	% Wet[a] gluten	% Dry gluten	Fermentation index[b]	Protein content	Ash content
Candeal blanco	21.00	7.30	26.50	9.44	1.88
Candeal	21.16	7.52	26.67	9.20	1.83
Manitoba (irrig.)	18.81	7.26	163.00	10.74	2.15
Manitoba (non irr.)	21.98	8.40	157.00	11.14	2.20
Empedrado	18.04	6.43	24.00	8.05	1.96
Hybrido L-4	16.15	5.98	31.00	8.37	1.76

[a] > 20% classified as 'glutinoso', < 20% as 'amiloso'.
[b] Fermentation index according to Pelshenke: < 15 = low, 20–40 = medium, 40–50 = good, 50–100 = very good, > 100 = exceptional.
Source: Anonymus (1940).

Official annual classification of wheat after 1945 was carried out by the Servicio Nacional de Productos Agrarios (SENPA), which defined four quality groups:

I: 'mejorantes y de fuerza' (improver and strong quality)
II: 'finos' (fine quality)
III: 'comunes' (regular quality)
IV: 'bastos' (poor quality)

No minimum values for protein and sedimentation were given. The limiting values for exterior characters were determined as: minimum hectoliter weight 75 kg, maximum humidity 15% and maximum impurity 5%. Only 5% of the varieties analysed by SENPA in 1980 were classified in Group I, 45% in

Groups II and III, and 50% in Group IV. The price difference between wheat in Groups I and IV was 8.3% (IV = 100) in 1980 (Strzybny, 1982). After the introduction of free market pricing of wheat in 1983, the amount of wheat classified by SENPA fell dramatically and the above mentioned system of classification ceased to be used (Garcia et al., 1989).

Chopin alveogram and farinogram values were introduced as parameters of bread-making quality in the 1970s. At that time, a dough with a W of Chopin of 300 was considered to have good baking quality (Villena, 1975). As in France and Italy, Chopin values still play an important role in the quality standard of wheat in Spain (Bolling, 1989).

In 1981 Lopez Bellido suggested the use of the following quality characteristics for the classification of wheat:

- grain characteristics: hectolitre weight, vitrosity (especially for durum wheat), ash content (rarely used in Spain) and protein content;
- flour characteristics:
  * Alveogram values (W > 300 strong; 200 < W < 300 good; W < 200 weak)
  * Zeleny sedimentation index (< 18 poor; 18–28 good; 28–38 very good, > 38 improver wheat)
  * Hagberg falling number
  * Mixogram (1:very poor – 8:very strong)
  * Brabender farinogram
  * Pelshenke index (> 100 strong; 60–100 good; < 60 poor)
  * Baking test

Since the 1980s, the quality of the Spanish wheat harvest has been registered by the Spanish Association of Cereal Technicians (AETC), in collaboration with the ministry of Agriculture (MAPA). The quality of the 1997 harvest was described with the following parameters: hectolitreweight, protein content; Hagberg falling number; Zeleny sedimentation value; alveogram values P, G, L and W, and the percentage of degradation of these Chopin values.

The use of Zeleny sedimentation values was studied by Garcia et al. (1989), who analysed the quality of the most commonly grown wheat varieties cultivated in the province of Salamanca (central western Spain) with the aim of providing a basis for the progressive selection of varieties that were most suited to the region. The results are discussed in Section 4.5.

A modified SDS sedimentation index, used at the INIA from 1984 onwards, was described by Silvela et al. (1993). The analysis can be carried out on 600 mg of flour, implying that a single ear is enough, thus leaving sufficient remnant seed for crosses, multiplication, etc. Compared to the conventional SDS index, the soaking period was extended (overnight instead of 20 minutes). The authors found that this method gave a more accurate determination of quality differences between lines.

4.5. Yield and quality figures of Spanish wheat varieties

The following description of the quality of Spanish breadwheat in 1949 was given by Silvela, at the Centro de Cerealicultura in Madrid: 'The bread-making quality of the Spanish traditional varieties in general is not very high in the sense of gluten quality. There are exceptions like for instance the wheats grown at high altitude in Guipúzcoa with a Pelshenke index of 200 (extremely high), followed by wheat grown in the mountains of Bergantiños (La coruña) with an index of 60. Also in the Candeales types have been found with good indices.' Silvela suggested that quality improvement was possible by selection in these groups of wheat, which had the advantage that they were well adapted to local circumstances. Moreover he stated that most of the traditional Spanish wheats (except for some very primitive and red types) had high yields of white flour with a relative small amount of bran. The quality of wheat from Argentina (in those days the main source of imported wheat) was in terms of flour strength better than Spanish wheat. However, the flour was darker and the flour yield generally lower. The content in wet gluten was 30% higher than that of the traditional Spanish varieties, and the quality of the gluten was 25% higher. Commercial varieties showed a great variation in physical, chemical and fermentoscopic characters. The dry gluten content varied from 5% in L-4 to 10% in Manitoba; thus the protein level of the bread produced from this latter wheat was twice as high as that of the former.

Garcia et al. (1989) studied the bread-making quality of the most commonly grown varieties in the province of Salamanca, which is a semi-arid region in central western Spain. The aim of the study was to provide a basis for selecting varieties most suited to the region. The varieties studied were Splendeur, Astral, Pané, Capitole, Barbilla, Talento, Candeal, Champlein, Anza, Caton, Marius, Mocho, Alcazar, Arcole, Golo and Mistral. Many were foreign varieties, mostly from France, Anza from Mexico, and Ariana from Tunesia. Only Pané, Barbilla, Candeal and Mocho were Spanish, together covering 21% of the area under wheat. The results of the analyses of Zeleny sedimentation values and protein content are presented in Table SP.5.

From these data Garcia et al. concluded that the protein content of the wheats harvested in Salamanca was lower than those in France in the same year, higher than in Ireland and lower than in Germany (FRG). Forty percent of the wheats analysed were above the 11% value stipulated by the EU market for bread-making wheat, and 66% of the samples were within the 10–12% range. The Zeleny values of the wheats harvested in the study region were relatively low. Only 9% of the samples surpassed the value of 25, which was the minimum EU value for bread-making wheat. With a Zeleny value of 23.10, Pané was clearly better than the other varieties.

Table SP. 5. Average levels, ranges and variation coefficients of Zeleny sedimentation values and protein content of wheat varieties grown in the province of Salamanca in 1984–1985

Variety (nr. of samples)	Protein content (%)			Zeleny value (ml)		
	average	range	CV	average	range	CV
Splendeur (70)	10.69	8.83–13.91	9.73	15.37	9–32	27.12
Astral (39)	11.06	9.42–13.16	7.32	18.72	13–29	17.46
Pané (25)	11.02	8.99–14.33	11.16	23.10	16–29	15.50
Capitole (12)	10.15	9.57–11.20	4.75	17.29	14–20	11.98
other (37)	10.98	8.86–15.30	12.93	15.11	6–43	45.00
total (183)	10.84	8.83–15.30	10.27	17.17	6–43	30.58

Source: Garcia et al. (1989).

Based on the criteria proposed by the ONIC (Office National Interprofessionel des Cereales in France), 40% of the varieties under study could be classified as suitable for bread-making purposes, 38% were of mediocre quality; only 4% of excellent quality; 1% of high quality and useful for improving bread quality; while the rest, 16%, was of poor grade. The authors pointed out that the quality level of bread wheat harvested in the south of Spain is generally higher than that of Salamanca. For this reason, flour manufacturers from Salamanca sometimes purchase wheat from southern areas to obtain mixtures suitable for bread-making.

Michelena et al. (1995) studied the stability of grain quality and yield in bread wheat. Twenty-five bread wheat varieties were grown at sixteen sites in northeast Spain for two years. All the analyzed characters showed significant differences between environments and years. Alveogram values tenacity (P) and strength (W) had a main genetic component. Test weight, protein content, extensibility (L) and yield were influenced mainly by year x site interaction. Within environments, L and W values had a significant correlation with protein content. Genotype × environment interaction was studied by AMMI anlysis but it was impossible to find a general explanation for geographical areas or varietal groups.

5. The HMW glutenin composition of Spanish wheats

Several extensive studies on HMW glutenin subunits in Spanish wheat varieties have been carried out since the late 1980s. Payne, Carrillo et al. (1988)

Table SP. 6. HMW glutenin subunit composition and *Glu-1* score of the 10 most widely grown varieties in Spain in 1985

	HMW subunits			*Glu-1*
	1A	1B	1D	score
Aragon 03	N	20	4 + 12	3
Candeal de Arévalo	N	20	4 + 12	3
Pané 247	N	7 + 8	2 + 12	6
Anza	N	7 + 8	2 + 12	6
Astral	N	7 + 8	2 + 12	6
Estrella	N	7 + 8	2 + 12	6
Marius	N	7 + 9	4 + 12	4
Cajeme	1	17 + 18	5 + 10	10
Rinconada	1	7 + 8	5 + 10	10
Yecora	1	17 + 18	5 + 10	10

Source: Carrillo et al. (1988)

calculated the correlation of the *Glu-1* score of 33 Spanish-grown wheat varieties with various measures for seed quality (alveograph W value, Zeleny sedimentation volume, protein content, wet gluten content, thousand kernel weight and falling number). They found that approximately 69% of the variation in the Zeleny values and 79% of the variation in the Chopin W values was accounted for by variation in *Glu-1* quality score and protein content. The variation in *Glu-1* score was caused primarily by variation in the 1D encoded subunits. The 1BL/1RS wheat rye translocation, causing dough to be undesirably sticky and weaker in mixing, was not present in any of the 33 varieties studied.

Carrillo et al. (1988) analysed the HMW glutenin subunit composition of 110 registered Spanish varieties. The composition of the most widely grown varieties in 1985 is listed in Table SP.6.

The highest score was found in Yecora, Cajeme and Rinconada, three varieties of Mexican origin, with almost similar HMW glutenin composition. Surprisingly, the varieties Estrella (French), Anza (Mexican) and Pané 247 (Spanish) had the same composition, even though they are not genetically related. In Candeal and Aragon 03, both traditional Spanish land varieties, subunit 20 (*Glu-B1*) was identified, which (at that time) had not yet been associated with bread-making quality.

Michelena & Sanchez (1990) evaluated 185 Spanish land varieties (from the INIA Germplasm Bank in Madrid) by measuring grain quality parameters, morphological characters of the spike and the HMW glutenin composition. They found that the frequency of several poor quality alleles (N, 20 and 2 + 12) was high, which in their opinion was not surprising, as selection for bread-making quality had never been applied to land varieties.

Silvela et al. (1993) estimated the contribution of HMW subunits to the variation in flour quality in two different wheat populations, using SDS sedimentation as an index. The first population was a synthetic of 10 bread wheat cultivars known for their good quality and selected under forced random mating for high SDS sedimentation. The second population was the selfed progeny of a cross between Sibereño, a very poor quality bread wheat with subunits (N, 7 + 8, 2 + 12) and 7681, a very good quality bread wheat with subunits (2*, 7 + 9, 5 + 10). Silvela et al. concluded that slightly more than half the phenotypic variance in the populations was genetic and that more than half of this was accounted for by HMW composition. They estimated that over half of the maximum possible advance in quality in heterogeneous populations similar to the ones used in their study should easily be achieved in two years of marker-assisted selection.

The composition of 174 Spanish land varieties (Germplasm Bank INIA, Madrid) was analyzed by Rodriguez-Quijano & Carrillo (1994). The HMW glutenin subunit composition was compared with the gluten strength as determined by SDS sedimentation test. Significant correlations were found between high sedimentation volume and HMW subunits 1 and 2* (*Glu-A1*), 13 + 16 and 7 + 8 (*Glu-B1*) and 5 + 10 (*Glu-D1*), and between low volume of sedimentation and HMW subunits N (*Glu-A1*), 6 + 8 and 20 (*Glu-B1*) and 2 + 12, 3 + 12 and 4 + 12 (*Glu-D1*). A summary of the results is presented in Table SP.7.

Table SP. 7. Frequencies of HMW glutenin patterns, SDS sedimentation values and *Glu-1* quality score in 174 Spanish land varieties of wheat

HMW subunit			Cultivar		SDS sediment	*Glu-1* score[a]	
Glu-A1	*Glu-B1*	*Glu-D1*	n	%	± s.d.	Payne	Pogna
1	20	2 + 12	15	8.6	48.83 ± 14.8	6	6
1	13 + 16	2 + 12	5	2.8	69.00 ± 11.1	8	11
1	7	2 + 12	5	2.8	55.07 ± 6.2	6	7
2*	20	2 + 12	67	38.3	52.53 ± 15.6	6	8
2*	13 + 16	2 + 12	11	6.3	62.68 ± 7.0	8	13
2*	20	4 + 12	15	8.6	34.30 ± 12.2	5	7
N	20	2 + 12	11	5.7	33.00 ± 16.3	4	5
Other patterns			45	25.8	–	–	–

[a] Quality score calculated according to Payne et al. and Pogna et al. (see Table SP.9).
Source: Rodriguez-Quijano & Carrillo (1994).

In addition to the *Glu-1* score calculated according to Payne et al. (1984), a quality score according to Pogna et al. (1986) was calculated (for a comparison of these scores the reader is refered to Table SP.9.), which allowed a better differentiation of the quality in Spanish wheats.

The most frequent HMW glutenin pattern in this collection of land varieties apeared to be (2*, 20, 2 + 12), of which the latter two subunits are associated with poor bread-making quality. Subunit 13 + 16, associated with good bread-making quality, appeared in higher frequencies than in other land-variety collections. Spanish land varieties with this subunit may be a useful source for quality breeding.

Ruiz et al. (1995) studied the relationship between HMW glutenin subunits, different quality scores and various indices of quality in a collection of 65 Spanish-grown bread-wheat varieties. Zeleny sedimentation and Chopin W value showed significant positive correlations with subunits 1 and 2*, 17 + 18 and 5 + 10, and negative associations with 6 + 8, 20 and 4 + 12. Protein content was found to be independent of rheological quality but it influenced the balance of P/L ratios, and in varieties with the same HMW glutenin composition it influenced the dough strength positively. Quality scores showed high significant correlation with dough strength, thus providing a useful method for selecting HMW glutenin compositions with good quality. The most common patterns found in these Spanish wheat varieties are listed in Table SP.8.

Table SP. 8. Frequencies of HMW glutenin patterns in 65 bread wheat varieties grown in Spain and their quality score

HMW subunits			Varieties		Score[a]	
Glu-A1	*Glu-B1*	*Glu-D1*	n	%	Payne	Pogna
N	7 + 8	2 + 12	9	14	6	8
2*	17 + 18	5 + 10	5	8	10	17
N	7 + 9	2 + 12	5	8	5	9
N	7	2 + 12	4	6	4	6
N	17 + 18	2 + 12	4	6	6	10
Other patterns			38	58		

[a] Quality score calculated according to Payne et al. and Pogna et al. (see Table SP.9).
Source: Ruiz et al. (1995).

Comparing Tables SP.7 and SP.8 it appears that there is a remarkable difference in HMW patterns between Spanish land varieties and commercial wheat varieties. Obviously, this is related to the fact that most wheat varieties grown in Spain are either foreign or derived from foreign progenitors. The

Table SP. 9. Quality scores assigned to HMW glutenin subunits by Payne et al. (1984, 1987), Pogna et al. (1986) and Branlard et al. (1992)

HMW subunit	Payne	Pogna	Branlard
Glu-A1			
1	3	3	15
2*	3	5	30
N	1	2	0
Glu-B1			
7	1	2	8
7 + 8	3	4	18
7 + 9	2	5	20
6 + 8	1	1	2
20	1	1	2
13 + 16	3	6	32
17 + 18	3	6	18
Glu-D1			
2 + 12	2	2	7
3 + 12	2	2	9
4 + 12	1	1	5
5 + 10	4	6	30
Range	3–10	4–17	2–92

Source: Ruiz et al. (1995)

glutenin patterns common in Spanish land varieties, such as (2*, 20, 2 + 12), do not occur in the commercial varieties.

Ruiz et al. (1995) compared three different *Glu-1* scores and their relationship with various quality measurements in the collection of wheat varieties of Table SP.8. They compared the score according to Payne et al. (1984), Pogna et al. (1986) and Branlard et al. (1992). The scores assigned to the different HMW subunits by these three groups of authors are presented in Table SP.9. The best characterization of the Spanish varieties for their dough strength was achieved by applying the *Glu-1* score according to Pogna et al. (1986).

Rodriguez-Quijano & Carrillo (1996) showed that the gluten strength of wheat was not only influenced by allelic variation of the *Glu-1* loci, but also by allelic variation of the *Gli-1*/*Glu-3* loci, with genes that code for low molecular weight (LMW) glutenins (*Glu-3* loci) and the closely linked genes that control the majority of gliadins (*Gli-1* loci). Nieto-Taladriz et al. (1998) ascribed a significant positive effect of subunit d_4 at the *Glu-B3* locus on the gluten strength of wheat. The amount of Spanish wheat material that has been analyzed for HMW glutenin composition since 1988 is impressive. Several wheat breeders, both private and public, apply the available knowledge in the choice of the crossing parents and the selection of the progeny in wheat-breeding programmes.

Several authors have given clear suggestions on units that may be interesting for Spanish wheat breeders. The combination of subunits 2*, 17 + 18 and 5 + 10, resulting in the highest *Glu-1* score, may be a big step away from the composition of Spanish land varieties. The combination of subunits 2* (*Glu-A1*) and 13 + 16 (*Glu-B1*), however, is frequently present in Spanish indigenous material. Both subunits are associated with good bread-making quality and seem to be an interesting source for quality breeding in wheat.

D.A.D.

Acknowledgments

The author wishes to thank Prof. Dr J.M. Carrillo (Agricultural University, Madrid) for his valuable comments on this chapter.

References

Anonymus (1940) Metodos de analisis de trigos y harinas por el Centro de Cerealicultura. Madrid: Ministerio de Agricultura.

Anonymus (1960) Progressive wheat production. Geneva: Centre d'Etude de l'Azote (ed), 338 pp.

Barber S, Benedito de Barber C, Llácer MD (1982) Componentes de la fibra dietética en trigo, harina y salvado. Revista Agroquímica Tecnologia Alimentaria 22(3): 431–440.

Bolling H (1989) Qualitätsbewertung in den EG-ländern. Bericht über die 14. Getreide-Tagung, Detmold, pp. 108–121.

Branlard G, Pierre J, Rousset M (1992) Selection indices for quality evaluation in wheat breeding. Theor Appl Gen 84: 57–64.

Broekhuizen S (1969) Atlas of the cereal-growing areas in Europe. Agro-ecological Atlas of Cereal Growing in Europe, vol II, 157 pp. Wageningen: Prudoc; Amsterdam, London, New York: Elsevier Publishing Company.

Carrillo JM (1998) personal communication, Wageningen.

Carrillo JM, Rodriguez de Quijano M, Galiano A, Hamie B, Vasquez JF, Orellana J (1988) Composición en gluteninas de alto peso molecular de variedades de trigo blando registradas en España y su relación con la calidad panadera. Anales Aula Dei 19(1–2): 239–250.

Gadea M (1949) Los trigos Españoles. Agricultura; Revista Agropecuaria 18: 197–202.

Gadea M (1954) Trigos Españoles. Madrid: Instituto Nacional de Investigaciones Agronómicas, 451 pp.

Garcia A, Garcia B, Blanco M (1989) Variability and relationship between protein and Zeleny sedimentation value in wheat varieties. Agrochimica 33(6): 424–433.

Garcia-Olmedo F, Carbonero P (1980) Chromosomes, genes, proteins and wheat endosperm. Proc 3rd Intern Wheat Conf, Madrid, pp. 664–670.

Lopez Bellido L (1981) Calidad del trigo. Agricultura 583: 45–51.

Lupton FGH (1992) Changes in varietal distribution of cereals in central and western Europe. Agro-ecological Atlas of Cereal Growing in Europe, vol IV, 161 pp. Wageningen: Pudoc.

Manrique LR, Gorge JL (1981) Estudio de los factores nutricionales que modifican la calidad de la proteina en el grano de trigo; (I) Influencia de la cantidad de fertilizacion nitrogenada aplicada en cobertera y de la variedad empleada. Anales de edafologia y Agrobiologia 40(3–4): 625–635.

Michelena A, Martin-Sanchez JA (1990) Evaluación de la calidad en collecciones de germoplasma autóctona de trigo harinero. Anales Aula Dei 20(1–2): 183–188.

Michelena A, Romagosa I, Martin JA, Lopez A (1995) Estabilidad de la calidad y el rendimento del trigo harinero. Invest Agr (prod y prot veg) 10(2): 203–212.

Molina RT, Peña MA (1952) Los Trigos de la Ceres Hispanica de Lagasca y Clemente. Madrid: Instituto Nacional de Investigaciones Agronomicas, 516 pp.

Nieto-Taladriz MT, Rodriguez-Quijano M, Carrillo JM (1998) Biochemical and gentic characterisation of a D glutenin subunit encoded at the *Glu-B3* locus. Genome 41: 215–220.

Payne PI, Holt LM, Krattiger AF, Carrillo JM (1988) Relationships between seed quality characteristics and HMW glutenin subunit composition determined using wheats grown in Spain. J Cereal Sci 7: 229–235.

Plaza Perez S de la (1971) Aproximacion de un modelo mecanico y de un modelo electrico para representar la relajacion de esfuerzos de la masa de harina de trigo. Anales INIA (general) 1: 183–194.

Pogna NE, Mellini F (1986) Alla ricerca delle basi biochimiche e genetiche della qualità del glutine. Inform Agr 42: 65–66.

Rodriguez-Quijano M, Vázquez JF, Carrillo JM (1990) Variation of high molecular weight glutenin subunits in Spanish landraces of *Triticum aestivum* ssp. *vulgare* and ssp. *spelta*. J Gen Br 44: 121–126.

Rodriguez-Quijano M, Carrillo JM (1994) Relacion entre subunidades de gluteninas de alto peso molecular y fuerza del gluten en cultivares indigenas españoles de *Triticum aestivum ssp. vulgare*. Invest Agr (prod y prot veg) 9(3): 327–339.

Rodriguez-Quijano M, Carrillo JM (1996) Relationship between allelic variation of *Glu-1* and *Gli-1/Glu-3* prolamin loci and gluten strenth in hexaploid wheat. Euphytica 91: 141–148.

Ruiz M, Rodriguez-Quijano M, Carrillo JM (1995) Relaciones entre la subunidades de gluteninas de alto peso molecular (HMW), distintas puntuaciones *Glu-1* y varios parametros de calidad en trigo harinero. Invest Agr (prod y prot veg) 10(1): 15–29.

Sanchez Monge E (1949) El punto de partida en la mejora de trigos. Agricultura 18: 506–509.

Sanchez Monge E (1957) Catalogo Genetico de Trigos Españoles, Madrid: Ministerio de Agricultura, 920 pp.

Sanchez Monge E (1959) Cereal breeding problems in Spain. Proc 2nd Eucarpia Congress, Köln.

Sanchez Monge E (1977) Plant breeding in Spain. Proc 8th Eucarpia Congress, Madrid.

Silvela F (1947) Sobre la variabilidad de la composición de las harinas de trigo de 'extracción legal'. Boletin INIA 17: 105–124.

Silvela F (1949) Sobre la mejora de la calidad del pan. Agricultura, revista agropecuaria 18: 244–248.

Silvela L (1992) La epistasia y los caracteres cuantitativos. Informaci6n Tecnica Econ6mica Agraria (ITIA) 88A(1): 57–60.

Silvela L, Ayuso MC, Gil-Delgado LG, Salaices L (1993) Genetic and environmental contributions to bread-wheat flour quality using the SDS sedimentation test as an index. J Appl Gen 86: 889–894.

Strzybny X (1982) Der Spanische Getreidemarkt. Berichte über Landwirtschaft 60: 140–149.

Vavilov NI (1935) The origin, variation, immunity and breeding of cultivated plants (transl. Starr Chester K) Chronica Botanica 13 (1949), 364 pp.

Villena LM (1975) Trigos enanos Españoles. (I) Primeros resultados. Anales de Aula Dei 13(1–2): 122–139.

Italy

1. Introduction

With a wheat area of around 2.5 M ha and a production of some 8 Mt of wheat, Italy (I) is among Europe's leading wheat producers. Almost two-thirds of the Italian wheat area and more than half of the total wheat production consists of durum wheat,[1] used for the production of semolina and pasta products for export. The area under wheat in Italy has decreased considerably since the beginning of the 20th century. Whereas wheat represented more than half the arable crops grown in Italy in 1956, its share was less than one-third forty years later in 1996. Notably, the decrease concerned bread wheat more than durum wheat. The area of durum wheat has remained relatively stable, as may be seen in the figures presented in Table I.1. In spite of the considerable reduction of the area under wheat since the beginning of the century, the yield per hectare has tripled over the same period. A major step in this development was the 'Grain Battle', proclaimed by Mussolini in 1925. Against a background of poor long-term cereal production and a dramatic reduction of wheat imports during and after World War I, the aim of the Grain Battle was to achieve self-sufficiency in cereals rapidly. Under the wings of a permanent National Committee, every province was given a 'Commissione per la Propaganda Granaria' and contests were organized between provinces.

Table I.1. Wheat area (1000 ha) and production (1000 t) in Italy between 1910 and 1997

Year	Area	Production	Year	Area		Production	
1910–1922	4 655	4 648					
1924–1925	4 645	5 591					
1934–1938	5 040	7 254		bread	durum	bread	durum
1948–1950	4 704	7 004					
1961–1965	4 398	8 857	1956	3 508	1 386		
1969–1971	4 089	9 756	1967	2 651	1 360		
1979–1981	3 373	8 989	1988	1 094	1 699		
1989–1991	2 800	8 312	1992	988	1 539		
1995–1997	2 447	7 746	1995	859	1 620	3 875	4 166

Source: IIA and FAO statistics and Statistiche economiche e finanziare.

[1] As durum wheat is beyond the scope of this book, this chapter only deals with bread-wheat breeding in Italy.

With the help of research and breeding activities, wheat yields, which had been stable at around 1 t/ha since the beginning of the century, increased in a few years time to 1,5 t/ha in 1935 (Anonymus, 1937). Wheat imports decreased from 2.58 Mt just after the war to 0.79 Mt in the years 1931–1935. The production increase continued after World War II, albeit in a somewhat more liberal atmosphere.

Practically all wheat in Italy is grown as a winter crop. Bread wheat is mainly grown in the northern and central districts of Italy, whereas durum wheat is generally grown in the central and southern districts (Table I.2). North of the Appenines, bread wheat is mainly grown in the Po Valley with Piemonte, Lombardia, Veneto and Emilia Romagna as the most representative districts. South of the Appenines, Toscana, Umbria, Marche, Abruzzo and Campania are the most important districts for bread wheat. Italy has

Table I.2. Area (1000 ha) of bread and durum wheat for North, Central and South Italy in the year 1992

	Bread	Durum	Total wheat
North	470	86	556
Central	469	447	916
South	43	997	1 040
Italy	982	1530	2 518

Source: Statistiche economiche e finanziare.

a mediterranean climate, characterized by a limited availability of water and thermal stress during grain filling, which may cause large fluctuations in grain yield but also in grain protein content and composition (Borghi et al., 1997). Generally speaking, the Italian climate is well suited to the production of high quality wheats.

2. Wheat breeding before 1960

Wheat breeding in Italy during the first half of the 20th century was principally associated with the names of two outstanding Italian breeders, Todaro and Strampelli (Vavilov, 1935). These two breeders have been called the antipodes of breeding: whereas Todaro worked principally by individual selection from local populations, Strampelli on the contrary, used the method of hybridization, considering that new forms had to be created. Both breeders produced a number of valuable varieties which have occupied a considerable proportion of the wheat acreage in Italy and which have largely contributed to win the Grain Battle.

Todaro considered that the Italian local wheat populations had a high potential from which valuable types could be selected. This resulted in varieties such as Gentilrosso 48 and Rieti 11. He also made selections in foreign material such as the Canadian variety Coronation and the French variety Hâtif inversable, resulting in Inallettabile (= 'non-lodging') 95 and Inallettabile 96 respectively (D'Amato, 1989).

Todaro was a professor at the university of Bologna and played a decisive role in the foundation of a private seed company in Bologna, the Società Produttori Sementi, still one of Italy's wheat breeding companies today. After 1921, Todaro directed the Institute of Cereal Crops Breeding in Bologna until his retirement in 1935. He was succeeded by Bonvicini, who followed another strategy than Todaro and obtained several varieties by hybridization, among which Fortunato, Funo and Fiorello have been the most successful.

Strampelli, in a way is considered to be the Italian rediscoverer of Mendelian laws, as he was the first scientist to systematically observe the segregation of inherited traits in planned crossing activity (Maliani, 1979). He made his first cross in 1900 (with no sound theoretical basis) between Noé, a French selection from a Russian wheat, and Rieti, a local variety noteworthy for its universality and adaptation to different conditions. In combining these two varieties Strampelli aimed to combine the straw quality of Noé with the rust resistance of Rieti; at the time lodging and/or rust were responsible for poor Italian wheat harvests. By 1905 Strampelli had already made 105 crosses, using mainly Rieti. In the same year his first variety, Carlotta Strampelli, named after his wife and main collaborator, was released.

Strampelli founded the Experimental Station for Grain Culture in Rieti in 1907 and the National Institute of Genetics for Cereal Research in Rome in 1920. The latter was named Istituto Sperimentale per la Cerealicoltura (Experimental Institute for Cereal Research) in 1967, with its headquarters in Rome and several breeding stations throughout the country. Nowadays, the sub-station of San Angelo Lodigiano in Milan is the most important centre for bread-wheat-breeding research in Italy.

During the 1920s Strampelli took the important step of crossing Italian varieties with the Japanese variety Akagomuchi, with the aim of introducing the semi-dwarf characteristics of the latter into Italian wheat. From a complex hybridization involving Rieti, Akagomuchi and the Dutch variety Wilhelmina (including an English squarehead type in its pedigree), several well known Italian varieties such as Ardito, Mentana and Villa Glori, have resulted.

Ardito was particularly valuable for its stiff short straw, earliness and yield. Its main defect was shattering of the grain, which is a characteristic of Japanese wheats. From Italy, Ardito spread to Argentina, Chile and France. Eventually it found its way back to Europe in the famous Russian variety

Bezostaya-1 (Ardito/Klein 33), which in its turn was used widely as a quality parent by several European wheat breeders.

Mentana was a non-shattering variety with good adaptation from the Po valley to Sicily, as well as many other countries around the world. Moreover, Mentana was described as a variety with 'technological quality', although no specification of this quality has ever been reported. Mentana was one of the varieties that played an important role in winning the Grain Battle and was still widely grown in Italy around 1950.

Villa Glori was later used as a parent of the varieties San Pastore and Jacometti (see Table I.6). San Pastore, released after World War II, was a variety with some resistance to cold temperature, performing well under various conditions. Apart from being grown in Italy for more than 25 years, it spread to Yugoslavia, where it contributed considerably to the increase in wheat production (Vallega, 1974).

Ardito, Villa Glori and Mentana were introduced into China in 1930, where they have been grown on vast areas. Mentana, grown under the Chinese name Nanda 4219, covered an area of 4 660 000 ha in 1962. Together with other varieties introduced in subsequent years, Italian varieties have had a large impact on Chinese wheat breeding (Zheng, 1993).

Strampelli was called 'a forerunner in green revolution' by Maliani (1979), who stated that 'the first green revolution has been accomplished in Italy forty years earlier than the Mexican one, even with much less financial, scientific and technical support'. But although the yielding capacity of Strampelli's varieties was widely recognized, those varieties were considered to be qualitatively poor, in spite of the fact that Strampelli was amongst the first to introduce laboratory technology and quality testing in plant breeding, both in Rieti and in Rome. In 1904, he used a small manual mill; two years later he was testing flour milled with a more advanced mill, technological instruments and an experimental oven.

Another breeder who has had quite a large impact on Italian wheat breeding during the first three-quarters of the 20th century was Michahelles. He started his work at a private institute in Frassineto and later continued on his own. Important varieties resulting from his work were Frassineto 405, Autonomia, Impeto, Mara, Marzotto and Mec. These high yielding semi-dwarf varieties have been grown on vast areas both in Italy and abroad. The last variety from this breeder was Nobel, released in 1985. Michahelles & Son is still one of the leading wheat-breeding companies of Italy.

3. Breeding for bread-making quality since 1960

Compared to the first half of the 20th century, the number of Italian breeders working on bread wheat has increased considerably and due to their efforts practically all the wheat varieties grown in Italy nowadays are home-bred.

Although the assessment of quality parameters was part of wheat breeding programmes during the first half of the century, bread-making quality was not a breeding issue in itself. As long as it was easy to find high quality wheats on the international market at prices lower than those of homegrown wheats, breeders were not encouraged to pay much attention to quality. Assessments of quality parameters were made in the final stages of the breeding programme, only in advanced material ready for trial.

It was not until 1961 that Samoggia, working with Bonvicini in Bologna, advocated a systematic approach to bread-making quality in wheat. 'Now that we have passed the stage of winning the Grain Battle, Italy should make use of its favourable conditions to produce high quality wheat with regard to the economic possibilities offered by the European market.' From then on intensive work has been carried out in order to achieve progress in the field of bread-making quality. Three major incentives in this 'quality battle' were:

- the development of a technological basis for quality assessment and for the classification of wheat varieties,
- the introduction of germplasm, and
- the improvement of selection methods.

3.1. Technological investigations

The most important techniques used at the Institute for Cereal Research in Bologna to measure gluten quantity and quality in wheat were, until the 1960s, a Chopin alveograph, a farinograph, a fermentograph and a Brabender amilograph (Samoggia, 1958). The Pelshenke test, which had been introduced in Italy by Borasio in 1934, was used sporadically.

Based on his comparative study of different quality assessment techniques, Kökény (1959), a co-worker of Bonvicini in Bologna, advised breeders to use a Zeleny test in the early stages of selection for quality. Samoggia & Kökény (1962) found that the W value of a Chopin alveogram, the sedimentation value according to Zeleny and the valorimetric index of the farinogram according to Brabender lead to the same order in baking quality. They based their conclusion on data of three years of analyses of the most important wheat varieties grown in Italy in the 1950s. Some of these data are presented in Table I.3.

Brunori et al. (1984) investigated the possibilities of breeding for higher protein content in bread wheat. It was suggested that by crossing genotypes

Table I.3. Quality parameters of some important Italian bread-wheat varieties harvested in 1956

Variety	N (%)	dry gluten (%)	Alveogram W	P	L	Zeleny sediment
Funo	1.83	10.81	87.54	41.50	22.23	–
Fortunato	1.54	8.24	102.06	52.50	20.48	27
Fiorello	1.61	9.31	106.14	50.00	20.90	26
Mara	1.82	9.87	117.56	52.40	20.87	30
Generoso	1.89	10.89	107.92	48.40	22.63	30
Produttore	1.58	8.90	77.26	32.20	22.86	27
San Pastore	1.77	11.95	53.39	32.40	22.47	–
Inallettabile	1.89	12.23	52.25	28.60	21.73	–
Average	1.79	10.66	86.79	41.41	21.65	29

Source: Samoggia & Kökény (1962).

with high rates of N accumulation and genotypes with a long duration of accumulation, recombinant lines with transgressive amounts of N per grain might be obtained. Crosses of the Indian line Pusa 5–3 and the Romanian line F 26–70, both sources of high protein, with the high yielding, low protein variety Kalyan Sona resulted in lines with short stature, high grain yield and high protein percentage.

The greatest gain in quality, however, was achieved through the selection of varieties with improved protein composition, with special attention being paid to the gliadin and glutenin fraction in the protein.

Arcioni et al. (1974) working at the Agricultural Institute of Perugia, were the first Italians to describe the use of electrophoregrams to characterize the gliadin fraction of the endosperm. Ten years later the method was further elaborated by researchers such as Dal Belin Peruffo and Pogna.

Dal Belin Peruffo et al. (1984) established gliadin electrophoregrams for the wheat varieties on the Italian variety list (Registro delle Varietà) and described a key for cultivar identification based on gliadin patterns.

The study of HMW glutenin subunits to characterize the quality of wheat material was undertaken by Pogna from 1985 onwards. Together with his co-workers, he described the HMW subunit composition of hundreds of Italian wheat varieties. Their work will be discussed in Section 4.

3.2. The classification of wheat varieties

The Experimental Institute for Cereal Research, which has been carrying out the national variety trials since the 1970s, used to describe wheat varieties on the basis of productivity, hectolitre weight, thousand kernel weight, Pelshenke test, fresh gluten content, Chopin alveogram values, mixogram, farinogram and a (micro and EEC) baking test (Boggini et al., 1979). Bread was baked according to AACC Method 10–10, with some modifications to make the procedure similar to the one traditionally used in Italy. Baking tests for bread wheat have been standardized according to EU rules since 1975.

The first official classification system for bread wheat in Italy was introduced in 1979. The classification, in four quality classes, was based on the criteria listed in Table I.4.

A new classification system proposed in 1996 by the Millers' Association was based on the criteria listed in Table I.5. In addition to these characteristics, the Italian list of recommended varieties in 1996 described hectolitre weight, gliadin and HMW glutenin composition of the protein, gluten content, SDS sedimentation index and bread volume. A note concerning grain hardness appeared for the first time in 1996.

Table I.4. Classification criteria for bread wheat as proposed in 1978

Category	Alveogram		Farinograph	Dry	Protein
	W	P/L	index	gluten (%)	content
Optimal	>180	<0.6	>3	>11.5	>12.5
Good	140–180	<0.6	2–3	10.5–11.5	11.5–12.5
Mediocre	100–140	0.6–0.8	1–2	<10.5	11.5–12.5
Biscuit	<100				

Source: Annal dell' ISC (1979).

Table I.5. Classification criteria for bread wheat as proposed by the Millers Association in 1996

Category	Alveogram		Farinograph	Protein	Falling
	W	P/L	index	content	number
Strong wheat	≥300	≤1.0	≥15	≥14	≥250
Superior breadm. wh.	≥220	≤0.6	≥10	≥13	≥220
Bread-making wheat	≥160	≤0.6	≥ 5	≥11.5	≥220
Biscuit-making wheat	≤115	≤0.5	–	≤10.5	≥240

Source: Borghi (1997).

Moreover, four marketing classes were used for breadwheat, based on Chopin alveogram values in combination with protein content (Borghi, 1995):

- Class 1: improver wheat ($W > 260 \times 10^{-4}$ J)
- Class 2: wheat for direct bread-making (W 155–180 $\times 10^{-4}$ J)

- Class 3: ordinary wheat (W 120 $\times$ 10^{-4} J)
- Class 4: wheat for biscuit (W $<$ 120 $\times$ 10^{-4} J)

As in France and Spain, in Italy Chopin alveogram values play a larger role in the quality assessment of wheat than in many other European countries. Although there is a great variety of Italian bread types, they may be generally characterized as small breads with hard crusts, having a relatively dry consistancy and small pores.

3.3. The introduction of foreign germplasm

The material introduced into Italian wheat breeding programmes during the first half of the 20th century was related to the breeding objectives in that period: yielding capacity, straw length and disease resistance. Strampelli used germplasm of French, British, Dutch and Japanese origin. Material from Australia, Canada, Mexico and other countries was introduced later into breeding programmes by other breeders.

Considering the pedigree of the most important Italian varieties released before 1970 (Table I.6), one will find North and South American varieties such as Manitoba, Saytana and Norin. By the use of the Canadian variety Manitoba around 1940, good bread-making characteristics were introduced into Italian wheat by chance. This became apparent fifty years later through the analyses of HMW units (Borghi et al., 1991a).

Some of the varieties listed in Table I.6. were grown until 1990, e.g. Irnerio, Conte Marzotto and Autonomia B (a selection from Autonomia, released in 1969). Libellula was a very popular variety in Hungary in the 1970s and 1980s.

The use of high quality parents from abroad with the direct aim of improving the bread-making quality of Italian wheat started after 1960. Benvenuti (1965) mentioned the value that Russian varieties like Bezostaya-1 could have in this respect. A large number of crosses has been made since, between high yielding Italian varieties and varieties known to have good bread-making quality such as the Russian varieties Bezostaya-1 and Novosadska-1933, the Australian varieties Mendos and Gabo, the Mexican variety Lerma Rojo and the American variety Atlas 66, the latter being used especially for its high protein content; but also with Italian varieties such as Conte Marzotto and Jacometti 49, the quality of which derives from its parent Manitoba.

Breeders were faced with the fact that with the introduction of baking quality from foreign material, other unwanted plant characteristics were automatically introduced. Especially the high quality varieties from America were largely unsuitable for Italian conditions. To obtain varieties with the typical Italian short straw, the crosses made with foreign material were generally backcrossed with the local varieties.

Table I.6. The origin of the most important Italian bread wheat varieties released before 1970

Variety	Year	Pedigree	Breeder
Ardito	1916	Rieti/Wilhelmina//Akagomuchi	Strampelli
Villa Glori	1918	Rieti/Wilhelmina//Akagomuchi	Strampelli
Autonomia	1930	Frassineto 405/Mentana	Michahelles
San Pastore	1940	Balilla/Villa Glori	Strampelli
Jacometti	1948	Manitoba/Villa Glori	Soc. Sisforaggera Bologna
Mara	1947	Autonomia/Aquila	Ist. Cerealic. Frassineto
Produttore	1955	Salto//Saytana 27/Quaderna	Soc. Prod. Sementi Bologna
Glutinoso	1957	Ardito//Mottin/Norin 2	Soc. Prod. Sementi Bologna
Marimp 8	1959	Mara/Impeto	Ist. Sper. Cerealicoltura
Conte Marzotto	1959	Mara/Impeto	Michahelles Firenze
Libellula	1962	Tevere/Giulliani//San Pastore	Ist. Gen. Lonigo
Funo	1969	Inallettabile 96/Rieti//Damiano	Ist. Allev. Veg. Bologna
Demar 4	1969	Demeter/Mara	Ist. Sper. Cerealicoltura
Fiorello	1969	Cologna 188/Damiano	Ist. All. Veg. Bologna
Irnerio	1970	Produttore S6/Manitoba	Soc. Prod. Sementi Bologna

Sources: Ann.ISC (1974), Boggini et al. (1979), Samoggia (1985), Lupton (1992), Canevara et al. (1994).

The pedigrees of the most important varieties with bread-making quality released since 1970 (Table I.7.), clearly show that especially Russian material has influenced Italian varieties to a large extent.

Table I.7. Origin of the most important varieties with good bread-making characteristics released after 1970

Variety	Quality group	Year	Pedigree
Alpe		1974	Bezostaya-1/Strampelli
Mec	2	1974	Marzotto/Combine
Saliente	1	1979	Marimp 3/Bezostaya-1
Centauro	2	1983	Irnerio/Strampelli
Manital	1	1983	Marzotto/Mendos
Pandas	1	1983	Orso//Bezostaya-1/S1/3/Generoso7/Marzotto
Salmone	1	1985	Bezostaya-1/Glutinoso
Brasilia	1	1985	Osjecka-20//Libellula/Bezostaya-1

Sources: Samoggia (1985), Lupton (1992), Perenzin et al. (1996).

Several of the varieties in Table I.7. have been grown on considerable areas (Lupton, 1992), especially Mec, Centauro and Pandas. The variety Centauro, released by the Società Produttori Sementi in Bologna, has been the most commonly grown bread wheat variety since 1988, in 1995 it still covered one third of the area under wheat.

Salmone was described by Borghi et al. (1985) as 'the first Italian strong wheat'. It had very good bread-making characteristics, a high Zeleny index, good values for other technological characteristics and a high hectolitre weight. Moreover, it had short straw, cold- and lodging resistance and resistance to stripe rust. However, it lacked yielding capacity and was never grown to an important extent.

3.4. Selection methods

From 1970 onwards, increasing attention has been given to the use of early selection for bread-making quality. At the ISC San Angelo Lodigiano, from 1972 onwards the traditional bulk selection procedure was modified into one year of intensive selection on F_2 spaced plants, with the intention to establish segregating populations of high agronomical and bread-making quality (Borghi et al., 1975). Based on the high heritability of Pelshenke test and sedimentation value, a rapid fixation of genes for quality was expected in segregating generations. From the first results it appeared that at least two selection cycles were necessary to fix these genes (Corino et al., 1975).

After the selection scheme had been used for six years, it was clear that the populations obtained from the plants with the highest Pelshenke values had superior technological qualities, and that those remained unchanged throughout the generations (Borghi et al., 1978).

Based on these results, the quality assessment in the ISC breeding programme was adapted:

F_2 plants	Pelshenke
F_3-F_5 lines	Zeleny
F_6-etc.	Zeleny, mixogram, gluten
Preliminary variety trial	gluten, mixogram, farinogram, alveogram

The material in second year variety trial was submitted to a baking test, carried out according to EU standards.

The analysis of the protein composition of the wheat grain has been integrated gradually into breeding programmes since the 1980s. It is applied for the selection of crossing parents and before registration of a new variety.

3.5. Results of quality breeding in bread wheat

The progress made between 1940 and 1975 in combining higher productivity with better bread-making quality may be discussed with the data presented in Table I.8; they have been selected from the results of variety trials carried out at ISC in 1978. The varieties are listed in order of release and compared on their productivity and loaf volume.

The productivity of the old variety San Pastore and the loaf volume of Marzotto were still far above the average of the varieties in trial in 1978. The most productive variety in trial was Orso, released in 1972 and classified as a biscuit quality wheat.

With regard to the combination of productivity and high loaf volume, the varieties Reno (with a loaf volume and a yield of 9% and 2% above the average, respectively) and Valle d'Oro seem interesting. These varieties have, however, never been grown on a large scale. The varieties Alpe and Mec, apparently the best possible combination at that time, have become very popular.

Results of baking tests, carried out according to EU standards (Boggini & Pogna, 1979), showed that most of the varieties grown in Italy in the 1970s could meet minimum criteria for kneadability.

A classification of varieties in 1978, based on five years results of rheological data, the EU baking test and a micro-baking test (AACC), showed that 20% of the Italian wheat production was suitable for bread baking, while 10–

Table I.8. Loaf volume and productivity of varieties in trial in 1978

Variety	Year of release	Loaf[a] volume	Production (t/ha)
San Pastore	1940	549	4.81
Marzotto	1959	643	4.60
Demar 4	1969	569	4.54
Orso	1972	523	5.44
Alpe	1974	647	4.37
Mec	1974	638	4.55
Valle d'Oro	1975	620	5.20
Reno	1975	652	4.82
Average of 25 varieties	602	4.70	

[a] Based on 100 g flour of 14% humidity.
Source: Annali ISC (1978).

15% was of intermediate quality, 45% was suitable for biscuit making, and the rest for mixing and for home use (Annali ISC, 1978).

Some 20 years later, based on the production of certified seed in 1996, 31% of the Italian wheat varieties were classified as improver wheat (Class 1), 64% as suitable for bread making (Class 2) and 5% as biscuit wheat (Class 4) (Perenzin et al., 1996). It shows that substantial progress in this respect has been made since the 1970s.

An extensive survey of the evolution of bread wheat varieties bred in Italy since 1900 was carried out by Canevara et al. (1994). Thirty-four representative Italian cultivars, grouped in seven generations, were evaluated on morphological, physiological, agronomical and qualitative traits. The cultivars, ranging in age from Rieti (1900) to Centauro (1983), were grown under both classical and modern husbandry conditions for 3 years.

Analyses of protein content in the grain revealed a gradual decrease from the oldest cultivars (Group 0) up to the second last generation (Group 5); this decrease was mainly associated with the increased yield potential and harvest index of modern varieties. The most recent generation of cultivars (Group 6) had an increased protein content compared to Group 4 (1947–1955) and Group 5 (1959–1974); apparently this is the result of breeders' efforts to raise the protein content in combination with higher yielding levels.

The general decrease in protein content was compensated for by the improved protein composition of the modern varieties over the older ones. The average quality score (according to Pogna et al., 1989), based on the HMW glutenin composition of the protein, ranged from 5.5 in Group 0 to 10.9 in Group 6. The improved quality score was confirmed by increased alveograph values W, P and P/L ratio. The SDS sedimentation test did not confirm the evolution of protein compostition; it only indicated a significantly lower quality of the cultivars in one group.

4. Gliadin and glutenin composition of Italian wheats

An impressive amount of research in this field has been carried out by Pogna and his co-workers at the ISC. As in many other countries, it marked an important step forward in breeding for bread-making quality.

The bread-making quality of Italian bread wheat may be evaluated on the basis of the analysis of the HMW glutenin subunit composition of 232 Italian cultivars carried out by Pogna et al. (1989). The compostition of 30 of these varieties is presented in Table I.9.

Twenty-three of the 232 cultivars analysed had a composition that appeared to be characteristic for Italian germplasm. The analyses revealed eight novel alleles that had not been previously described. For example, the variety

Table I.9. HMW glutenin subunit composition and quality score of 30 new and old Italian bread wheat varieties after Pogna et al. (1989)

Variety	Year of release	HMW glutenin subunit			Quality score[a]	*Glu-1* score
		Glu-A1	*Glu-B1*	*Glu-D1*		
Rieti	1900	N	20	3+12	5	4
Ardito	1916	2*	7+9	2+12	12	7
Mentana	1918	N	7*+8[b]	2+12	6	6
Villa Glori	1918	N	20	2+12	5	4
Frassineto 405	1927	1	7	2+12	7	6
Autonomia	1930	1	7*+8	2+12	9	8
San Pastore	1940	1	20	2+12	6	6
Impeto	1930	1	7*+8	2+12	9	6
Mara	1947	N	7	2+12	6	4
Jacometti	1948	1	7*+8	5+10	13	10
Produttore	1955	N	7*+8	2+12	8	6
Glutinoso	1957	N	20	2+12	5	4
Marimp 8	1959	N	7	2+12	6	4
Conte Marzotto	1959	1	20	2+12	6	6
Libellula	1962	1	20	2+12	6	6
Funo	1969	1	20	2+12	6	6
Demar 4	1969	N	6+8	2+12	5	4
Fiorello	1969	1	7*+8	5+12[c]	9	8
Irnerio	1970	1	7*+8	5+10	13	10
Orso	1972	1	20	2+12	6	6
Alpe	1974	2*	7*+8	2+12	11	8
		2	7*+8	5+10	15	10
		2	7+9	2+12	12	7
Mec	1974	1	7	2+12	7	6
Valle d'Oro	1975	1	6+8	2+12	6	6
Saliente	1979	1	7+9	5+10	14	9
Salvia	1981	1	20	2+12	6	6
Centauro	1983	1	7*+8	5+10	13	10
Manital	1983	2	17+18	2+12	13	8
Pandas	1983	1	7+9	2+12	10	7
Salmone	1985	1	7+9	2+12	10	5
Brasilia	1985	N	7+9	2+12	9	5

[a] Score according to Pogna et al. (1986); for comparison of this score (ranging from 4–17) with the Payne score (ranging from 3–10), the reader is referred to chapter Spain (Section 5, Table SP.9).
[b] The subunit pair 7*+8 has been given the same value as 7+8.
[c] The subunit pair 5+12 has been given the same value as 2+12, based on SDS sedimentation test.

Fiorello contains the unique subunit pair 5+12 encoded at the *Glu-D1* locus. Moreover, an intense band 7∗, with a faster mobility than band 7, was found in several older varieties in combination with subunit 8, i.e. as the subunit-pair 7∗+8.

Twenty-five cultivars, both old and new ones, were heterogeneous for HMW glutenin subunit composition. Pogna et al. suggested that this heterogeneity could provide an opportunity to improve the bread-making quality of a cultivar by choosing the type with good quality subunit composition for foundation seed production. Previous analyses of gliadin and glutenin patterns (Pogna et al., 1982) had shown that different genotypes with different gliadin patterns occured within the varieties Saliente and Alpe, both derived from crosses with the Russian variety Bezostaya-1.

A quality score based on the HMW glutenin subunits was given to each of the analysed cultivars. The score was based on the effect of individual bands on gluten quality as determined by the alveograph test. The score of a variety ranged from a minimum of 4 to a maximum of 17. Pogna et al. (1989) calculated that the average score of the Italian bread wheats that were registered in 1987 was 8.4. They considered the average Italian score to be rather low compared the mean score of cultivars grown in Canada (12.7), Australia (11.6), Finland (10.6) and England (7). (N.B. These scores must not be confused with the *Glu-1* scores calculated according to Payne et al. (1987), which are discussed in other chapters of this book). Pogna et al. (1989) suggested that the very common occurrence of the poor quality subunit pair 2+12 and the scarcity of good quality subunits such as 2∗ and 17+18 partly accounted for the low Italian average score. Whereas subunit 2∗ was present in old cultivars such as Ardito and Strampellino, subunits 17+18 (*Glu-B1*) have recently been introduced into Italian wheats from the Australian cultivar Mendos and the Indian cultivar Kalyan Sona, both of which inherited the allele from the Australian cultivar Gabo.

Pogna et al. suggested that in order to improve bread-making quality of future Italian wheat cultivars, it could be advantageous to select for good quality subunits 5+10, 17+18, 7+9, 7∗+8, 2∗ and/or 1. Moreover, it would be useful to introduce subunit 13+16, associated with good bread-making quality, into Italian germplasm.

The data in Table I.9 show that:

- Both registered and old Italian cultivars are characterized by the very common occurence of subunits 2+12. In contrast, most new cultivars contain the allelic subunits 5+10.

- Other *Glu-D1* encoded subunits, such as 3+12, 4+12 and 2+11, are very rare or absent, a situation which favours bread-making quality.

- The highest score was found in the variety Alpe, apparently consisting of a mixture of lines; other high scores were encountered in the old varieties Ardito, Jacometti and Irnerio, and in the more recent varieties Saliente, Centauro and Manital. Surprisingly, the score of Salmone, described as 'the first Italian strong wheat' by Borghi et al. (1985) is not very high.
- A common subunit combination 1,20,2+12 is encountered in the varieties San Pastore, Marzotto, Libellula, Funo, Orso and Salvia. Surprisingly Marzotto was famous for its excellent bread-making quality, whereas Orso is a high yielding biscuit quality wheat. In this respect the influence of genetic background, discussed in the following paragraph, is considered to play a decisive role.

Brunori et al. (1989) and Brunori (1991) studied the correlation between different quality parameters in the progenies of the crosses Kalyan Sona × Pusa 5–3 and Kalyan Sona × F26–70. Their data suggested that it is feasible to obtain high dough strength (W of Chopin) by concentrating on favourable combinations of HMW subunits of glutenin. These combinations must, however, be present 'in a genetic background capable of expressing high glutenin/gliadin ratios'. Furthermore they stated that, in the presence of high W values, high protein content of the flour was necessary to maintain the P/L ratio within the range 0.4–0.6 required for bread making.

The authors suggested that instead of selecting for increased W values, wheats should be developed with sufficient dough strength W to suit bread-making, combined with a high glutenin/gliadin ratio and high protein content, thus minimizing the fluctuation in bread-making quality, particularly the P/L ratio.

A summary of the change in protein quality achieved by Italian breeders was given by Borghi et al. (1991a), who analysed the frequency of different HMW glutenin subunits in Italian varieties bred since 1900. The varieties were divided in seven groups, representing seven generations of breeding and a reference group, 0. The results are presented in Table I.10.

Comparing the ratio 'good' to 'poor' subunits in 1900 and 1991, the progress that has been made is evident. Apparently the range and distribution of HMW glutenin subunits in Italian wheats was rather limited at the beginning of the 20th century. The variation in bread-making quality was mainly based on subunits present on chromosome 1A.

The good quality subunit 2* was already present in Ardito and the introduction of other good quality subunits 1, 7+9, 17+18 and 5+10 with varieties from USSR, Canada, Australia and Central America marked the noticeable improvement of bread-making quality of modern Italian bread wheat varieties. As it has been said before, the subunit 5+10 was introduced in the cultivar Jacometti (released in 1949) from the Canadian variety Manitoba.

Table I.10. Frequency (%) of HMW glutenin subunits and quality score in seven generations of bread wheat cultivars bred in Italy since 1900

Group	Year	Chromosome 1A encoded subunits		Chromosome 1B encoded subunits		Chromosome 1D encoded subunits		Average quality score
		Poor	Good	Poor	Good	Poor	Good[a]	
0	1900	50	50	100	–	100	–	5.5
1		86	14	57	43	100	–	5.6
2		50	50	25	75	100	–	8.1
3		43	57	57	43	96	4	7.2
4	1960	40	60	70	30	93	7	
5		36	64	67	33	94	6	7.3
6	1979	35	65	63	37	79	21	8.4
7	1991	21	79	22	78	59	41	10.9

[a] Subunits classified as 'poor' are N on *Glu-1A*; 20, 7 and 6+8 on *Glu-1B*; and 2+12 on *Glu-1D*; classified as 'good' are 1 and 2* on *Glu-1A*; 7+8, 7+9 and 17+18 on *Glu-1B*; and 5+10 on *Glu-1D*.
Source: Borghi et al. (1991a).

5. Concluding remarks

Looking upon the evolutionary trend of bread wheat cultivation and production in Italy, it appears that now, at the end of the 20th century, an increasing amount of Italian wheat is marketed according to its technological value and farmers are paying more attention to quality as a consequence of the higher reference price for high quality wheats. The only way for a bread wheat crop to survive in Italy is linked to the possibility of obtaining grain with qualitative characteristics substantially better than those from Central or Northern Europe (Borghi et al., 1991b).

However, although Italy may be favoured from a quality viewpoint due to its climate, it has difficulty to compete with Central European countries with cool, humid summers as far as productivity in terms of yield is concerned. After substantial efforts have been put into the improvement of bread-making quality of Italian wheat, its yielding capacity may be in need of simultaneous attention. This may be one of the reasons why some Italian breeders have put relatively much energy into the research of hybrid wheat varieties. These may contribute to circumventing the negative correlation between a high level of quality and good yielding capacity. Borghi et al. (1989) made a study of the combining ability in 100 F_1 hybrids of bread wheat and Perenzin et al. (1992) reported the results of a study on the combining ability for bread-making quality in wheat. One of their conclusions was that 'although further information on the economic implications of the introduction of hybrid wheat into practical agriculture is needed, it seems that satisfactory bread-making properties combined with high yields can be obtained with F_1 hybrids from crosses between high yielding cultivars and good-quality cultivars.'

D.A.D.

Acknowledgment

The author wishes to thank Prof. Dr Basilio Borghi (Istituto Sperimentale per la Cerealicoltura, San Angelo Lodigiano) for his valuable comments on this chapter and for his help in providing additional literature on wheat breeding in Italy.

References

Anonymus (1937) Notes sur l'agriculture italienne et sur l'organisation agricole cooperative. Report 17th International Agricultural Congress in The Hague, Netherlands. Rome: S.A. Arte della Stampa.

Annali del' Istituto Sperimentale per la Cerealicoltura 1970–1992.

Arcioni S, Piano E, Briganti Giannoni MG, Lorenzetti F (1974) Proteine e forza della farina in relazione ai genotipi della pianta e dell'endosperma in varietà e incroci di *Triticum aestivum* L. Genetica Agraria 28: 177–190.

Benvenuti A (1965) Contributo al progresso della cerealicoltura nazionale. Sementi Elette 11(4): 313–316.

Boggini G, Bonali F, Borghi B, Stanca M (1973) Valutazione agronomica e tecnologica di nuove linee di grano tenero. Annali ISC 4: 123–132.

Boggini G, Pogna NE (1979) Prove di panificazione secondo la metodica communitaria. Molini d'Italia 30(11): 487–491.

Boggini G, Cardillo F, Zangara L (1979) Prove tecnologiche su frumenti teneri. Sementi Elette 25(4): 11–16.

Borasio L (1934) Il metodo Pelshenke per l'esame rapido della qualità dei frumenti. Plant Breeding Abstracts 5: 114.

Borghi B, Boggini G, Corino L (1975) Breeding for quality in common wheat, I: Early selection in F_2 and F_3 generations. Cereal Res Comm 3(3): 205–214.

Borghi B, Boggini G, Cattaneo M, Corbellini M, Corino L (1978) Effect of early selection for bread-making quality on subsequent generations in soft winter wheat. Proceedings 5th International Wheat Genetics Symposium, New Delhi, pp 836–844.

Borghi B, Boggini G, Cattaneo M, Corbellini M, Pogna NE, Dalli A, Bianchi A (1985) Salmone: il primo grano di forza italiano. Sementi Elette 31(6): 3–6.

Borghi B, Perenzin M, Nash RJ (1989) Combining ability estimates in bread wheat and performance of 100 F1 hybrids produced using a chemical hybridising agent. J Gen Br 43: 11–16.

Borghi B, Corbellini M, Perenzin M, Pogna NE (1991a) Breeding for high quality bread wheats in southern Europe: results and perspectives. Proc Eucarpia Congress 20: 278–283.

Borghi B, Cattaneo M, Corbellini M, Perenzin M, Pogna NE (1991b) The present and the future of wheat production in Italy. Proceedings of III days Kromeriz Cereals, Kromeriz, Czechoslavakia.

Borghi B, Giordani G, Corbellini M, Vaccino P, Guermandi M, Toderi G (1995) Influence of crop rotation, manure and fertilizers on bread making quality of wheat (*Triticum aestivum* L.). Eur J Agronomy 4(1): 37–45.

Borghi B (1997) Personal communication, Wageningen.

Borghi B, Corbellini M, Minoia C, Palumbo M, di Fonzo N, Perenzin M (1997) Effects of Mediterranean climate on wheat bread-making quality. Eur J Agronomy 6: 145–154.

Brunori A, Figueroa A, Micke A (1984) Strategy of breeding for high protein content in *Triticum aestivum*. In Cereal-grain-protein-improvement. Vienna: International Atomic Energy Agency, pp 321–331.

Brunori A, Galterio G, Zannettino C, Pogna NE (1989) Bread-making quality indices in *Triticum aestivum* progenies; implications in breeding for better bread wheat. Pl Br 102: 222–231.

Brunori A (1991) Breeding for breadwheats in Italy. Plant Breeding Abstracts 61: 1336.

Canevara MG, Romani M, Corbellini M, Perenzin M, Borghi B (1994) Evolutionary trends in morphological, physiological, agronomical and qualitative traits of *Triticum aestivum* L. cultivars bred in Italy since 1900. Eur J Agronomy 3 (3): 175–185.

Ceccarelli S, Lorenzetti F., Briganti Giannoni GM (1971) Analisi della variabilità genetica per la 'forza della farina' in sette varietà di *Triticum aestivum* L. Genetica Agraria 25: 1–16.

Corino L, Boggini G, Bonali F, Borghi B (1975) Breeding for quality in common wheat II: Response to selection in early generations. Cereal Res Comm 3 (3): 137–147.

D'Amato F (1989) The progress of Italian wheat production in the first half of the 20th century: the contribution of breeders. Agr Med 119: 157–174.

Dal Belin Peruffo A, Pogna NE, Pallavicini C, Pegoraro E, Mellini F, Bianchi A (1984) Diagrammi elettroforetici delle gliadine e chiave di identificazione delle varietà di grano tenero iscritte nel Registro delle Varietà. Sementi Elette 30: 1–29.

Kökény B (1959) Ricerche sul metodo di valutazione della qualità del frumento secondo Zeleny; confronto fra i metodi Pelshenke e Zeleny. Sementi Elette 5 (1): 28–42.

Lupton FGH (1992) Changes in varietal distribution of cereals in central and western Europe. Agro-ecological Atlas of Cereal Growing in Europe, vol IV, 161 pp. Wageningen: Pudoc.

Maliani C (1979) Nazareno Strampelli, a forerunner in green revolution. Genetica Agraria 33: 1–14.

Perenzin M, Pogna NE, Borghi B (1992) Combining ability for bread-making quality in wheat. Can J Pl Sc 72: 743–754.

Perenzin M, Minoia C, Borghi B (1996) Scelta delle varietà di frumento tenero. L'Informatore Agrario 33: 4–11.

Pogna NE, Dal Belin Peruffo A, Boggini G, Corbellini M (1982) Analysis of wheat varieties by gliadin electrophoregrams; (II) Nature, origin and quality of biotypes present in six Italian common wheat varieties. Genetica Agraria 36: 143–154.

Pogna NE, Mellini F (1986) Alla ricerca delle basi biochimiche e genetiche della qualità del glutine. Inform Agr 42: 65–66.

Pogna NE, Mellini F, Bretta A, Dal Belin Peruffo A (1989) The high molecular weight glutenin subunits of common wheat cultivars grown in Italy. J Gen Br 43 (1): 17–24.

Samoggia S (1985) Panorama varietale del frumento tenero e duro in Italia. Sementi Elette 33 (4): 3–10.

Samoggia V (1958) Ridimensionare, ma migliorare la qualità del frumento. Sementi Elette 4 (5): 32–36.

Samoggia V, Kökény B (1962) Ricerche sui metodi di analisi qualitativa dei frumenti teneri. Sementi Elette 8 (3): 12–28.

Scalfati MC, Leonori-Ossicini A (1971) Superfici e produzioni di frumento dal 1966 al 1971. Annali dell'ISC 2: 75–90.

Vallega J (1974) Historical perspective of wheat breeding in Italy. Extr Proc 4th FAO/Rockefeller Foundation Wheat Seminar, Tehran (Iran), pp 115–126.

Vavilov NI (1935) The origin, variation, immunity and breeding of cultivated plants (transl. Starr Chester K) Chronica Botanica 13 (1949), 364 pp.

Zheng D (1993) Use of Italian wheat varieties in China. Gen Res Crop Evol 40 (3): 137–142.

Switzerland

1. Introduction

Located between the mountain chains of the Alps and the Jura, only 10% of the total area (4 100 000 ha) of Switzerland (CH) can be used for arable cropping. More than half of this area consist of cereals. Although the area under cereals and wheat (Table CH.1) represents a very small part of the total area in Europe, Swiss wheat breeding deserves special attention with regard to bread-making quality. Breeders in Switzerland have given high priority to quality in their programmes since the early 20th century. The production of high quality wheat is stimulated by the government by price guarantees for farmers. Nevertheless, bread prices in Switzerland are the highest in Europe. Switzerland is practically self-supporting in providing bread wheat for its inhabitants.

Most of the wheat grown in Switzerland is sown as a winter crop. Spring wheat (10 000 ha in 1985) is mainly grown at higher altitudes in the Jura and the lower Alps. Spelt is also used for bread-making purposes and is still grown on considerable areas in parts of central Switzerland.

Land varieties have survived in Switzerland for a relatively long time, due to the combination of several valuable characteristics specific for Swiss conditions. Squarehead types did not find their way easily into Swiss varieties, mainly due to a lack of bread-making quality. Especially for winter wheat, for

Table CH.1. Area (1000 ha) and production (1000 t) of cereals and wheat (including spelt) in Switzerland since 1917

Year	Cereals		Wheat	
	area	production	area	production
1917	117		39	
1934–1938	122	277	85	196
1948–1952	172	446	98	260
1961–1965	177	582	109	355
1969–1971	173	683	100	378
1979–1981	172	843	88	409
1989–1991	210	1331	99	604
1995–1997	205	1548	103	665

Sources: Volkart & Wagner (1945); FAO statistics.

a long time foreign varieties could not compete in terms of quality, winter-hardiness and earliness with the traditional local varieties (Volkart & Wagner, 1945). Earliness and sprouting resistance are two important features required for wheat in Switzerland, especially for higher locations, where the weather in August may be wet and cold.

2. Organisation

The emphasis that is put on the quality of bread wheat in Switzerland is directly linked to the 'Getreidegesätz' (= wheat-law), dating from 1932. According to this law, which aims to stimulate the production of local bread wheat, the Swiss government guarantees the availability of seed of quality varieties, as well as a price for quality wheat that makes its production economically feasible. Moreover, the government guarantees to buy a certain amount of Swiss bread wheat.

Swiss varieties on the variety list are classified in four price categories:

I = excellent bread-making quality
II = medium – good quality
IV = biscuit making quality and
V = inferior quality

The difference between the price of Group I and Group V was 18% in 1980 and 10% in 1994 (Weilenmann, 1994).

The 'Getreidegesetz' is implemented through a close involvement of the state in breeding activities. As a matter of fact, two state breeding institutions, one in the French- and one in the German-speaking part of Switzerland have the monopoly on Swiss wheat breeding. They are:

- the 'Station Fédérale des Recherches Agronomiques' (RAC) in Lausanne (Mont Calme), founded in 1898; the main wheat-breeding activities of this station are undertaken at the Domaine de Changins in Nyon (Canton Vaud);
- 'Die Eidgenössische Forschungsanstalt für Agrarökologie und Landbau (FAL) founded in 1878 in Oerlikon and in 1968 moved to Reckenholz, close to Zürich.

These two stations have worked together closely in the field of wheat breeding since the early 20th century. Nevertheless, separate variety lists existed for the French- and German-speaking parts of Switzerland until World War II.

3. Wheat breeding before 1950

Wheat breeding during the first half of the century was restricted to winter wheat. Wheat-breeding activities in Lausanne started around 1900, where Martinet, the director of the research station, played a vital role in the initial years. The aim of his programme was to increase yielding capacity and to improve the straw of the traditional land varieties without, however, conceding on their quality, the high level of which he was well aware.

Martinet mainly used land varieties from the western part of Switzerland, e.g. Blanc du pays and Petit rouge du pays, but also Erlacher Landweizen. A selection of Erlacher Landweizen – Mont Calme 22 –, introduced in 1920, became an important variety and was frequently used as a parent by Martinet. From 1910 onwards he started to make systematic crosses, mainly with French cultivars. In time this resulted in Mont Calme 245 (MC 22/Blé hâtif inversable) and Mont Calme 268 (Carré vaudois/MC 22//Vuiteboeuf), two high yielding varieties that covered the major part of the Swiss winter-wheat area between 1928 and 1950 (Oehler, 1950). The aim to maintain the quality level of the old land varieties was not achieved, however, as both varieties were of medium quality. In order to obtain a product with a satisfactory quality level, they were often grown in mixture with high quality land varieties. Mont Calme 245, for example was commonly sown together with Bisnachter Weizen, a local variety with very good quality.

In the German-speaking part of Switzerland, Dr Volkart, the director of the FAP in Oerlikon, started in 1917, in close cooperation with local farmers, the selection of promising individual wheat plants from land varieties. As a result, 17 winter-wheat and two spring-wheat varieties were on the market in 1927. Like his colleague Martinet in Lausanne, Volkart was well aware of the risk of loosing quality while trying to increase quantity. During the 1920s he made several crosses with the local variety Plantahof as a parent. Plantahof had a remarkably high quality level, comparable to that of the Canadian spring-wheat variety Huron.

An analysis of the quality of Plantahof and Mont Calme 245, compared to Vilmorin 27, was carried out by Volkart & Wagner (1945). The French variety Vilmorin 27, yielding better than Plantahof and MC 245, was widely grown around Basel in those years. Vilmorin 27 and MC 245 had Hâtif inversable as a common parent. The dough yield, considered a measure of the output for the bakery, was lower for Vilmorin 27 and MC 245 than for Plantahof, caused by low gluten content in Vilmorin 27 and low gluten quality (Quellzahl) in MC 245 (Table CH.2). Despite its low gluten content, Vilmorin 27 gave higher bread volumes than Plantahof. However, bakeries valued Plantahof more than Vilmorin because it was easier to handle in practice ('The baker does not like

Table CH.2. Quality parameters of three winter-wheat varieties; average of 3 years (1939–1941) trial in one location

Quality parameter	Winter-wheat variety		
	Plantahof	MC 245	Vilmorin 27
Gluten content (%)	25.0	27.8	23.9
Quellzahl	8.8	1.1	9.3
Dough yield (%)	161.2	155.9	155.3
Loaf volume (cm^3)	603	465	620

Source: Volkart & Wagner (1945).

a difficult flourtype, which he has to monitor with a watch and thermometer in his hand'). Still, Vilmorin 27 could be considered as a valuable source for breeders as it combined high yields with good gluten quality. In this respect it was comparable to the German variety Tassilo.

The first important variety resulting from Volkart's crosses was registered in 1934. It was the winter-wheat variety Alpha, a cross between Plantahof and the Swedish variety Sol. Alpha was never grown on a large scale. A real breakthrough was the variety Probus, a cross between the Bavarian land variety Trubilo and Plantahof. Probus, introduced in 1948, was grown on 80–90% of the Swiss winter-wheat area between 1950 and 1970, and it remained on the Swiss variety list until 1990. The long-standing success of this variety was based on its combination of good quality characteristics and satisfactory yields. Huron, the most important spring-wheat variety grown in Switzerland until the 1950s, was used as a reference for outstanding gluten quality (Table CH.3).

Table CH.3. Characteristics of the most important winter-wheat varieties grown in Switzerland around 1950

	Plantahof	MC 245	MC 268	Probus	Huron
Yield	3.2	3.9	3.4	3.8	2.5
Wet gluten M	29.8	34.7	32.8	31.8	35.0
Pelshenke value (min.)	44.9	23.2	22.9	62.9	100.0
Brabender valorimeter	54.8	33.6	30.5	63.5	64.1
Hectoliterweight (kg)	79.8	81.3	81.5	81.0	82.0
1000 kernel weight (g)	45.2	49.2	43.2	45.9	37.0

Source: Oehler (1950).

Table CH.4. Quality characteristics (average of 1985 and 1986 harvests) and yield (average of 1984–1986 harvests) of some important Swiss winter-wheat varieties and Partizanka

	Protein content	Falling number	Wet gluten	Quellzahl		Farinogram		Extensogram		Yield	
				Q0	Q30	H_2O	KA[a]	DW/DB	cm^2	kg/ha	CHF/ha
Probus	14.8	310	32.1	16	8	76.2	123	1.0	56.0	5230	5857
Zenith	13.7	374	29.3	15	8	63.7	110	1.1	64.8	6060	6242
Eiger	14.7	379	33.2	14	6	68.5	108	1.0	55.7	6225	6660
Arina	14.6	343	32.2	13	7	68.7	105	1.0	59.2	6270	6625
Partizanka	14.4	328	30.1	21	14	66.3	98	2.3	73.4	5975	6390

[a] Konsistenzabfall (= dough softening).
Sources: Achermann & Rudin (1987); Achermann & Tièche (1987).

4. Wheat breeding after 1950

4.1. Winter wheat

Probus covered more than half of the Swiss winter-wheat hectarage between 1952 and 1975. During the long-lasting success years of Probus, breeders focused somewhat more on disease resistance, mainly for *Septoria nodorum* but also for rusts. Sprouting resistance as well received considerable attention, especially in the 1970s.

It was not until 1968 that a serious replacement for Probus was introduced in the form of the variety Zenith, a cross between Heine VII and C3842-3663, a Canadian winter-hardy spring wheat. Zenith was an improvement on Probus as far as yielding ability was concerned but not its quality: Zenith was classified in Group II.

A cross of Zenith with the Russian variety Bezostaya-1 resulted in the varieties Zenta, Eiger and Sardona, released in the years 1979–1980, all with improved baking quality compared to their father Zenith. All three varieties were very high yielding and classified in Group I but they never covered such vast areas as Probus, and Zenith.

The release of the variety Arina (Moisson/Zenith) in 1982 meant the end of the Zenith era. With a protein content comparable to Probus and Eiger, i.e. higher than Zenith and producing higher yields than the previous varieties (Table CH.4), Arina covered more than half of the Swiss wheat area for a period of 10 years.

It is typical for the Swiss situation that the winter-wheat varieties released since the beginning of the century until the 1980s have continued to be grown for many years and that they have covered vast areas. In other words; the market has been dominated by a limited number of varieties, as may be seen

Table CH.5. Proportions of the winter-wheat area covered with the main varieties in Switzerland between 1930 and 1994

	1930	1940	1950	1960	1970	1980	1984	1990	1994
MC 22	48	11							
Plantahof	23	18	1						
MC 245	6	37	59	7					
MC 268	2	24	19	2	3				
Probus			20	89	68	19	1		
Zenith					23	67	22	4	
Arina							45	71	52
other	21	10	1	2	6	14	32	25	48

Sources: Lupton (1992); Weilenmann (1994).

in Table CH.5. The larger number of varieties in use since the 1980s is related to a wider variation in quality required by industry.

For an adequate interpretation of the data in Table CH.5, note that the use of mixtures of varieties is quite common in Switzerland. The variety list even has a seperate recommendation list for mixtures.

Recent first-class quality winter-wheat varieties are Tambo (irradiation of line derived from Probus/Bankuti//Hoeser 52) released in 1986, Tamaro (Kormoran/NR72-837/Monopol), released in 1992, and Runal (Virtue//Hoeser 52/ Sturdy/3/Kavkaz//Zenith/Backa), released in 1995.

4.2. Spring wheat

A Swiss breeding programme for spring wheat was initiated in the 1950s. Until then, foreign varieties, such as the Canadian Huron and the Austrian Kärtner Frühweizen (very early – for higher altitudes) were grown. Later, other foreign varieties, such as Svenno, Walter and Kolibri, were also quite successful in Switzerland.

The first Swiss spring-wheat variety, released in 1963, was Relin, a mixture of 5 lines derived from a back-crossing programme of Lichti on Newthatch (N/L//L///L). Together with Huron, these two varieties resulted in several high quality spring-wheat varieties in succeeding years: Hinal, Ronega and Granat (Table CH.6).

After this first 'generation' of Swiss spring-wheat varieties, based on German and North American parents, the pedigrees of succeeding varieties included several South-American and Italian parents (Frontana, Redriver and Mentana). The most important Swiss spring-wheat varieties with first-class

Table CH.6. Characteristics of the first generation of Swiss spring-wheat varieties compared to Svenno (average of harvest 1967–1969)

	Yield	Protein content	Zeleny value	Bread volume[a]
Svenno	36.2	13.4	68	366
Relin	37.3	12.3	45	369
Hinal	34.4	12.6	42	370
Ronega	35.1	12.6	53	389
Granat	34.8	13.0	53	358
Average	35.6	12.7	52	370

[a] In a micro baking test.
Source: Weilenmann (1970).

Table CH.7. Pedigree of the most successful spring-wheat varieties that have been classified as first-class quality wheats

Variety	Year	Pedigree
Relin	1963	Lichti / Newthatch
Hinal	1963	Lichti // Huron / Newthatch
Ronega	1967	Heines Koga // Huron / Newthatch
Granat	1970	LNL / Garant // Huron / Newthatch
Lita	1972	Fasan B114
Tano	1972	Probat / Kentana 54B // Probat
Albis	1984	Hermes / 90847 // Kentana 54B / B564
Remia	1987	selection of MG75Z1OO (bred by use of male sterility)
Lona	1991	Redriver / Walter
Greina	1994	94096/93964
Balmi	1994	93884/93940

Sources: Mitteilungen für die Schweizerische Landwirtschaft 1965–1988; Landwirtschaft Schweiz 1988–1993; Agrarforschung 1994–1995.

quality are listed in Table CH.7. The most commonly grown spring-wheat variety in 1995 was Lona, frequently used in mixtures with Greina and Balmi.

Although spring-wheat varieties with winter-hardiness have replaced some winter wheat during the 1970s, apart from that period spring wheat has never become an important crop in Switzerland (Weilenmann, 1994). Farmers have been warned against the sowing of spring wheat in autumn or early winter (to obtain an early crop and higher yields) because of the negative effect on bread-making characteristics (Müller, 1977).

Since the 1980s, androgenetic doubled haploids have been commonly used in spring-wheat breeding. (In winter-wheat breeding this method was introduced at a somewhat later stage.) With respect to quality characteristics, such as protein level and grain hardness, doubled haploid lines appeared to have a slight advantage over lines selected in a pedigree system (Winzeler et al., 1987). For most of the agronomical characteristics there was no such advantage reported, apart, of course, from the gain in time.

5. Quality assessment

To illustrate the importance of quality, the number of analyses carried out in Switzerland for the classification of wheat varieties, compared to Germany, Austria, the United Kingdom and France is presented in Table CH.8. Although there have been several changes since 1981, the comparison is still representative of the situation in the 1990s. The Quellzahl (swelling number)

Table CH.8. Importance of quality characteristics used for the classification of bread wheat in Switzerland compared to Germany (FRG), Austria (A), the United Kingdom (UK) and France (F)

Quality characteristics	CH	FRG	A	UK	F
Protein	••	••	o	•	•
Zeleny	•	o	o		o
Wet gluten	••		••		
Quellzahl Q0	•		••		
Q30	•				
Farinogram:					
water absorption	••	o	o	o	o
dough characteristics	•	o	o	o	o
Mixogram				o	
Extensogram	•	o	o	o	
Alveogram					••
Amylogram	o	o	o	o	o
Baking test in a tin	••	o		••	
Baking test on floor	••			••	
Rapid mix test	••	••	••		
Mixed baking test		••	o		

o = analysis carried out; • analysis used for classification.
Source: Brönnimann (1981).

is analysed as a measure for gluten quality: Q_0 is the Quellzahl of gluten

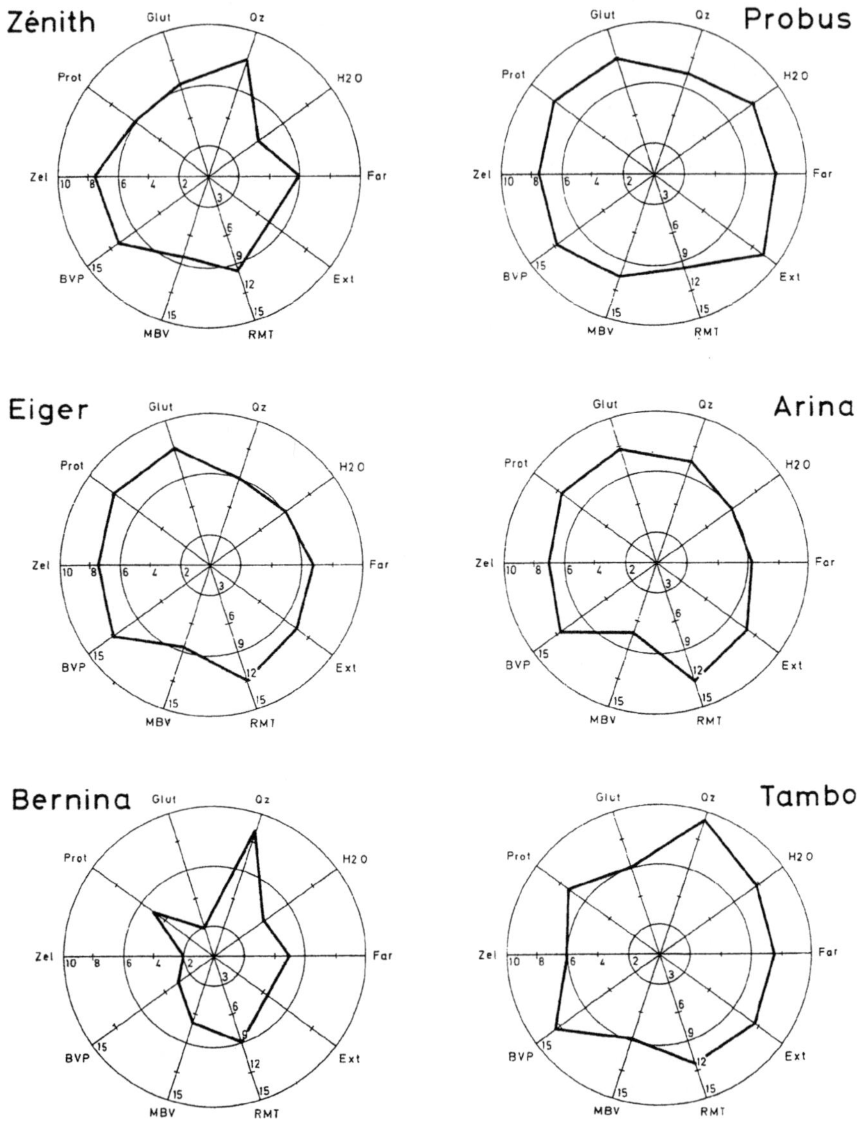

Figure CH.1. Graphical representation of the quality characteristics of six Swiss winter-wheat varieties; twelve characters are represented in each diagram; the distance to the centre is positively correlated to the value of the character.
Zel = Zeleny test; Prot = protein content; Glut = wet gluten; Qz = Quellzahl; H_2O = water absorption; Far = farinogram; Ext = extensogram; RMT = rapid mix test; MBV = baking test in a tin; BVP = sheath baking test CIBM Pully.
Source: Fossati et al., 1986.

that has been washed out immediately after dough preparation; Q_{30} is the Quellzahl of gluten that has been washed out 30 minutes after preparation of the dough.

A new classification system was introduced in 1990. Compared to the analyses presented in Table CH.8, the amylogram value and the falling number are included in the calculation of the quality score; the mixed baking test was skipped. Material with a falling number lower than 180, is excluded as quality wheat.

Based on the analyses in the laboratory and the results of different baking trials, wheat varieties are given a score with a maximum of 200 points. A detailed description of this quality classification system was given by Saurer et al. (1991).

Swiss varieties are described by a graphical presentation of their quality characteristics. Most of the parameters are represented as a score between 1–10. Only the rapid mix test (RMT), the baking test in a tin (MBV) and the sheath baking test (BVP) have scores between 1–15 (Saurer et al., 1982). As an illustration, the quality diagrams for five bread-making varieties (Probus, Zenith, Eiger, Arina and Tambo) and one biscuit wheat variety (Bernina) are presented in Figure CH.1.

Analyses of HMW glutenins and gliadins have been carried out at the FAP in Reckenholz (Tätigkeitsberichte 1986/7) but varietal data have not been published.

D.A.D.

Acknowledgment

The author wishes to thank Dr A. Brönnimann (Swiss Federal Research Station for Agroecology and Agriculture, FAL, Zürich-Reckenholz) for reviewing this chapter.

References

Achermann J, Rudin P (1987) Qualität der Weizenernte 1986. Die Grüne 115(2): 9–13.

Achermann J, Tièche JD (1987) Neue Getreidesorten im Streifenanbau. Die Grüne 115(49): 9–11.

Brönnimann A (1982) Möglichkeiten und Aussichten zur Optimalisierung der Erträge. Schweizerische Landwirtschaftliche Monatshefte 60(2): 25–48.

Fossati A, Paccaud FX, Weilenmann F, Winzeler H, Saurer W, Fried PM, Jaquiery R (1986) Tambo, a bread wheat variety produced by mutation. Revue Suisse d'Agriculture 18(5): 267–271.

Müller H (1977) Gefährdet der Einsatz von ertragreichen, aber qualitativ schlechten Weizensorten das zukunftige Anbauprogramm von Brotgetreide? Schweizerische Landwirtschaftiche Monatshefte 55(2): 33–45.

Oehler E (1950) Die Züchtung der Getreidearten und die Produktion und Anerkennung von Getreidesaatgut in der Schweiz. DLG-Nachrichten für Pflanzenzucht, Lausanne.

Saurer W, Rudin PM, Achermann J, Pouly E (1981) Ein neues Bewertungsschema für die Beurteilung der Backqualität von Brotgetreidesorten; Resultate der Hauptversuche 1981. Mitt Schweiz Landw 30(9): 185–199.

Saurer W, Achermann J, Tièche JD, Rudin PM, Mändli K (1991) Das Bewertungsschema '90 für die Qualitätsbeurteilung von Weizenzüchtungen. Landwirtschaft Schweiz 4(1/2): 55–57.

Volkart A, Wagner S (1945) Getreide Züchtung: Qualitäts- oder Quantitätszucht? In Späti W (1945) 25 Jahre Saatzuchtgenossenschaft bei der Basel, Liestal, pp 30–38.

Weilenmann F (1970) Ergebnisse der Sortenversuche bei Sommerweizen. Mitt Schweiz Landw 18(9): 153–160.

Weilenmann F (1984) Zur Züchtung und Stellung des Brotgetreides in der Schweiz. Mitt Schweiz Landw 32(4): 81–88.

Weilenmann F (1994) Sortenentwicklung und Sortenstrategie bei Getreide in der Schweiz. Bericht über die 45. Arbeitstagung der Arbeitsgemeinschaft der Saatzuchtleiter im Rahmen der 'Vereinigung Oesterreichicher Pflanzenzüchter', BAL Gumpenstein, 22–24 November 1994.

Winzeler H, Schmid J, Fried PM (1987) Field performance of androgenetic doubled haploid spring-wheat lines in comparison with lines selected by the pedigree system. Plant Breeding 99: 41–48.

Austria

1. Introduction

The written history of the Austrian baking industry goes back to the beginning of the 19th century, when the Viennese bakery had a special reputation in Europe. This reputation was partly based on the technical perfection of the mills built in the Austrian-Hungarian empire (Glattes, 1978). An important Austrian contribution to baking technology was the introduction of the 'Presshefe' in 1850. The Austrian brewer Mautner was the first to produce a yeast for use by the bakery that took the place of beer yeast which had been used until that time. The industrial production of this pressure yeast spread from Vienna to the rest of the world under the name 'Wiener Verfahren' (Vienna Process).

The production of wheat was of minor importance in the territory of Austria (A) at the beginning of the 20th century compared to the vast amounts grown in the Hungarian part of the Austrian-Hungarian empire. With their separation into two countries after World War I, Austria had to increase its wheat production in order to become self-sufficient. The expansion of the wheat area took place mainly at the expense of rye and oats, which were the two main cereals at the beginning of the century. Moreover, considerable progress was made in increasing yields per hectare, due to improved husbandry and cultivars. A further increase of the production of high quality bread wheat was enhanced by a pricing policy, initiated in 1953, with a strong price differentiation between bread wheat and rye and other feed cereals. Whereas in the early 1960s Austria was an importer of wheat, the country has become more than self-sufficient (170% in 1993) and now exports quality wheat. As a result of a well organised production system, based on contracts between wheat growers and the wheat industry, the quality of Austrian wheat is among the highest in Europe. On the other hand, average wheat yields are somewhat lower than in many Western European countries, due to a relatively high proportion of quality wheats and a modest application of N fertilizers, growth regulators and fungicides (Oberforster & Werteker, 1995).

The production figures of wheat presented in Table A.1 show that average yields have tripled over the past fifty years. The area under barley is similar to that of wheat, varying between 250 000 ha and 300 000 ha. Wheat covered 17% of the arable land or 30% of the total cereal area of Austria, in 1996. Rye and oats together accounted for less than 100 000 ha in the same year. Most of the wheat grown in Austria is winter wheat; the area of spring wheat

Table A.1. Wheat acreage (1000 ha) and annual production (1000 t) in Austria

Year	Wheat	
	area	production
1924–1925	195	261
1934–1938	250	417
1948–1952	204	384
1961–1965	276	704
1969–1971	279	912
1979–1981	271	1 025
1989–1991	276	1 381
1995–1997	255	1 277

Sources: IIA and FAO statistics.

varies between 1% and 5% of the total wheat area. In addition some 10 000 ha of durum wheat has been grown annually since the 1960s.

Two-thirds of Austria is dominated by the Alps, so that wheat production is limited to land bordering the mountains in the eastern part of Austria, encompassing three regions each with a different micro-climate. The most important area for wheat growing is the Pannonic region[1] in the northeast, where almost half of the Austrian wheat is produced. Traditionally, this semi-arid region has produced wheats with a high quality level. Adam (1993) even suggested that the 'roots' of the Canadian variety Marquis, which has been used as a quality parent all over the world, through its parent Red Fife, were to be found in this region. The second most important wheat producing region lies in the upper Austrian foreland of the Alps, where the climate is influenced by the Baltic sea. The third most important wheat growing area are the plains and hills of lower Austria, including the province Burgenland, with its more humid climate.

Wheat breeding in the Pannonic region has obviously been greatly influenced by Hungarian wheats, whereas German varieties have had more impact on wheat breeding in the Baltic region. Yugoslav and Italian wheats (Sava and Libellula, respectively) have been grown on the southeastern plains and hills of Austria. Efforts to breed wheat varieties equally suitable for all Austrian cropping sites have not been successful so far (Oberforster & Werteker, 1995).

[1] Named after the Roman province Pannonia which covered eastern Austria and Western Hungary, bounded by the Danube, the foothills south of Sava and a line drawn south from about Vienna (The Oxford Atlas, 1951).

2. Wheat breeding before 1945

The earliest years of organised wheat breeding in Austria are related to the person of von Tschermak, who acquired world fame as a botanist and as one of the rediscoverers of Mendel's laws at the turn of the century. He was a professor of plantbreeding at the 'Hochschule für Bodenkultur' in Vienna between 1906 and 1941. The plant breeding department of this university, which was integrated into the Institute of Agronomy and Plant Breeding in 1948, had a large impact on Austrian wheat breeding during the first half of the 20th century. The same accounts for the Vienna Station of Seed Control founded in 1881, named Bundesanstalt für Pflanzenbau und Samenprüfung in 1895 (and changed into Bundesamt und Forschungszentrum für Landwirtschaft in 1996).

Through combination breeding von Tschermak created several wheat varieties, such as Tschermak's weiszer begrannter Marchfelder and Tschermak's Burgenländer Winterweizen. Many of the crosses carried out by von Tschermak are presumed to be based on the Hungarian Banat wheat (Adam, 1993).

On the whole, hybridization was not applied on a large scale by Austrian wheat breeders before 1945, and most of the wheat consisted of improved land varieties. The characteristic land varieties grown in the Pannonic region were early maturing, strong tillering, tall wheats with a high level of cold and drought tolerance. Fifty percent of the Pannonic wheat hectarage in the 1940s and 1950s was covered with the cultivar Austro Bankut, selected from the Hungarian variety Bankut. Other cultivars grown in this part of the country were Kadolzer and Tschermaks Marchfelder. Typical land varieties in the Baltic region were Manker Kolben and Marienhofer Kolben, and later the improved cultivar Ritzlhofer (Hron, 1981).

3. Wheat breeding after 1945

3.1. The 'Qualitätsweizenaktion'

The Qualitätsweizenaktion, hereafter referred to as the Quality Action, was initiated by the Austrian government in 1954 to stimulate the production of quality wheat. Wheat growers in the Pannonic region were contracted to grow certain varieties under specific conditions and would receive a premium for the resulting product. Only those wheat varieties that met a fixed standard of quality were included in the action. Seed of these varieties was provided by the seed firm for a reduced price. From 1958 onwards the Quality Action was extended to include a system of separate storage of wheat batches.

The definition of quality was based on the Wertzahl (W). This parameter was calculated according to the formula: $W = 2K + 3Q_0$ where K is the gluten content and Q_0 is the swelling index (Quellzahl). Compared to the Gütezahl used in Germany ($G = 25K + 100Q_0 + 50P$), the Wertzahl did not include the Pelshenke value. Moreover the relative importance of the gluten content was higher in the Wertzahl than in the Gütezahl. Hänsel (1963) pointed out that this meant that breeders could not confine themselves to selecting for a high swelling index (Q_0), a characteristic which he considered to be relatively easy to achieve and which was less influenced by environmental variation than the gluten content. The quality classes defined under the Quality Action of 1954 are included in Table A.2.

Initially Austro Bankut was the most important variety that contributed to the Quality Action. From the mid-1950s on, the place of Austro Bankut was taken by varieties with a lower gluten content but higher gluten quality, such as Stamm 101 and Record. These were replaced in the 1960s by Erla Kolben and Probstdorfer Extrem, cultivars with improved straw quality and disease resistance. Since then a large number of cultivars have been available for the Quality Action, which has continued to be effective until today.

Under the influence of a new market economy, adapted to EU regulations (Austria joined the EU in 1995), a new quality classification system was introduced in 1994. The classification of wheat cultivars within this system was based an a large number of parameters (see footnote Table A.2). Moreover, wheat for export was analysed by a Chopin alveograph (Oberforster et al., 1994). Within the new system the production of quality wheat was no longer limited to the Pannonic region but could take place in any region as long as its quality met a fixed standard.

The premium paid for wheat under the Quality Action has varied widely over the years, from 10 to 40% compared to wheat of medium to low quality, respectively (Hänsel et al., 1994). In 1980, for example, the price of quality wheat was 19% above the price of normal milling wheat; in 1995 the compensation for quality was far less attractive (Oberforster & Werteker, 1995).

3.2. Breeding for improved bread-making quality

3.2.1. The semi-arid (Pannonic) region

By far the most important wheat-breeding company for this part of Austria is the Probstdorfer Saatzucht, located in the neighbourhood of Vienna. Another private company working in the same region is Neuhof Rohrau Pflanzenzucht. The work of these companies is supported by research carried out at the Federal Office and Research Centre of Agriculture (Bundesanstalt für Pflanzenbau) and the Agricultural University in Vienna.

Table A.2. Definition of Austrian quality wheat classes in different periods of time

	Wertzahl	Minimum value set:		
		wet	swelling index	
		gluten (%)	Q_0	Q_{30}
before 1954				
Quality wheat			25	18
1954–1962				
A1	≥126	25	10	35
A2	115–125	25	10	35
A3	104–114	25	10	35
B	76–103	20	6	–
C	≤75			
1962–1994				
A1	≥118	28	14	[a]
A2	108–117	26	12	
After 1994[b]				
7–9	–	31.2	–	

[a] The gluten degradation (Kleberabbau) between Q_0 and Q_{30} was not allowed to exceed 35%.

[b] The classification system introduced in 1994, including 9 quality groups with the highest quality in group 9, was based on four indirect parameters (protein content, wet gluten content, Zeleny sedimentation value and falling number), three rheological characters (quality number from a Brabender farinogram, water absorption and dough area from a Brabender extensogram) and the results of a rapid mix test; swelling index was no longer included in the classification.

Sources: Fuchs (1954); Hänsel (1960, 1963); Oberforster et al. (1994).

Starting in the mid-1950s, Austro Bankut was gradually replaced by cultivars with a medium gluten content but better gluten quality. Although the gluten content of Austro Bankut was very high, its low swelling index, which was exacerbated by increased levels of N application, sometimes resulted in a Wertzahl that was not high enough to meet the standard set by the Quality Action (Meinx, 1977). Moreover its yielding capacity needed improvement.

Stamm 101, a cross between a Hungarian land variety of the Tisza type and the French cultivar Bleu Dome, was released in 1955. Apart from a higher yielding potential than Austro Bankut, it also achieved higher quality levels. The somewhat lower gluten content was compensated by a much higher swelling index than that of Austro Bankut. Still, it needed improvement in frost tolerance and straw strength. Professor Hänsel at the Probstdor-

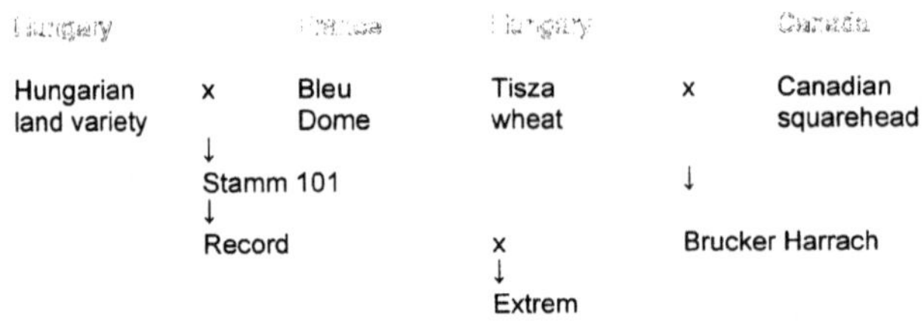

Figure A.1. The pedigree of Probstdorfer Extrem, a Class I winter-wheat cultivar released in 1968.

fer Saatzucht selected a line of a similar quality level but improved frost and lodging resistance that was released in 1958 as the cultivar Record. Its popularity spread rapidly, such that between 1960 and 1970 it covered 80% of the austro-pannonic wheat area. The second important variety was Erla Kolben, released in 1962 by Kärntner (see Section 3.2.2).

With the introduction of the use of CCC, the emphasis in breeding shifted from straw characteristics to yielding capacity, whilst preserving at least the same quality level. When the production of Record was at its height, in 1968 Probstdorfer Saatzucht released the variety Extrem, which became even more popular. It covered more than half the Austrian wheat hectarage in the mid-1970s and was grown for a period of 30 years. The high yields obtained with Extrem demonstrated that it was possible to combine a high quality and yielding level within one variety (Hänsel, 1970). The pedigree of Extrem (Figure A.1) demonstrates that its Hungarian 'blood' is inherited from two sides.

Probstdorfer Extrem has been used frequently as a parent in quality breeding programmes, so it may be found in the pedigree of several high quality varieties. The Probstdorfer cultivars Karat and Perlo, both selected from a cross between Extrem and Bezostaya-1, belonged to the first group of Austrian semi-dwarf varieties. Perlo was grown on one-third of the total Austrian wheat area during the 1980s (Lupton, 1991). More recent varieties derived from Extrem are Martin and Josef (both from Probstdorfer Saatzucht), and Brutus (from Saatbau Linz). The pedigree, year of release and quality characteristics of these cultivars are presented in Table A.3.

Oberforster et al. (1995) pointed out that Bezostaya-1, Extrem and Pokal were the most successful parents of modern Pannonic wheat varieties. Probstdorfer Pokal (Diplomat/Purdue 5517//Diplomat) was known for its outstanding quality and was grown both in Pannonic and in Baltic regions, albeit not on a large scale.

3.2.2. The Baltic region

During World War II a national institute in Admont, directed by Isenbeck, was entrusted with the breeding of wheat for alpine conditions. The material

Table A.3. Austrian-bred wheat varieties with their year of registration, pedigree, HMW glutenin composition and quality score

Cultivar	Year	Pedigree	BMQ score	HMW glutenin composition A1	B1	D1	*Glu-1* score
Austro Bankut	1948	selection from Bankuti 1202					
Lassers Dickkopf	1950	Thatcher/Svalöfs Kronen					
Stamm 101	1953	Hungarian l.v.(Tisza)/Bleu Dome					
Admonter früh	1953	Thatcher/Svalöfs Kronen					
Record	1958	selection from Stamm 101					
Erla Kolben	1962	Admonter früh/Stamm 101	9	1	7+9	5+10	9
Extrem	1968	Record/Brucker Harrachweizen	7	N	7+9	5+10	7
Karat	1977	Extrem/Bezostaya-1	7	N	7+9	5+10	7
Perlo	1979	Extrem/Bezostaya-1	8	2*	7+9	5+10	9
Agron	1981	Neuhof nr.I/Artemovka/Bezostaya-1	8	2*	7+9	5+10	9
Martin	1983	HP 35719/Extrem	8	N	7+9	5+10	7
Hubertus	1986	Tassilo/Carsten//Tassilo 502	5	1	6+8	5+10	8
Amadeus[a]	1986	Pokal[b]/Kavkaz	8	2*	7+9	5+10	9
Capo	1990	Martin/Pokal	7/8	1	7+9	5+10	9
Leopold	1993	Pokal/Karat	7	N	7+9	5+10	7
Josef	1994	Extrem/HP35719//Pokal/3/Perlo	7	2*	7+9	5+10	9
Brutus	1994	Agron/Extrem	7	2*	7+9	5+10	9
Alidos	1995	Arkos/Hadmerslebener 914.76	8	N	17+18	5+10	8
Exquisit	1995	Pokal/Agron	8	1	7+9	5+10	9
Spartakus	1995	Perlo/Extrem/Bezostaya-1	8	2*	7+9	5+10	9

[a] Amadeus was noted to have a 1B/1R rye translocation.
[b] the pedigree of Pokal is Diplomat/Purdue 5517//Diplomat.
Source: Hänsel et al. (1994), Oberforster & Werteker (1995); Gröger et al. (1997b, 1998).

from this institute later formed the basis of the Kärntner wheat breeding programme of Dr Lasser. The plant breeders' annual meeting, first held in Admont and later in Gumpenstein, has been an important exchange-platform for breeders within Austria, as well as surrounding countries, since the early 1950s. Today, the most important private wheat-breeding station in this region is Saatbau Linz.

During the first half of the 1950s, 50% of the Baltic region was cropped with the Bavarian variety Tassilo. It was gradually replaced by the cultivars Lassers Dickkopf, Admonter früh, Hubertus and Drauhofener Kolben (Hron, 1981). Lassers Dickkopf and Admonter früh were both selected by Dr Lasser in Admont from a cross between the North American spring wheat Thatcher (with Marquis as a quality parent) and the Swedish cultivar Kronen. A cross between Admonter früh and Stamm 101 resulted in the cultivar Erla Kolben, released in 1962. This variety with a high quality level comparable to that of Probstdorfer Record, but shorter straw and improved resistance to stem rust, was grown mainly in the Pannonic region. Erla Kolben remained a standard variety for quality and was the cultivar with the highest quality score on the 1995 variety list of Austria.

In the 1960s the German cultivars Jubilar and Diplomat and the Swiss cultivar Probus were widely grown in the Baltic region. They were replaced by the Austrian cultivars Multiweisz and Linzerbraun, semi-dwarf varieties with improved straw quality but an unsatisfactory baking quality (Hron, 1981). A combination of Diplomat and Multiweisz resulted in the cultivars Danubius and Önus, released in the 1970s. They were succeeded by Ikarus, released in 1984. The quality level of all these varieties was moderate. Recent cultivars released by Saatbau Linz are Brutus (1994, A7 quality), Silvius (1995, A6 quality) and Spartakus (1996, A7 quality).

4. Analyses of glutenin and gliadin composition

Lelley & Gröger (1993) investigated the effect of HMW glutenins, gliadins and rye translocations on the bread-making quality of wheat in Austria. For the glutenin composition, they stated that a high *Glu-1* score should be considered as an indicator for the quality potential of a wheat, rather than a measure of its actual quality. A cultivar with the 'wrong' alleles was very unlikely to give good baking results. Gröger et al. (1998) described a new Probstdorfer wheat variety, Achat, with HMW glutenin subunits 1,6+8,2+12 and very good bread-making characteristics.

The HMW glutenin composition of 66 Austrian-grown winter-wheat cultivars was investigated by Gröger et al. (1997a,b) and related to several bread-making quality parameters. Some 91% of the Austrian quality wheats (class 7–9) contained the *Glu-B1* allel 7+9 and the *Glu-D1* allel 5+10, whereas 44% of the feed wheats (Class 1–3) contained the *Glu-B1* allel 6+8 and the *Glu-D1* allel 2+12. Gröger et al. recommended using SDS-PAGE for the selection of favourable HMW glutenin subunits in the potential crossing partners and in early generations. Subsequently, a Zeleny sedimentation test may be used to assess the quality of the preselected material.

Data from Hänsel et al. (1994) show that the subunit 2+12 was present in Austro Bankut and Austro Kolben. Austrian breeders, who have mainly used high Zeleny sedimentation values as a selection criterion to improve the quality of their material, appear to have succeeded in retaining the subunits 7+8, 7+9 and 5+10 and to eliminate the subunits 2+12. Based on a study of the pedigree of 22 winter-wheat varieties, the conclusion of Hänsel et al. is that the high quality of these wheats originates from very few sources.

Cerny et al. (1992) compared the gliadin and glutenin alleles in two sets of wheat selections from Freising in Bavaria (n=76) and Probstdorf in Austria (n=67). A score predicting baking quality based on gliadin (*Gld*) and glutenin (*Glu*) blocks was correlated to the Zeleny sedimentation value. In the Bavarian set, highly positive correlations were found for the prediction value of *Gld* blocks, *Glu* blocks and *Gli+Glu* blocks. In the Austrian set, however, no positive correlation was found for the prediction value of *Gld* blocks. This was compensated for by a highly positive correlation for *Glu* blocks, which resulted in a positive correlation for the summary prediction of *Gld+Glu* blocks. The discrepancy of the share of both protein systems in the baking quality formation in the Bavarian and the Austrian sets was related to the different selection pressure in the two regions, related to their specific agro-ecological conditions.

D.A.D.

Acknowledgment

The author wishes to thank Prof. P. Ruckenbauer (Institute of Agrobiotechnology, Tulln) for reviewing this chapter.

References

Adam J (1993) Der Banater Weizen – ein Genreservoir für die Züchtung von Qualitätsweizen; historischer Rückblick. Bericht über die Arbeitstagung 1993 der Arbeitsgemeinschaft der Saatzuchtleiter im Rahmen der Vereinigung Österreichischer Pflanzenzüchter, Gumpenstein, pp 29–35.

Anonymus (1923) Le blé; production, industries de transformation, commerce. Compte rendu des travaux de la semaine nationale du blé, Paris.

Cerny J, Sasek A, Zimmermann G, Seibert L (1992) An evaluation of Bavarian and lower Austrian new varieties of winter wheat by means of genetic protein markers of baking quality. Scientia Agriculturae Bohemoslovaca 24(2): 113–123.

Fuchs H (1954) Qualität als Zuchtziel bei den vier Hauptgetreidearten. Bericht über die Arbeitstagung der Arbeitsgemeinschaft der Saatzuchtleiter im Rahmen der Vereinigung Österreichischer Pflanzenüchter, Admont, pp 57–67.

Glattes H (1978) Austria's contribution to solving international problems in milling and baking technology (abstr). Cereal Foods World 23(8): 456.

Gröger S, Oberforster M, Werteker M, Grausgruber H, Lelley T (1997a) Use of HMW glutenin subunits as markers for bread-making quality in wheat. In Application of marker aided selection in cereal breeding programs. Eucarpia Section Cereals Meeting 1997, September 22–23, Tulln, Austria, pp 50–51.

Gröger S, Oberforster M, Werteker M, Grausgruber H, Lelley T (1997b) HMW glutenin subunit composition and bread-making quality of Austrian grown wheats. Cereal Research Communication 25(4): 955–962.

Gröger S, Löschenberger F, Hetzendorfer H (1998) Breeding for bread-making quality in wheat at Probstdorfer Saatzucht. Proceedings 16th ICC conference, Vienna, 66 pp.

Hänsel H (1960) Beziehung zwischen Ertrag und amtlicher Qualitätseinstufung der in Österreich gebauten Winterweizen. Bericht über die Arbeitstagung der Arbeitsgemeinschaft der Saatzuchtleiter im Rahmen der Vereinigung Österreichischer Pflanzenzüchter, Gumpenstein, pp 35–45.

Hänsel H (1963) Grenzbereiche der Kombination von Ertrag und 'Qualität' und anderer Merkmale in der Winterweizenzüchtung Österreichs, insbesondere im pannonischen Gebiet. Bericht Arbeitstagung in Gumpenstein / Irdning, Ver Österr Pfl Züchter, 58–82.

Hänsel H (1970) Die Züchtung eines ertragreichen Winter-Qualitätsweizens: Probstdorfer Extrem. Bericht über die Arbeitstagung der Arbeitsgemeinschaft der Saatzuchtleiter im Rahmen der Vereinigung Österreichischer Pflanzenzüchter, Gumpenstein, pp 45–68.

Hänsel H (1974) Mahlfähigkeit als Zuchtziel bei Weizen. Berichte der 8. Tagung der Internationalen Gesellschaft für Getreidechemie (ICC), Wien, pp 34–39.

Hänsel H, Seibert L, Gröger S & Lelley T (1994) Pedigree and HMW glutenin subunits of 22 winter-wheat varieties bred for high bread-making quality and increasing yield potential in the semiarid region of Austria since 1948. In Brönnimann A, Keller B, Winzeler H (eds), Proceedings of the Eucarpia Cereal Section Meeting 'Prospectives of cereal breeding in Europe', Landquart, Switzerland.

Hron R (1981) Getreideanbau und Getreidezüchtung in Österreich. In Festschrift '100 Jahre Bundesanstalt für Pflanzenbau und Samenprüfung in Wien', 1881–1981, pp 35–58.

Lelley T, Gröger S (1993) Einfluss der HMW- und LMW glutenine, der Gliadine und der 1BL/1RS Translokation auf der Backqualität des Weizens. Bericht über die 44. Arbeitstagung der Arbeitsgemeinschaft der Saatzuchtleiter im Rahmen der Vereinigung Österreichischer Pflanzenzüchter, BAL Gumpenstein, pp 13–19.

Lupton FGH (1992) Changes in varietal distribution of cereals in central and western Europe. Agro-ecological Atlas of Cereal Growing in Europe, vol IV, 161 pp. Wageningen: Pudoc.

Meinx R von (1966) Der Anbau und die Züchtung von Weizen in Österreich Die Bodenkultur (17. Sonderheft): 97–114.

Meinx R von (1977) Der Aufbau einer qualitätsorientierten Weizenproduktion in Österreich. Die Bodenkultur 28: 423–428.

Oberforster M, Schmidt L, Werteker M (1994) Bewertungsschema '94 der Technologischen Qualität von Weizensorten (Weichweizen). Jahrbuch 1993 der Bundesanstalt für Pflanzenbau, Wien.

Oberforster M, Werteker M (1995) Wheat breeding and bread-making quality in Austria. Sjemenarstvo 12(6): 413–425.

Hungary

1. Introduction

The quality of Hungarian wheat has been renowned for centuries, even long before wheat breeding was consciously practiced. Wheat from Hungary (H) traditionally commanded a good price on European wheat markets. Over three hundred years ago, in 1686, a deputy of the English Royal Medical Society, traveling through Europe, wrote that the bread made from Hungarian wheat flour was better than all other European loaves (Ellinger et al., 1934). At the end of the 19th century Hungary was the third largest wheat exporting nation in the world after the USA and Russia. The main buyer was Austria, its partner in the Austrian-Hungarian empire until 1919. Until World War II, the main goal of Hungarian wheat breeders was to maintain and possibly improve the quality of Hungarian wheats (Lelley, 1994). Two Hungarian investigators acquired international fame in the field of wheat quality: Hankóczy, who developed the farinograph in Budapest in 1929, and Zeleny, who developed the Zeleny test in Canada in 1947.

Wheat is the most important cereal in Hungary, covering 20–25% of the arable cropping area (more than 30% until the 1960s). Practically all wheat grown in Hungary is winter wheat. Wheat production is concentrated on the Great Plain of Hungary (Alföld) and the best areas for wheat production in terms of quality and quantity are in the middle and lower part of the Tisza

Table H.1. Area (1000 ha) and production (1000 t) of cereals and wheat in Hungary

Year(s)	Cereals		Wheat	
	area	production	area	production
1924–1925			1 421	1 677
1934–1938			1 589	2 220
1948–1952	3 818	5 647	1 385	1 909
1961–1965	3 265	6 765	1 078	2 009
1969–1971	3 132	9 057	1 289	3 410
1979–1981	2 878	13 001	1 187	4 800
1989–1991	2 818	14 592	1 207	6 249
1995–1997	2 829	11 822	1 182	4 508

Sources: IIA and FAO statistics.

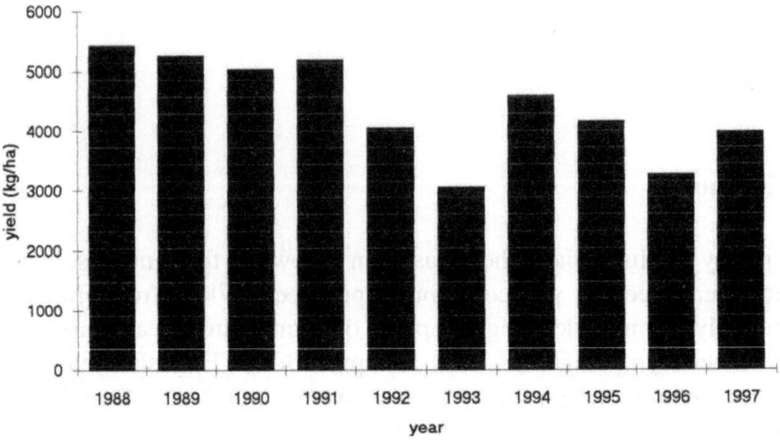

Figure H.1 . Average grain yield of wheat (t/ha) in Hungary, 1988–1997.

Basin (Szabo et al., 1994). The river Tisza (Theiss) is a branch of the Danube, running through the country from north to south. The area under wheat (Table H.1) has gradually decreased from 1 500 000 ha to 1 100 00 ha from the start to the end of the 20th century.

After a production increase shortly after World War I, yield levels in Hungary remained stable until the late 1960s, whereas yields in surrounding countries had already started to increase earlier. Presumably this was a reflection of the emphasis that Hungarian breeders continued to put on the maintenance of the high quality level. A major step forward in the level of grain yields of wheat was made during the 1970s and 1980s, resulting in production figures of 6–7 Mt (7.4 Mt in 1984) of wheat. Since 1990 however, production figures have not increased and have even been very low in certain years. This was not only due to a smaller area under wheat but also to low grain yields, especially in years with dry summers. Whereas Hungary was able to export 2 Mt of wheat in 1988, the margin for export decreased in the following years. Nevertheless, production still met home consumption in every year, including 1.5–1.6 Mt for human consumption and 3.0–3.5 Mt for animal feed.

Average wheat production per hectare increased from 1.22 t/ha in 1900–1950 and 1.86 t/ha in 1961–1965, to 4.65 t/ha in 1981–1988 (Balla et al. 1989). With an average of 4,34 t/ha for 1989–1997 (FAO statistics) the growing trend came to a halt. In part this was due to social and political circumstances, including a reduction in the use of fertilizers (Balla, 1998).

Nevertheless, annual variation in wheat yields is rather large in Hungary, as illustrated in Figure H.1. Climatic conditions play a major role in this fluctuation. One of the major constraints for Hungarian wheat production is drought at critical periods of the growing season, especially in the booting stage, between the end of April and the first half of May. Summer heatwaves prior to ripening are frequent and may influence the quality of the harvest. Under these circumstances the gluten stability of old Hungarian varieties appears to be better than that of the newer ones (Bedö et al., 1998). Generally speaking, Hungary has a natural potential for the production of excellent quality wheat.

2. Wheat breeding before 1950

Wheat breeding in the late 19th and early 20th century has been undertaken by private breeders whose names may be recognized in some of the older Hungarian varieties, such as Székács and Fleischmann. Locations of breeding stations, such as Bánkut, Fertöd and Kompolt, may be recognized in other names. Early breeding activities consisted mainly of mass and individual selection in local populations. Organized breeding started in 1891 at the governmental breeding institute Magyaróvár (Grabner, 1930).

The most important wheats grown at that time were populations of the Banatka and Tiszavideki type, consisting of awned, red-grained types with prevalently white ears. The names of these land varieties were derived from the Banat region[1] and the river Tisza.

Banatka and Tiszavideki land varieties were characterized by high levels of winter-hardiness and drought resistance combined with excellent bread-making properties. In comparison with Western European awnless wheats their productivity was low and they had a tendency to lodge (Vavilov, 1935), which is not surprising as some varieties grew up to 140 cm. Banatka and Tiszavideki wheats were reputed for their bread-making quality and were grown in neighbouring countries, including Poland and Germany (Jakubziner, 1962). Selections from Banatka and Tiszavideki populations have been used in many breeding programmes as a source of bread-making quality. Through the Danube countries and southern Russia these wheats found their way to America and to Australia (Adam, 1993).

The export of Hungarian wheat has had to meet several challenges around the turn of the century. The first time was at the end of the 19th century, when the introduction of more productive foreign varieties in certain parts of the country induced a decline in the overall level of quality. The second time was around World War I, when selection on productivity without regard

[1] The Banat region lies in the south of the Karpat Basin and is part of the Great Hungarian Plain (Alföld). It is delimited by the rivers Maros, Tisza and Danube (north, west and south respectively) and by the Karpathian mountains (east).

to quality aspects led to the release of some varieties with a lower quality. The establishment of a State Control and Testing Station and the foundation in 1928 of the Royal Hungarian Institute for Investigations on Cereals and Flour helped prevent a further decline. It was at this institute that Hankóczy developed the farinograph for the quality assessment of wheat samples.

One of the first varieties obtained by hybridization was Fleischmann-481, released in 1928 and named after its breeder. It was the result of a cross that included Rumai-244 (selected in a population from Romania), Scékács-1 and the Caucasian variety Kanred. During the same period, the work of László Baross at an estate in Bánkut resulted in two very successful cultivars: Bankuti 1201 and Bankuti 1205, both released in 1931. They were obtained by hybridization of a Tiszavideki wheat with the North American spring wheat Marquis. Due to their high protein content and excellent bread-making characteristics, Bankuti 1201 and Bankuti 1205 acquired fame all over the world. Bankuti 1201 won an award at the International Wheat Exhibition in Canada in 1933 as a variety with an excellent quality. Indeed, under certain conditions it could even surpass that of Manitoba (Ellinger et al., 1934). Bankuti 1201 covered 50% of the Hungarian wheat area until after World War II and Bankuti 1201 and 1205 together covered 50–65% of the Hungarian wheat area between 1950 and 1961. Bankuti 1201 remained on the variety list until 1970. (The variety Bankuti 178, from the same breeding station, was used as a parent in winter-wheat breeding in Sweden for its gluten quality; see pedigree Kosack, Figure Sw.2).

3. Wheat breeding after 1950

The restoration after World War II of the multiplication and propagation of certified seed was undertaken by the Növénytermelési Hivatal (Plant Cultivation Office). The seed distributed during the 1950s consisted mainly of Bankuti 1201, Bankuti 1205, Beta Bankuti and Fleischmann-481.

With the nationalization of private estates and the institution of large agricultural concerns, the use of combine harvesters and fertilizers increased. The old Hungarian varieties, which had excellent quality characteristics, were susceptible to lodging under these circumstances and combine harvesting was difficult. Due to low yields, Hungary became a wheat importing country between 1960 and 1964. New cultivars suitable for intensive production were urgently needed but not available.

At this point, the initiative of Hungarian breeding programmes was needed. One institute that has played a leading role in Hungarian wheat breeding during the second half of the century is the Agricultural Research Institute at Martonvásár. It was established by the Ministry of Agriculture in 1949 and operated under the Hungarian Academy of Science from 1953 onwards. Two breeders who left their traces on wheat breeding at this institute were Sandor Rajki (1958–1968) and László Balla (1968–1984).

Other institutes that have greatly contributed to Hungarian wheat breeding are the Cereal Research Institute (GKI) at Szeged, the Agricultural Research Institute and the Fleischmann Rudolf Institute (University of Gödöllö) at Kompolt and the Agricultural and Horticultural Research Station at Fertöd, as well as several others. Two smaller programmes in Bánkút Karcag and Lovászpatona combined their activities under the name Bánkút. Breeding work on quality aspects has been supported by the Research Institute of the Baking Industry in Budapest.

Until 1957, the cultivars Fleischmann-481, Bankuti 1205, Bankuti 1201 and Beta Bankuti covered most of the Hungarian wheat area. The variety Fertödi-293 was released in 1957 and became quite popular in following years. Resulting from a cross between Bankuti 1201 and an American wheat, this variety was noted for its bread-making quality.

Breeders at Martonvásár took the initiative to introduce and test French, Italian and Soviet wheats under Hungarian conditions. During the winters of 1962/63 and 1963/64 it appeared that the Italian varieties did not have sufficient frost resistance. The Soviet variety Bezostaya-1, however, passed the test and became very popular in the following ten years. In 1968, 75% of the Hungarian wheat area was sown with Bezostaya-1. After 1980, Yugoslav varieties became popular and Italian varieties were introduced at regular intervals. Although the cultivation of these varieties never lasted for more than a few years, they were valuable breeding material. This is reflected in the pedigree of the Hungarian varieties that began to appear on the market in the 1970s. The genetic background of these new varieties was totally different from that of the old Hungarian ones as, apart from Fertödi-293, none of the older local varieties was used as a parent (see Table H.2).

3.1. Wheat breeding at Martonvásár

Wheat breeding at Martonvásár was from the onset carried out with the aim of combining high yield with tolerance to drought and improved bread-making quality (Rajki, 1971). Eventually the aims of suitability for machine harvesting and resistance to rust and powdery mildew were included. From 1951 onwards the selection on quality was supported by a flour-quality laboratory, where baking tests were carried out. Starting in 1960, many crosses were made between Hungarian and Russian material, as well as Russian cultivars among themselves (Balla, 1973).

The breeding programme at Martonvásár has been very successful: 38 winter-wheat varieties were released up to 1998 and Martonvásár varieties

Table H.2. Three generations of Martonvásár varieties (Mv) with their pedigree, year of release and quality characteristics

Variety	Year	Pedigree	Quality description	Farino-graph value	Quality group[a]
Mv-1	1971	Bezostaya-1/Mv-65-07	good	75	A2
Mv-2	1972	Bezostaya-1*2/Fertödi-293	excellent	88	A1
Mv-3	1973	Bezostaya-1/Fertödi-293	excellent	92	A1
Mv-4	1974	Mironovskaya 808/Bezostaya-1	excellent	71	A2
Mv-5	1974	Mir-808/Bez-1*2//Produttore/Bez-1	excellent	69	B1
Mv-6	1976	Bezostaya-1/Moisson	good	65	B1
Mv-7	1978	Bezostaya-1/0pal	good	69	B1
Mv-8	1978	Ranks. mutant/Bezostaya-1 mutant	fair	61	B1
Mv-9	1980	Mir-808/Bez-1//Bez-1*2/Produttore	excellent	71	A2
Mv-10	1981	Moisson/Arthur	good	67	B1
Mv-11	1982	Bezostaya-1/Purdue-4930	good	68	B1
Mv-12	1982	Mir-808/Bez-1//Bez-1*2/Produttore	good	69	B1
Mv-13	1983	Krasnodari/Zg-1477-69			
Mv-14	1985	Mir-808/Bez-1//Kavkaz/Rana-1// Zl.Dolina/Arthur	good	61	B1
Mv-15	1985	Kavkaz/Mir-808//Zl.Dolina/Kavkaz	excellent	70	A2
Mv-16	1987	Mv-4/Kavkaz//Purdue-4039/ Zl.Dolina/Arthur/Rubin	excellent	73	A2
Mv-17	1988	Slavia/Mv-Tf//Bananjka	good	68	B1

[a] Classification based on farinograph values.
Sources: Balla et al. (1989); Balla (1998).

have occupied 55–60% of the Hungarian wheat area since the early 1980s. Several Martonvásár varieties have been registered in Turkey, Ukraine, Yugoslavia and Italy. A comparison of three generations of Martonvásár varieties (Balla et al., 1989) is summarized in Table H.2. The dominance of Bezostaya-1 in the pedigree of these varieties is apparent. Obviously the bread-making quality of these 'Mv' cultivars was comparable to Bezostaya types, implying that their protein and gluten content was lower than the old Hungarian varieties but still sufficiently high (Balla, 1996).

Within the first generation (Mv-1 to Mv-5), the best bread-making characteristics were found in Martonvásár-2 and -3, but the most successful variety was Martonvásár-4. It was the result of a cross between Bezostaya-1 and Mironovskaya-808, that was backcrossed with Bezostaya-1. The cross between Bezostaya-1 and Mironovskaya-808 also resulted in the varieties Jubilejnaya-50 (Ukraine), Sadovo (Bulgaria) and Slavia (Czechoslovakia), all varieties combining good quality characteristics with a high yield potential. The second generation of Martonvásár varieties (Mv-6 to Mv-12) had a higher yield potential and lodging resistance than the first generation but it still lacked resistance to leaf and stem rust. Mv-8, Mv-9, Mv-10 and Mv-12, of which Mv-9 and Mv-12 had the best quality, were successfully grown during the 1980s. The third generation (Mv-13 to Mv-17) represented a further improvement of yield potential combined with improved resistance to stem and leaf rust.

The improved yield potential achieved in the Martonvásár varieties was the result of increased biological yield and improved harvest index. As a matter of fact the plant height decreased from 125 (Bankuti 1201) to 80–90 cm. Martonvásár cultivars contributed greatly to the production increase achieved in Hungary between 1970 and 1990.

Table H.3. Changes in the genetic background of the Martonvásár wheat breeding programme

Origin	Weight of parents in the pedigrees of the programme (%)								
of parents	1985	1986	1987	1988	1989	1990	1991	1992	1993
Hungary	34	32	36	31	41	49	54	56	56
Martonvásár	*26*	*25*	*28*	*25*	*31*	*38*	*44*	*48*	*49*
GKI Szeged	*7*	*7*	*6*	*5*	*8*	*10*	*10*	*9*	*7*
Yugoslavia	33	34	26	21	17	15	12	12	7
Novi Sad	*19*	*21*	*18*	*14*	*9*	*5*	*4*	*3*	*3*
Zagreb	*11*	*10*	*6*	*5*	*5*	*5*	*4*	*6*	*3*
Romania	4	6	10	8	8	6	6	5	7
Bulgaria	9	8	6	6	5	3	3	3	2
Czechoslovakia	3	3	4	3	6	7	9	9	7
Soviet Union	9	7	10	4	4	3	2	1	2
Western Europe	2	2	3	2	3	4	4	6	12
France	*1*	*1*	*1*	*1*	*2*	*2*	*2*	*3*	*6*
USA	2	2	2	1	2	1	1	1	1

Source: Lang & Bedö (1994).

To broaden the genetic basis of the Martonvásár material, the input of Western European 'blood' in the breeding programme was increased in the 1980s (Table H.3). The proportion of Russian and Ukrainian genotypes dropped, although they still represent valuable sources of winter-hardiness and quality. Increased priority given to quality criteria also led to a reduction in the proportion of Yugoslav varieties (Lang & Bedö, 1994).

3.2. *Wheat breeding at Szeged*

The current wheat breeding programme of the Cereal Research Institute (GKI) at Szeged is the result of a combination of programmes from different wheat breeders, for example Lelley (from Kompolt) and Beke (from Fertöd). A new programme at Szeged was initiated by Barabás in 1970. The Cereal Research Institute is located in a region with frequent droughts. During the starting period the wheat breeding programme emphasized the development of (semi)-dwarf varieties with a high level of disease resistance and winter-hardiness (Barabás, 1975). Improved bread-making quality was not a specific goal of the breeding programme, but the breeding material was submitted to a quality test from F_3 onwards. For this reason the Zeleny sedimentation test was mechanized, allowing the analysis of 12 000 to 14 000 seed samples within fifty days (Lelley, 1973). The first variety released by GKI that was grown on a considerable area was GK Kincsö, a variety with a high level of field resistance against several diseases (Barabás, 1989).

Major efforts have been made to improve the protein and lysine level of wheat varieties by focusing on genotypes in which the negative correlation between protein level and lysine content was absent (Belea & Sági, 1978). One of the high protein sources resulting from this selection was Tiszatáj. Derived from a cross between Bezostaya-1 and Fiorello, it had a high protein content and excellent baking and dough handling properties. GK Tiszatáj was released in 1977 and classified as A1-A2 quality. Although it could not compete with the existing high yielding varieties, it remained a standard variety for baking quality (Lelley, 1994). The quality of GK Tiszatáj was similar to that of Bánkuti 1201 (Balla, 1996).

Several successful varieties from Szeged were released in the 1980s (Table H.4). GK Zombor, GK Öthalom and GK Bence, all semi-dwarfs with a straw length of 75–100 cm, covered 6, 19 and 5%, respectively of the winter-wheat area in Hungary in 1990. GK Öthalom was still widely grown in 1996.

GK Delibáb, released in 1994, was the first Hungarian wheat cultivar obtained by biotechnological breeding methods, i.e. the production of doubled haploid plants from anther cultures. It was described as a very early maturing, high yielding variety with winter-hardiness and medium baking quality (Pauk et al., 1993).

Hungarian varieties released in the 1980s and 1990s were characterized by their improved yield potential and disease resistance (Lelley, 1994). However, they did not have the same quality level as the old Hungarian wheat varieties. Lelley stated that several decades of neglecting quality were to blame for this shortcoming. He advocated that both industry and breeders should pay more attention to wheat quality and should adopt the quality level of GK Tiszatáj as a standard. The next most important breeding aim in wheat, according to

Table H.4. Quality characteristics of seven winter-wheat varieties from the Cereal Research Institute at Szeged

Variety	Year	Pedigree	A	B	C	D
GK Kincsö	1983	Arthur/Sava	49	B2	C1	6
GK Öthalom	1985	GK Szeged^{*2}/Jubileynaya-50	69	B1	B1	9
GK Zombor	1985	Kavkaz/Produttore/Sava	60	B1	C1	4
GK Gereben		Rusalka/Rannyaya-12/Rubin/Mini Manó	–	B2	A2	–
GK Szeged		Strampelli/Marco Michahelles/Bezostaya-1				
GK Bence	1987	Arthur/Sava/Libellula	64	B1	C1	9
GK Délibáb	1994	Mini Manó//Jub-50/Sadovo Super/3/Mini Manó/Mv-12		B1	B2	

A: Brabender farinograph value.
B: Classification based on farinograph value: $A1>90$; $70<A2<90$; $55<B1<70$; $45<B2<50$; $30<C2<45$.
C: Quality group based on German classification system.
D: *Glu-1* score.
Sources: Balla et al. (1989); Balla (1996); Pollhamer (1994); Karpati et al. (1990).

Lelley should be drought resistance, an aspect that faded into the background in Hungary in a series of years with rainy summers. The dry summers of 1992 and 1993, however, resulted in a yield reduction of 21% and 41%, respectively, for Hungary and 45% and 50%, respectively, on the Hungarian plain. This amply demonstrated the need for continued emphasis on the selection of drought resistant genotypes.

4. Glutenin composition of Hungarian wheats

The HMW glutenin subunit composition of 72 wheats grown in Hungary was analysed by Karpati et al. (1990): the average *Glu-1* score of the Hungarian varieties (n = 53) was 7.2. This value is somewhat lower than the score calculated by Sontag et al. (1986) for a group of 35 Finnish varieties (average score 8.0), but it is higher than the average score calculated for cultivars grown in the UK (5.2) and in Germany (5.8) by the same authors.

Of the 72 varieties investigated by Karpati et al., 21 were released from GKI Szeged, 29 from Martonvásár and 8 from Zagreb (Yugoslavia). The remaining 14 genotypes were released from various breeding stations. The average *Glu-1* quality score calculated for each of the three groups of genotypes varied:

- 6.3 for the material from Szeged (n = 21);
- 7.8 for the material from Martonvásár (n = 29);
- 5.4 for the material from Zagreb (n = 8).

Table H.5. Proportion (%) of the different alleles in each of the 1A, 1B and 1D loci as found for three groups of genotypes

	1A locus			1B locus					1D locus		
	N	1	2*[a]	7	7+9	7+8	6+8	20[b]	2+12	4+12	5+10
Genotypes from Szeged (n = 21)											
absolute	11	8	1	8	10	1	1	1	12	1	9
relative	55	40	5	38	48	5	5	5	$54\frac{1}{2}$	$4\frac{1}{2}$	41
Genotypes from Martonvásár (n = 29)											
absolute	7	5	17	6	20	1	1	1	7	1	21
relative	24	17	59	21	69	3	3	3	22	3	74
Genotypes from Zagreb (n = 8)											
absolute	3	2	3	1	1	–	6	–	5	2	1
relative	$37\frac{1}{2}$	25	$37\frac{1}{2}$	$12\frac{1}{2}$	$12\frac{1}{2}$	–	75	–	$62\frac{1}{2}$	25	$12\frac{1}{2}$

[a] This column includes the allele 2, found in one Szeged variety.
[b] This column includes the allele 20+9, found in one Mv and two other varieties.
Source: calculated from data of Karpati et al. (1990).

These figures may be considered to reflect the intensity of breeding for bread-making quality and the parental material used.

Further analysis of the HMW glutenin subunit composition of the genotypes within the same three groups points to differences in the proportion of the different alleles on the 1A, 1B and 1C loci, which are presented in Table H.5. In summary, for each locus the following observations can be made. On the 1A locus:

- half of the genotypes from Szeged have a nullisome here;
- nearly 60% of the genotypes from Martonvásár have a 2∗ allele, which undoubtedly improves the *Glu-1* score and the baking quality;
- for the genotypes from Zagreb, there is a more or less well balanced distribution over the three alleles.

On the 1B locus:

- among the genotypes from Szeged, as well as those from Martonvásár, there is a majority with the 7+9 or the 7 alleles, whereas the 6+8 allele is in the majority in the genotypes from Zagreb.
- the presence of a subunit 20 is rather exceptional in European wheats

On the 1D locus:

- for the genotypes from Szeged, there is a more or less balanced distribution over the 2+12 and 5+10 alleles;
- in the material from Martonvásár the 5+10 allele is prevalent;
- in the material from Zagreb the 2+12 allele is prevalent.

The authors determined the association between the HMW subunit composition and various quality parameters for the 72 genotypes. The highest lineair correlation coefficients were obtained for the relation between *Glu-1* score and valorigraph value ($r = 0.690$) and for the relation between *Glu-1* score and baking-test loaf volume ($r = 0.573$). The correlation coefficient for the relation between valorigraph value and loaf volume was $r = 0.537$. The valorigraph was used in Hungary before the farinograph became available. It measured the same quality characteristics and its values are comparable to farinograph values.

The 2+12 (*Glu-D1*) subunit is of major importance for Hungarian bread-making quality. Most of the old Hungarian varieties with good bread-making properties, including Bankuti 1205, Beta Bankuti, Bankuti 1201 and Székacs-1242, possess this subunit (Bedö et al., 1995, 1998). Both the Hungarian Bankuti cultivars, as well as their Austrian derivates, have 2+12 or 3+12 glutenin subunits on the 1D chromosome but are heterogenous for the *Glu-A1* and *Glu-B1* locus. Populations consisting of sublines with various glutenin components provide the possibility for greater plasticity with respect to quality characters, as well as other agronomic traits.

J.M. & D.A.D.

Acknowledgments

The authors wish to thank Dr László Balla (Budapest) for his valuable comments on this chapter and for providing additional information on wheat breeding in Hungary.

References

Adam J (1993) Der Banater Weizen – ein Genreservoir für die Züchtung von Qualitätsweizen; historischer Rückblick. Bericht über die Arbeitstagung 1993 der Arbeitsgemeinschaft der Saatzuchtleiter im Rahmen der Vereinigung Österreichischer Pflanzenzüchter, Gumpenstein, pp 29–35.

Anonymus (1927) La distribution du froment dans le monde. Rome: Imprimerie de l'Institut International d'Agriculture, 46 pp.

Balla L (1973) Wheat breeding for quality and yield at Martonvásár. Proc 4th Intern Wheat Genetics Symp, Colombia (Missouri, USA) pp 483–488.

Balla L, Bedö Z, Lang L (1989) The aims and results of wheat breeding in Hungary. Sveriges Utsädesförenings Tidskrift 99: 109–116.

Balla L (1996) Personal communication, Rhenen.

Balla L (1998) Personal communication, Wageningen.

Barabás Z (1975) Wheat research of the Cereal Research Institute in Szeged. Cer Res Com 3(3): 245–254.

Barabás Z (1989) Breeding and biotechnology in the Cereal Research Institute, Szeged, Hungary, 1883–1988. Sveriges Utsädesförenings Tidskrift 99: 87–91.

Bedö Z, Kárpáti M, Vida G, Krammarik-Kissimon J, Láng L (1995) Good bread-making quality wheat (*Triticum aestivum* L.) genotypes with 2+12 subunit composition at the *Glu-D1* locus. Cer Res Com 23(3): 283–289.

Bedö Z, Vida G, Láng L, Karsai I (1998) Breeding for breadmaking quality using old Hungarian wheat varieties. In Braun HJ, Altay F, Kronstad WE, Beniwal SPS, McNab A (eds), Wheat: prospects for global improvement. Dordrecht, Boston, London: Kluwer Academic Publishers, pp 245–248.

Belea A, Sági F (1978) Evaluation of common wheats and interspecific Triticum hybrids for protein and lysine content. Cer Res Com 6(1): 15–19.

Ellinger T, Schandl L, Siegescu M (1934) Die Landwirtschaft Ungarns. Budapest: Königlichen Ungarischen Ackerbau-Ministeriums, 192 pp.

Grabner E (1930) La situation actuelle de l'amélioration des plantes en Hongrie, notamment des plantes fourragères et le commerce des graines ameliorées. In Production et Commerce des Semences en Hongrie. IV Congrès International du Commerce des Semences. Budapest: Société des Marchants Grainiers de Hongrie, pp 22–30.

Jakubziner MM (1962) Wheat species and varieties as resources in plant breeding. In Symposium on Genetics and Wheat Breeding. Martonvásár: Agricultural Research Insitute of the Hungarian Academy of Sciences, pp 405–422.

Karpati EM, Lasztity R, Kuroczi LG, Czirak L (1990) Relationship between high molecular weight subunits of glutenin and bread-making quality of Hungarian wheats. In Bushuk W, Tkachuk R (eds), Gluten Proteins, St Paul (MN) AACC, pp 245–252.

Kovacs I (1971) Plant Breeding Research at Martonvásár: wheat breeding. Martonvásár: Agricultural Research Institute of the Hungarian Academy of Sciences, pp 70–72.

Kovrig J (1942) Die ungarische Landwirtschaft. Prague: Berichte über Landwirdtschaft 156 (Sonderheft), 170 pp.

Lang L, Bedö Z (1994) Genetic background of the Martonvasar wheat breeding programme. In Balfourier F, Perretant MR (eds), Evaluation and exploiting of genetic resources: pre-breeding. Proceedings of the genetic resources section meeting of Eucarpia, Clermont-Ferrand (France), pp 117–122.

Lelley J (1973) Einzelpflanzauslese und Mechanisierung in der Weizenzüchtung. Zeitschrift für Pflanzenzüchtung 69: 129–134.

Lelley J (1994) Stand und Perspektive der Weizenzüchtung in Ungarn. Bericht über die 45. Arbeitstagung der Arbeitsgemeischaft der Saatzuchtleiter, Gumpenstein (Austria).

Lupton FGH (1992) Changes in varietal distribution of cereals in central and western Europe. Agro-ecological Atlas of Cereal Growing in Europe, vol IV, 161 pp. Wageningen: Pudoc.

Pauk J, Mesterházy A, Kertész Z, Matuz J (1993) Eine neue Winterweizensorte 'GK Délibáb' mit hervorragender Backqualität, herrgestellt durch biotechnologische Methoden. Bericht über die 44. Arbeitstagung der Arbeitsgemeinschaft der Saatzuchtleiter, Gumpenstein (Austria).

Pollhamer S (1994) Die Qualität der Winterweizensorten in Ungarn. Getreide, Mehl u Brot 49(2): 3–8.

Rajki S (1971) The first twenty years of Martonvásár. Martonvásár: Agricultural Research Institute of the Hungarian Academy of Sciences, 329 pp.

Sági F, Barabás Z, Pongor S (1980) Evaluation of winter-wheat lines for protein, lysine quality and yield. Cer Res Com 8(2): 385–391.

Sontag T, Salovaara H, Payne PI (1986) The high-molecular–weight glutenin subunit compositions of wheat varieties bred in Finland. J Agric Sc Finland 58: 151–156.
Szabo M, Angyan J, Szalai T (1994) Biometric analysis of climatic conditions in Hungary with a view to the yield and quality of winter wheat. Acta Agronomica Hungarica 42(1–2): 83–93.
Vavilov NI (1935) The origin, variation, immunity and breeding of cultivated plants (transl. Starr Chester K) Chronica Botanica 13 (1949), 364 pp.

Yugoslavia

1. Introduction

The Federal Republic of Yugoslavia (YU), which was proclaimed in 1991, has been a part of the larger Federal Republic of Yugoslavia since 1945 and part of the Kingdom of Serbia-Croatia-Slovenia since 1918. This chapter covers the breeding work carried out on Yugoslav territory, as it existed between 1918 and 1991. Unless mentioned otherwise, in this chapter Yugoslavia covers Slovenia, Croatia, Bosnia-Herzegovina, Macedonia and the Federal Republic of Yugoslavia with Montenegro and Serbia, including its autonomic provinces Kosovo and the Vojvodina.

The country comprises a coastal belt bordered by the Adriatic Sea in the west and covers a considerable part of the Pannonic Plain in the northeast. Between these two areas, with Mediterranean and Danube characteristics respectively, the Dinaric Alps form a central mountain barrier. A transitional zone between the mountains and the plain is flooded by the rivers Morava and Vardar. Several important rivers, such as the Danube, the Sava and the Drava, run through the country, providing an excellent infrastructure for transport (Figure YU.1). In this complex pattern of natural regions, the most productive area for wheat growing is Vojvodina, in the northeast of the country, where it shares the Pannonic Plain with Hungary and a small part of Austria.

The territory of Yugoslavia was one of the earliest areas of wheat cultivation in Europe (Jost & Cox, 1989). Forms and populations of wheat that are highly adapted to this region have resulted from many centuries of agricultural practice.

Before 1939, Yugoslavia was a well known exporter of high quality wheat of the Banat type. Most of it was exported by way of the Danube to Austria and Hungary, later to Czechoslovakia and Germany. The majority of the Yugoslav population at that time consumed principally maize. After 1945, due to an increasing urban population and a rising standard of living, Yugoslavia became an importer of wheat, mainly of American quality wheats. The amount of imported wheat varied between 400 000 and 1 500 000 t between 1959 and 1967. To eliminate these imports, serious efforts were made to increase wheat production. Apart from measures to stimulate use of mineral fertilizers, improvement of soil tillage and other treatments, much attention was paid to improving wheat varieties. As a result, wheat yields increased from 1.25 to 2.9 t/ha between 1950 and 1970, and Yugoslavia has been self-sufficient for wheat since 1968. This increase in wheat production, however,

Figure YU.1 . Location of wheat breeding stations in Yugoslavia. (Design: Lucas Janssen)

brought forward the urgent need to improve the baking quality of at least part of the national harvest (Martinic & Zanic, 1973). This chapter reviews the results of these breeding efforts.

More than half the total area under arable crops in Yugoslavia is covered by cereals; in 1938 this was 82%, and in 1969 about 75%. Wheat and maize are the two main cereal crops, maize being by far the largest crop with almost 2 000 000 ha in 1995 (Table YU.1). Practically all wheat grown in Yugoslavia is winter wheat.

2. Wheat breeding during the first half of the 20th century

The establishment of a plant breeding station within the Royal Farming School at Krizevci (near Zagreb) in 1885 is considered to be the beginning of organized selection and plant breeding research on the territory of Yugoslavia (Jost & Cox, 1989).

Up to 1914, most of the the wheat grown on the Balkan Peninsula consisted of land varieties of *T. aestivum*, and wheat breeding was limited to mass selection in native populations. These populations were characterized

Table YU.1. Area (1000 ha) and production (1000 t) of wheat in Yugoslavia between 1934 and 1997

Year	Area	Production
1934–1938	2 150	2 455
1948–1952	1 821	2 171
1961–1965	2 006	3 599
1969–1971	1 928	4 760
1979–1981	1 475	4 624
1989–1991	1 507	6 186
1995–1997	1 188	3 898[a]

[a] Total of Bosnia-Herzegovina, Croatia, Macedonia, Slovenia and Yugoslavia.
Source: FAO statistics.

by good winter-hardiness, tall, thin straw, late maturation and low productivity but good baking quality (Borojevic & Potocanac, 1966).

Some populations and varieties were introduced from Hungary and Austria. Sirban Prolifik, a high yielding variety with good rust resistance and a very high hectolitre weight that was introduced from Hungary in 1919 became very popular, especially in Croatia. Gradually, mass selection was replaced by individual selection. Thus, the 'Bohutinsky wheats', named after Professor Bohutinsky who worked in Krizevci, were the result of individual selection both in native and foreign populations, such as Sirban Prolifik. Later, Krizevci Prolifik and Maksimirski Prolifik 39 were also selected from this material.

The varieties Rumska Crvenka and Krusevacka-22, both selected from the old Banat population, were released in 1925 and 1935, respectively. They were mainly grown in the eastern part of the country (Serbia and Vojvodina), together with the very popular Hungarian varieties Bankut 1201 and 1205. Although Banatka wheats were renowned for their excellent bread-making quality, farmers were aware of their susceptibility to rusts and lodging.

Italian varieties (Mentana, Frassineto and Virgilio) were well adapted to conditions in Istria and Dalmatia and were widely grown in those regions. The variety Krusevacka-9 was obtained from the Italian variety Cologna. A cross between Strampelli (from Italy) and Marquis (from Canada) made at the Agricultural Institute in Osijek resulted in the variety Osjecka sisulja (= 'beardless'), called U-1 for short. This early, lodging resistant and highly productive variety took first rank in Yugoslav wheat production until 1958. Later, crosses with the North American varieties Marquis and Dakota resulted

Table YU.2. Yugoslav winter-wheat varieties bred and grown during the first half of the 20th century with their height (cm) and pedigree

Variety	Height	Pedigree	Breeding station	Period of production
Bohutinsky's wheats	150	sel. from Sirban Prolifik	Krizevci	1910–1925
Rumska. Crvenka	135	sel. from Starobanatska	Krusevai	1925–1958
K-9	150	sel. from Cologna	Krizevci	1926–1958
Maksimirska Prolifik 39	120	sel. from Sirban Prolifik	Zagreb	1929–1958
Krusevacka 22	135	sel. from Starobanatska	Krusevai	1937–1960
Maksimirska Brkulja 530	125	Maks.Prolifik 39/domestic awnless	Zagreb	1937–1959
Osjecka Sisulja (U-1)	125	Strampelli/Marquis	Osijek	1936–1960
Maksimirska Brkulja 24	110	Strampelli/domestic awnless	Zagreb	1939–1959

Source: Jost & Cox (1989).

in Novosadska-1439/3, Novosadska-1446 and Krusevacka-2217 (released in 1948–1950), all cultivars of reduced plant height that spread fairly rapidly.

The Yugoslav winter-wheat cultivars bred before 1940 (Table YU.2) were characterized by an average plant height of 135–145 cm, late maturity and susceptibility to rust and lodging, but with a satisfying winter-hardiness and good baking and flour quality (Potocanac, 1975).

After World War II, wheat breeding was continued with greater intensity, supported by the creation of three breeding centres in 1955:

- the Institute of Agricultural Research at Novi Sad (Vojvodina);
- the Institute for Breeding and Production of Field Crops at Zagreb (Croatia);
- the Institute for Small Grains 'Kragujevac' at Kragujevac (Serbia).

Apart from these three institutes, several other breeding stations have been active in wheat breeding, e.g. at Osijek and Skopje. The institutes in Novi Sad, Zagreb and Kragujevac formed a coordinating committee and worked out a programme for the development of new wheat varieties. The main objective was to increase the yielding capacity and straw stiffness of improved local varieties, mainly by crossing them with foreign varieties. To this end a large collection of foreign varieties with high yield potential was tested.

A three-year experiment carried out at different locations in Vojvodina showed that some Italian varieties outyielded Bankut 1205 and U-1 by 40% or more. San Pastore outyielded Bankut 1205 by 64% (Table YU.3). A limiting factor in the spread of Italian wheats was their poor winter-hardiness. After a radical selection in the severe winter of 1959/60, only a few varieties remained, amongst which San Pastore. A few years after its introduction in 1958 this variety covered more than 50% of the wheat area in Vojvodina.

Table YU.3. Average yield of the winter-wheat variety Bankut 1205 and some Italian varieties grown in Vojvodina between 1957 and 1959

Variety	Micro trials[a]		Macro trials[b]		Large scale		Quality
	t/ha	%	t/ha	%	t/ha	%	class[c]
Bankut 1205	4.39	100	3.00	100	2.70	100	I
Autonomia		123		128		154	II
San Pastore		131		135		164	III

[a] Average of 5 locations.
[b] Average of 8 locations.
[c] The quality classes were defined as:
I: sedimentation value > 40; crude protein > 13%;
II: sedimentation value 30–39; crude protein 11.5–13.0%;
III: sedimentation value 18–29; crude protein 10.0–11.5%.
Source: Borojevic & Potocanac (1966).

Although this and other foreign varieties (e.g. Etoile de Choisy) made great contributions to reaching self sufficiency in wheat production, they have also contributed to a decrease in quality levels of the Yugoslav wheat harvest.

3. Wheat breeding after 1955

3.1. The concept of creating high-yielding wheat varieties

The development of a new type of variety that would meet the requirements of intensive production under Yugoslav conditions was undertaken in a number steps (Borojevic & Potocanac, 1966). The ultimate goal was to obtain varieties suitable for high yet stable production with a genetic yielding potential of 8–10 t/ha (this aim was re-adjusted to 15 t/ha in 1978).

The first step was to improve the straw quality and winter-hardiness of the high yielding foreign (mainly Italian) varieties and combine that with early maturity, i.e. earlier than U-1 and Bankut 1205.

The second step of the breeding programme was to introduce resistance to stem and leaf rust into the improved material, if possible combined with good milling and baking quality. To this end, crosses were made with North American material, such as Thatcher, Selkirk and Purdue lines.

The third step, which started around 1963, involved the introduction of good quality combined with a further increase of yield and winter-hardiness by crossing or back-crossing mainly with the Russian cultivars Bezostaya-1 and Mironovskaya-808. The quality mentioned as a goal was described as 'very good milling and baking quality of group A and B'. The quality of

Table YU.4. Yugoslav winter-wheat varieties released between 1964 and 1967 with their pedigree and quality classification

Variety	Breeding station	Pedigree	Quality group[a]
Backa	Novi Sad	Heine VII/line 129[b]	B1
Panonya	Novi Sad	Heine VII/line 129	B1
Vuka	Zagreb	U-1 / Fiorello	C1-C2
Hybrid 013	Kragujevac	Etoile de Choisy/line 958[c]	B1-B2
Brkulja 4	Novi Sad	Mara / Funo	B2-C2
Crvena Zvezda	Novi Sad	U-1/Selkirk//San Pastore/Mara	B2-C2

[a] Based on farinogram value.
[b] This line, also referred to as '129 genus' was an early *T. vulgare* line from a wheat × rye cross made by M. Kump in Zagreb.
[c] Selection from Bulgarian material.

Yugoslav wheat varieties is indicated both by a quality class (I, II or III), based on sedimentation value and crude protein (see Table YU.3) and by a quality group (A1, B1-B2, C1) based on farinogram values.

The first varieties that resulted from the breeding concept initiated by Borojevic and his colleages, were released from 1964 onwards. The most important cultivars released between 1964 and 1967 are listed in Table YU.4. This first generation of improved Yugoslav varieties was not able to compete with the Italian and Russian varieties that covered most of the Yugoslav wheat area and almost 100% of Vojvodina in the 1960s. Although they were better adapted to Yugoslav conditions than their foreign parents, their yielding performance was unstable and varied greatly from region to region. From a quality point of view, some of them were better than the Italian wheats grown, but they could not compete with the quality of Bezostaya-1. The results of the analysis of their quality characteristics in a three-year trial (Borojevic & Potocanac, 1966) are presented in Table YU.5.

Average Yugoslav wheat yields increased from 1.25 t/ha (average 1950–1959) to 2.56 t/ha in 1967, and to 2.91 t/ha in 1971 (Borojevic, 1989). Once Yugoslavia had reached a wheat production level of self-sufficiency for home consumption, the emphasis on increasing yield levels could be shifted towards more attention for quality aspects, although attention was also needed for varieties that were suited to mechanized production and advanced crop-management techniques. The wheat breeding work carried out in three of the breeding centres will be discussed in the following paragraphs. The number of varieties released by the four main centres may serve as an illustration to their importance: of the 217 winter-wheat varieties that were released

Table YU.5. Quality characteristics (average of three years trial) of improved varieties bred in Yugoslavia compared to foreign standard varieties

Variety	HLW[a] (kg)	TKW[b] (gr)	Protein content	Gluten in % wet	Gluten in % dry	Quality value	Quality group	Output flour	Location of trial
Hybrid 013	78.9	40.4	–	40.0	13.2	53.3	B2	142	Kragujevac
San Pastore	79.3	39.3	–	36.0	11.0	24.9	C2	134	
Vuka	76.0	40.7	–	34.2	9.3		C1		Zagreb
San Pastore	77.2	41.4	–	33.9	9.3		C2	–	
Backa	79.7	39.6	16.9	36.3	12.5	58.4	B1	140	Novi Sad
Panonya	78.0	40.3	17.0	38.5	13.6	69.1	B1	138	
San Pastore	79.1	38.0	15.0	37.7	12.6	29.5	C2	139	Novi Sad
Bezostaya-1	80.1	40.6	15.5	39.8	13.6	87.1	A1	143	
Et.de Choisy	75.5	39.1	14.6	38.5	13.2	35.0	C1	135	

[a] Hectolitre weight.
[b] Thousand kernel weight.
[c] Output of bread from 100 g of flour in g.
Source: Borojevic & Potocanac (1966).

between 1964 and 1988, 88 were bred in Novi Sad, 36 in Zagreb, 21 in Osijek and 15 in Kragujevac. Of the 20 spring-wheat varieties released in the same period, 8 were bred in Novi Sad and 6 in Zagreb; the remaining varieties were bred at different institutes and state farms spread over the country.

3.2. Results of wheat breeding in Novi Sad

The breeding programme initiated by Borojevic was continued in cooperation with Misic and Mikic. The concept of developing wheat varieties included the following quality requirements (Misic & Mikic, 1975):

- raw protein content: > 13%
- wet gluten content: > 30%
- water absorption power: > 60%
- quality group: B_1, A_2 or A_1
- maltose content > 2%
- loaf volume > 450 ml (/100 g flour)
- milling ability: = Bezostaya-1
- hectolitre weight: > 80 kg

Table YU.6. Yield and quality of winter-wheat varieties with improved quality released from Novi Sad in the 1970s

	Relative yield[a]	Farinogram		Extensogram		Loaf volume
		absorbtion of H_2O	group	energy (cm^2)	resistance of ext./extension	
Partizanka	106.7	61	A1-A2	130	2.5	490
Nova Banatka	102.4	61	A1-A2	120	2.0	490
Vojvodanka	105.3	62	A2-A1	110	2.0	480
Podunavka	103.5	61	A1-A2	120	2.0	490
NS Rana-1	–	60	B1-A2	80	2.0	450
NS Rana-2	–	61	B1-B2	80	2.0	450
Bezostaya-1	100.0	61	A1-A2	110	2.0	490
Kavkaz	–	62	B1-A2	60	1.5	510
Aurora	–	62	B1-A2	60	1.5	520
Libellula	100.4	57	C1	50	1.0	390

[a] Average of three-year trial in five regions at 17 locations; Bezostaya-1 (100%) = 4695 kg/ha.
Source: Misic & Mikic (1975).

Indicators of milling and baking quality were included in three years of small-scale variety trials at the breeding station. The quality characteristics of the varieties that were admitted for variety approval were examined for another three years in the laboratories of the Institute of Agricultural Research and the Institute of Food Industry at Novi Sad.

The varieties Sava, Biserka and Drina, which were selected from a cross between Fortunato and Redcoat, were released in 1972 and 1973. They were mainly appreciated for their yielding capacity and as a result they spread rapidly. Together with Zlatna Dolina and Slavonka, bred in Zagreb and Osijek respectively, they replaced Libellula, an Italian variety with very poor quality. The milling and baking properties of the new group of Yugoslav cultivars were, however, not satisfactory (Misic & Mikic, 1975) (see also Table YU.8). The result of intensified selection on quality characteristics was more reflected in the varieties Partizanka, Nova Banatka, Vojvodanka and Podunavka, released in 1974. All these varieties were higher yielding than the widely spread Soviet varieties Bezostaya-1, Aurora and Kavkaz, and their milling and baking characteristics were close to or better than Bezostaya-1. Even more promising were the varieties Novosadska Rana-1 and Novosadska Rana-2, released in 1975. Their yielding capacity was comparable to that of Sava, Biserka and Drina, and their milling and baking quality was very good. Partizanka, Novosadska Rana-1, and especially Novosadska Rana-2, were widely grown during the 1970s and 1980s. Some yield and quality characteristics of the most important varieties mentioned here are presented in Table YU.6. The pedigrees of these varieties may be found in Table YU.8.

A further improvement of combined yield and quality potential was obtained in the varieties Balkan, Posavka 2, Partizanka Niska and Jugoslavija, released between 1979 and 1984. All these varieties were grown on considerable areas, especially in the province of Vojvodina. Looking at the pedigree of the Novi Sad varieties released between 1975 and 1985 (Table YU.8), it appears that practically all have Kavkaz, Aurora or Skorospelka as a parent. Jost & Cox (1990) stated that through these Russian varieties, and indirectly through their common German parent Neuzucht, the 1B/1R chromosome translocation, often responsible for sticky dough, was introduced into Yugoslav wheat varieties. It is therefore surprising that the cultivars Balkan, Posavka 2 and Partizanka Niska are all classified in Class I. Dimitrijevic (1997) suggested that the effect of the wheat-rye translocation might be influenced by the presence of the HMW glutenin subunits 2∗, 7+9, 5+10.

The varieties Italija, Europa and Francuska, released in 1989–1990, have been selected from a cross between the French variety Talent and Novosadska Rana-2. Europa is classified in Class I(II), and Italija and Francuska in Class II(I). Europa and Francuska are both low-input varieties.

Borojevic (1989) made an inventory of the winter-wheat varieties that contributed most to the yield increases (1.65 to 5.61 t/ha) in the province of Vojvodina between 1957 and 1988. On the basis of these data it is possible to compare the area grown with different quality classes of wheat throughout those years. From Figure YU.2 it is apparent that whereas up to the 1970s the wheat area was mainly covered with Class I and/or Class III varieties, after 1976 the importance of those two classes decreased in favour of Class II varieties.

From the pedigree of the Novi Sad varieties (Table YU.8), it is possible to make some remarks on the genetic sources for baking quality. It is evident that the varieties Bezostaya-1, Mironovskaya-808 and Bezostaya-4 have been important sources for baking quality, as well as the Yugoslavian variety U-1 (Strampelli/Marquis), and possibly the Canadian spring wheats Selkirk (via the lines NS-413, NS-422, NS-433 and NS-435) and Purdue 5369 (via the line NS-646).

3.3. The wheat-breeding programme in Kragujevac

Kragujevac is located in Serbia, along the Morava river, in the transitional area between the Dinaric Alps and the Danube Plain. The main objective of the wheat breeding programme at Kragujevac that was started in 1957/58 was

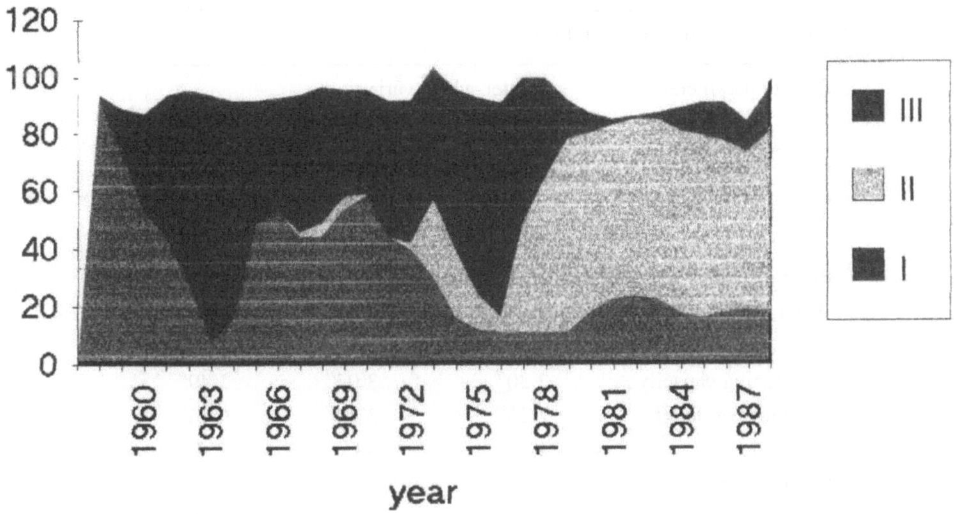

Figure YU.2 . The proportion of the total area of winter wheat in the province of Vojvodina grown with varieties of quality classes I, II and III between 1958 and 1988.

Notes to Figure YU.2.

1958–1963: Bankut 1205 (quality Class I) was replaced by San Pastore and Etoile de Choisy (both quality Class III).

1963: Introduction Bezostaya-1 (Iq), a quality improver variety.

1964–1971: Bezostaya-1 covered 50% of the wheat area; other important varieties: Backa, Aurora and Kavkaz (II), and Leonardo and Libellula (III).

1973–1978: Sava (III) 20% in 1973, 50–60% in 1974–1975.

1976: Introduction of NS Rana-1 and NS Rana-2 (II).

1977–1983: Partizanka (Iq) on about 10%.

1982: Introduction of Jugoslavija.

Source: Borojevic (1989).

to create high yielding semi-dwarf winter-wheat varieties of wide adaptability and good quality. During the first few years mainly Italian varieties were used as crossing parents. In the 1960s several crosses were made with Bezostaya-1. In order to increase the mildew resistance and yield capacity of Bezostaya-1, it was crossed with different Yugoslav and foreign cultivars. These crosses were backcrossed with Bezostaya-1, to maintain a high quality level in the offspring. Selection for quality was done in early generations with the Zeleny test. In advanced generations of more promising lines, quality was tested by farinograph and Chopin alveograph.

One of the most successful results of this breeding programme was the variety Kragujevac-56, from the cross Bezostaya-1/Halle Stamm//Bezostaya-1, released in 1975. This variety with large and vitreous kernels (thousand kernel weight 42.3 g) outyielded both Libellula (a standard for yield at that

Table YU.7. Yield and quality of the winter-wheat variety Kragujevac-56 compared to Bezostaya-1 and Libellula

Quality parameter	Winter-wheat variety		
	Kragujevac-56	Bezostaya-1	Libellula
Grain yield[a] (kg/ha)	5894	5339	5576
Yield (rel. to Bezostaya-1)	110	100	104
(rel. to Libellula)	106	96	100
Protein content (%)	14.5	15.7	15.1
Water absortion (%)	59.5	60.3	58.5
Development of dough:			
– stretching capacity	2′30″	3′07″	1′49″
– stability of dough	2′25″	3′30″	0′15″
– breakdown	42	38	132
Farinograph:			
– quality number	73.4	76.1	37.5
– quality group	A2	A2	C1
Loaf volume	349	365	302

[a] Average of three years (1972–1974) trials at 18 locations.
Source: Popovic & Popovic (1975).

time) and Bezostaya-1 whilst its quality level was close to that of Bezostaya-1 (Table YU.7).

Kragujevac-56 was selected by using the pedigree method up to F_4 and by progeny selection from F_5 onwards. The variety was characterized by a wide adaptability, which was probably the result of it consisting of several genotypes of similar phenotype (Milovanovic et al., 1990). After the registration of Kragujevac-56, this variability was maintained and only extreme genotypes were discarded. One of its sublines was released in 1992 as the variety Kragujevacka 56-s. It was classified in quality Class I, Group A2 (improver wheat), with a quality level similar to Partizanka and a yielding potential higher than Partizanka.

In the years 1968–1973, a collection of 1805 wheat entries were gathered in different parts of Yugoslavia and analysed in Kragujevac on form diversity, disease and quality characteristics (Jankovic, 1975). Of 200 *T. aestivum* entries, 27 were found to have a Zeleny sedimentation value of 50 or higher. Their origin was from Macedonia (11), Kosovo (8), Serbia (7) and Bosnia (1). The highest sedimentation values were found in samples from Macedonia. This region appeared to be especially rich in different forms and species. The possibilities that such a gene pool offers for incorporation of quality characteristics in modern varieties depends largely on how difficult it is to bridge the discrepancy in plant characteristics and yielding ability between the wild material and improved varieties.

The variety Lepenica, released in 1980, with subunit composition 2*, 7+8, 5+10 and *Glu-1* score 10, had the highest *Glu-1* score of the 36 winter-wheat varieties that were tested by Knesevic et al. (1993). The variety Studenica, released in 1990, was decribed as a high yielding, medium early cultivar with a potential grain yield of 11 t/ha. It was classified in quality Class II with values for loaf volume and bread crumb comparable to Partizanka and better than Super Zlatna (Kuburovic et al., 1991)

3.4. The wheat-breeding programme in Zagreb

Based in Croatia, the Institute of Breeding and Production of Field Crops in Zagreb faces climatic circumstances that are different from those in Novi Sad and Kragujevac. The natural conditions are not as favourable for growing high quality wheats as they are in the Pannonic Plain. Bread-making quality has not been as important an issue in the Zagreb wheat breeding programme as it has at the other two stations. However, varieties bred at the Zagreb institute are characterized by their suitability for a variety of agro-ecological conditions, which has allowed them to spread in neighbouring countries such as Hungary, Czechia, Bulgaria and Italy (Tomasovic & Koric, 1990).

The basic objective since the start of its wheat breeding programme in 1947 has been to create semi-dwarf varieties with increased production, wide adaptability, satisfactory quality of grain and flour, and resistance to the most important diseases. The Italian varieties that were widely grown in Croatia in the 1950s lacked winter-hardiness and bread-making quality. They were crossed with American cultivars in order to combine the good traits of both (Tomasovic et al., 1996).

The first varieties from Zagreb were marketed in 1967. The first variety grown on a large scale, both in Croatia, the rest of Yugoslavia and abroad, especially in Czechoslovakia, was Zlatna Dolina, which was registered in 1971. The success of this semi-dwarf variety was mainly due to its high yielding potential. It was the result of a double cross involving two Italian (Libero and Leonardo) and two North American parents (Dakota and Regent). Together with the variety Sanja, released in the same year, it contributed to the gradual replacement of Italian, Russian and French varieties in Croatia.

Some important winter-wheat varieties released from Zagreb between 1978 and 1982 were Superzlatna, Lonja, Sana, Baranjka and Zagrepcanka, of which the latter two were also registered in Hungary. In 1985, two-thirds of the Croatian wheat hectarage was covered with Zagreb varieties, mainly Super Zlatna and Baranjka. Most of the wheat varieties released from Zagreb before 1990 were classified in bread-making quality Class III, reflecting the emphasis that has been put on yielding potential. A marked improvement in the combination of yield and quality was achieved with the release of Marija, in 1990. Although it was graded in farinogram group A2, it was classified as a Class II variety. Two other varieties, released shortly after, were Marina and Olga, both in quality Class I, Group B1. Javor & Tomasovic (1995) stated that in the most recent varieties a definite progress had been achieved in terms of kernel and flour quality. Several of the new varieties had a higher level of protein content and sedimentation value combined with good extensograph and farinograph values, especially water absorption. Moreover, the thousand kernel weight had increased to 40–50 g for the new varieties, compared to 30–35 g for Zlatna Dolina.

3.5. Ancestors of Yugoslav winter-wheat cultivars

Jost & Cox (1990) studied the pedigree of 142 Yugoslavian winter-wheat cultivars and compared cultivars from the four major breeding institutions: Novi Sad, Zagreb, Kragujevac and Osijek. They found that the gene pool of the cultivars from Zagreb was the most genetically diverse and the most distant from the other three gene pools. In contrast Novi Sad and Kragujevac cultivars almost constituted a single gene pool. The most frequent ancestor of Yugoslavian wheat cultivars was the Japanese variety Akagomuchi, the source of the gene Rht8, which controls plant height. It was introduced in Yugoslav wheat breeding through various Italian varieties. Another prominent ancestor was the German variety Neuzucht, introduced in the 1960s through the varieties Kavkaz, Aurora and Skorospelka-35, and a source of the 1B/1R chromosome translocation.

4. Investigations on glutenin and gliadin composition of Yugoslav wheats

4.1. Glutenins

The HMW glutenin subunit composition and *Glu-1* score of the most important varieties mentioned in this chapter are presented in Table YU.8. It appears that the variation in HMW glutenin subunit composition is rather limited with a high frequency of the allele coding for subunits 7+9 on the *Glu-B1* locus and only two alleles on the *Glu-D1* locus. The subunits 2∗, 7+9 and 5+10, resulting in a very high *Glu-1* score (9) are commonly present in cultivars from Novi Sad.

This is in agreement with the findings of Vapa (1990), who compared the allelic variation on the *Glu-1* locus in cultivars from seven different breeding centres in Yugoslavia. She also found that the subunits 1, 6+8 and 2+12 were

Table YU.8. Yugoslav winter-wheat varieties released since 1970 with year of registration, breeding station (BS), pedigree, quality group, HMW glutenin subunit composition and *Glu-1* score

Variety	Year	BS[a]	Pedigree[b]	Qual. group	HMW glutenin subunits on			*Glu-1* score
					A1	B1	D1	
Sava	1970	NS	Fortunato^2/Redcoat	III	1	7+8	2+12	8
Zlatna Dolina	1971	ZG	Libero/Dakota//Regent/3/Leonardo	III	1	6+8	2+12	6
Biserka	1972	NS	Fortunato^2/Redcoat	III	1	7+8	2+12	8
Tena	1973	OS	Libellula/Bezostaya-1	I	1	7+9	5+10	9
Podunavka	1973	NS	Bezostaya-1/Argelato		2*	7+9	5+10	9
Partizanka	1973	NS	Bezostaya-1//Campodoro/Heine VII	Iq	2*	7+9	5+10	9
Nova Banatka	1973	NS	Bez-4/Argelato//Bez-1		2*	7+9	5+10	9
NS Rana-1	1975	NS	Bez-l/NS-262//Mir-808/3/NS-413/Kavkaz	II	N	7	2+12	4
NS Rana-2	1975	NS	Bez-l/NS-262//Mir-808/3/NS-413/Kavkaz	II	N	7	2+12	4
Kragujevac-56	1975	KG	Bezostaya-1/Halle Stamm//Bezostaya-1	Iq	2*	14+15	5+10	?
Super Zlatna	1977	ZG	Sanja/TP 114–1965A//Sanja	III	1	6+8	2+12[c]	
Baranjka	1979	ZG	Sanja/TP 114–196SA//Sanja	III	1	6+8	2+12	
Balkan	1979	NS	Backa/Bez-l//Mir-808/3/NS-433/4/Skor-35	I	2*	7+9	5+10	9[d]
Posavka-2	1980	NS	NS-646/Bez-l//Skorospelka-35	I	2*	7+9	5+10	9[d]
Zvezda	1982	NS	Bez-l/Sava//Mir-808/3/NS-413/Kavkaz	II(I)	N	7+9	5+10	7[d]
Zitnica	1982	NS	Bez-1/Backa//Sava	II(I)	1	7+9	2+12	7
Jugoslavija	1984	NS	Bez-l/NS-646//Aurora	I(II)	2*	7+9	5+10	9
Partiz.Niska	1984	NS	NS-646/Bez-l//Aurora/Partizanka	I	2*	7+9	5+10	9
Somborka	1986	NS	NS-2153/Aurora/Nova Banatka					
Marina	1989	ZG						
Italija	1989	NS	Talent/NS Rana-2	II(I)	N	7+9	2+12	7
Europa	1990	NS	Talent/NS Rana-2	I(II)	N	7+9	2+12	5
Francuska	1990	NS	Talent/NS Rana-2	II(I)	N	7+9	2+12	5
Pobeda	1990	NS	Sremica/Balkan	I				
Proteinka	1990	NS	NS-3726(3)/Macvanka-1	I	N	7+9	5+10	7
Rana Niska	1990	NS	Tobari-66/Kavkaz/Bacvanka-1/NS Rana-1	I	1	7+9	2+12	7
Zitarka	1990	OS		I(II)	1	7+9	5+10	9
Marija	1990	ZG		II	1	7+9	2+12	7
Studenica	1991	KG	Kavkaz/line 5393//Tena	II				
NS Rana-5	1991	NS	NS Rana-l/?/Tisza/?/Partiz./?/Macvanka-1	I				

[a] Breeding stations are abbreviated as ZG (Zagreb), NS (Novi Sad), KG (Kragujevac), OS (Osijek) and SK (Skopje).
[b] NS-646 = San Pastore^{*2} × Purdue 5369; NS-413, 433 = U1/Selkirk//San Pastore/3/Mara; NS-2 = Heine VII / genus 129 (wheat × rye).
[c] Soltes-Rak et al. (1991) found three different biotypes within this variety varying in the *Glu-B1* locus (7/6+8/7+9).
[d] These varieties were reported to have a 1B/1R translocation, which they inherited from Kavkaz, Aurora or Skorospelka (Jost & Cox, 1990).
Sources: Jost & Jost (1989); Vapa (1989); Knezevic et al. (1993); Semenarstvo (1991).

predominant in cultivars from Zagreb and that some cultivars from Kragujevac possessed the allele coding for subunits 14+15 on the *Glu-B1* locus. Vapa assumed that the source of this allele was the German line Halle Stamm, which is a parent of several Kragujevac varieties. She suggested that subunits 14+15 contributed to good bread-making characteristics. The average *Glu-1* score calculated by Vapa was 8.2 for Kragujevac, 7.8 for Osijek, 7.6 for Novi Sad and 6.8 for Zagreb. As the average scores are rather high compared to those from Knezevic et al. (1993), it is unfortunate that the number of varieties involved in this study was not mentioned. Knezevic et al. compared the allelic variation in 36 cultivars and lines from different breeding centres and calculated an average *Glu-1* score of 6.8 for 15 lines from Kragujevac, 7.9 for 11 cultivars from Novi Sad and 7.2 for 8 cultivars from Zagreb.

Vapa & Savic (1988) investigated the HMW glutenin subunits in 86 Yugoslav varieties. Twenty-seven percent of the Yugoslav varieties showed the HMW glutenin subunit composition (2∗,7+9,5+10), which is the composition of Bezostaya-1. The greatest variability was found on the *Glu-B1* locus. Fifty percent of the cultivars investigated had a *Glu-1* score of 8, 9 or 10. The average *Glu-1* score of Yugoslav wheat cultivars released between 1967 and 1987, calculated by Vapa (1989), was 7.3

The influence of *Glu-1* loci on the bread-making characteristics of wheat was investigated by Dencic & Vapa (1996) in a collection of 160 cultivars grown at Novi Sad. Their results confirmed the findings of Kolster et al. (1991) (see Chapter Netherlands and Belgium, section 4.1) that the quality of a cultivar depends far less upon the occurance of individual alleles than upon their combinations i.e. additive and epistatic effects. The null allele at the *Glu-A1* locus had the largest positive effect on most of the bread-making quality components, in combination with the presence of subunits 2+12 on *Glu-D1*. In cultivars with subunits 5+10 on *Glu-D1*, subunit 2∗ was superior to subunit 1 and the null allele.

4.2. Gliadins

Javornik et al. (1990) studied the gliadins (by APAGE) in a collection of 133 Yugoslav cultivars released between 1967 and 1987. They found that 105 had a unique gliadin electrophoretic pattern upon which identification of the cultivar was possible. Similar electrophoregrams were generally due to similar pedigrees. A 1BL/1RS wheat-rye translocation was present in 34 of the 133 cultivars.

Metakovski et al. (1991) investigated the gliadin pattern in 57 Yugoslav cultivars. They reported that about 40% of the varieties showed the presence of biotypes, which was twice as much as in Australian and British cultivars but was similar to Soviet wheats. The Osijek breeding centre was outstanding in the number of gliadin biotypes among its varieties, in contrast to the low number among Novi Sad varieties. The frequency of heterogeneity of HMW glutenin alleles in Yugoslav cultivars was only about 15%, which is similar to frequencies found in Canadian, Australian and Italian varieties.

The allele coding for the 1BL/1RS wheat-rye translocation was commonly present in Yugoslav varieties. Metakovski et al. suggested that selection of this allele might have been favoured due to its linkage to disease resistance genes. The frequency was higher in Yugoslav than in foreign varieties but it was absent in varieties from Osijek and had a very low frequency in varieties from Zagreb. In varieties from Novi Sad a noticeable decrease in frequency was observed in recent varieties.

In contrast, Vapa (1990) reported that the 1BL/1RS wheat-rye translocation was not frequent in the cultivars in her study. It had not been taken into consideration in calculating the *Glu-1* score as 'there are no complaints from Yugoslav bread-making industry on quality of wheats with translocation'.

D.A.D. & J.M.

Acknowledgments

The authors wish to thank Prof. Dr Slavko Borojevic (Faculty of Agriculture, Novi Sad) for his valuable comments on this chapter. We are grateful to Dr Marjan Jost and Dr Slobodan Tomasović for providing us with additional information on wheat breeding in Yugoslavia.

References

Anonymus (1985) Semenarstvo; wheat and barley varieties developed at the Institute of field and Vegetable crops of the Faculty of Agriculture, Novi Sad.

Borojevic S, Potocanac J (1966) The development of the Yugoslav programme for creating high-yielding wheat varieties. Contemporary Agriculture 14: 7–35.

Borojevic S (1989) Thirty years of growing high-yielding wheat cultivars in Vojvodina province. Proc Insitute for Field and Vegetable Crops, Faculty of Agriculture, University of Novi Sad, pp 213–224.

Borojevic S (1990) Genetic improvement in wheat yield potential. Contemporary Agriculture 38(1–2): 45–47.

Borojevic S, Misic T, Dencic S (1994) High yields of Novi Sad winter-wheat cultivars obtained on large farms in Vojvodina (Serbia) in 1993. Not published, personal communication, Wageningen.

Dencic S, Vapa L (1996) Effect of intra and inter-allelic variation in *Glu-A1* and *Glu-D1* loci on bread-making quality in wheat. Cer Res Com 24(3): 317–322.

Dimitrijevic M (1997) Presence and effect of wheat-rye translocation 1BL/1RS on the characteristics and grain yield of Novi Sad high yielding varieties (*Triticum aestivum* L.). Doctoral Thesis, Faculty of Agriculture, University of Novi Sad.

Jankovic M (1975) Properties of the local Yugoslav wheat population. Proc 2nd Intern Winter Wheat Conf, Zagreb, pp 145–152.

Javor P, Tomasovic S (1995) Achievements of Institute Zagreb in breeding of winter wheats. Sjemenarstvo 12(1): 39–46.

Javornik B, Sinkovic T, Metakovsky EV, Novoselskaya AY, Knesevic D (1990) Gliadins in Yugoslav winter-wheat cultivars. In Bushuk W, Tkachuk R (eds), Gluten Proteins, St Paul (MN) AACC, pp 595–602.

Jost M, Cox TS (1989) History of wheat breeding in Yugoslavia. Scientific Expert Journal Podravka 7(1): 1–15.

Jost M, Jost M (1989) Pedigrees of 142 Yugoslav winter-wheat cultivars released from 1967 till 1986. Scientific Expert Journal Podravka 7(1): 19–27.

Jost M, Cox TS (1990) Relative genetic contributions of ancestral genotypes to Yugoslavian winter-wheat cultivars. Euphytica 45: 169–177.

Knezevic D, Surlan-Momirovic G, Ciric D (1993) Allelic variation at *Glu-1* loci in some Yugoslav wheat cultivars. Euphytica 69: 89–94.

Kolster P, van Eeuwijk FA, van Gelder WJM (1991) Additive and epistatic effects of allelic variation at the high molecular weight glutenin subunit loci in determining the bread-making quality of breeding lines of wheat. Euphytica 55: 277–285.

Kuburovic M, Jankovic M, Pavlovic M (1991) Studenica, a new cultivar of winter wheat. Contemporary Agriculture 39(2): 53–56.

Martinic Z, Zanic D (1973) Anbau und Qualitätsmerkmale von Sommerweizen in Kroatien. Getreide, Mehl und Brot 27(3): 87–90.

Metakovsky EV, Knezevic D, Javornik B (1991) Gliadin allele composition in Yugoslav winter-wheat varieties. Euphytica 54: 285–295.

Milovanović MS, Lookhart GL, Knežević DS, Femić VM (1990) Investigation of sublines of Kragujevačka 56 variety of wheat (*Triticum aestivum* spp. *vulgare*). Genetika 22(2): 105–116.

Misic T, Mikic D (1975) Breeding for high yield and quality in winter wheat. Proc 2nd Intern Winter Wheat Conf, Zagreb, pp 393–401.

Popovic A, Popovic S (1975) Investigations of some hard red winter-wheat lines in regard to yield and quality. Proc 2nd Int Winter Wheat Conf, Zagreb, pp 372–378.

Potocanac J (1975) Effect of wheat breeding on wheat production in Yugoslavia. Proc 2nd Int Winter Wheat Conf, Zagreb, pp 2–23.

Soltes-Rak E, Kevresan S, Rajcan-Separovic E, Vapa L (1991) Relationship between the RLFP pattern at the *Glu-1* loci and the bread-making quality of the Yugoslav wheat cultivars. Periodicum Biologorum 93(3): 373–375.

Tomasovic S, Koric B (1990) Work on breeding winter-wheat varieties in the Institute for breeding and production of field crops of the Faculty of agricultural sciences (Zagreb); high yielding wheat varieties. Wheat Information Service 71: 32–35.

Tomasovic S, Javor P, Sesar B (1996) Contribution of the BC Institute Zagreb to the breeding and seed production of winter wheat in the Republic of Croatia. Acta Agronomica Hungarica 44(4): 397–402.

Vapa L, Savic M (1988) The HMW glutenin subunit composition of wheat varieties bred in Yugoslavia. Proc 7th Intern Wheat Genetics Symposium, Cambridge, pp 1019–1023.

Vapa L (1989) Genetic variability of HMW glutenins in Yugoslav winter-wheat cultivars. Scientific Expert Journal Podravka 7(1): 31–44.

Vapa L (1990) Allelic variation of *Glu-1* loci in Yugoslav wheat cultivars. In Bushuk W, Tkachuk R (eds), Gluten Proteins, St Paul (MN) AACC.
Vapa L, Dencic S, Soltes-Rak E, Kevresan S (1995) Genetic variation in *Glu-1* loci and bread-making quality in wheat. Cer Res Com 23(1–2): 161–166.

Romania

1. Introduction

At the beginning of the 20th century Romania (RO) was one of the main wheat exporting countries in the world, after Russia, Argentina, Canada and USA. The export was based on regular harvests of vast areas with favourable soil and climate for wheat production, combined with a good infrastructure through various waterways. Most Romanian wheat was exported by the Danube to Hungary, Austria and Germany (Loewé, 1923). Wheats from Romania were known for their outstanding milling and bread-making quality and used to be classified in the same category as strong wheats from Canada and Russia (Åkerman, 1929).

Nowadays, with a wheat area of nearly 2 500 000 ha and a production of more than 7.5 Mt, Romania is one of the most important wheat producers in Eastern Europe. Most of the wheat produced in Romania is used for human consumption. Historically, maize has played a large role in the nutrition of the population, and mixed wheat-maize bread was widely consumed. There is, however, a growing tendency towards the consumption of white wheat bread (Lasztity, 1995). Wheat and maize are the two major cereals; the production of rye and oats is very limited. The development of wheat, maize and total cereal production since 1925 is summarized in Table R.1.

Table RO.1. Acreage (1000 ha) and production (1000 t) of wheat, maize and cereals in Romania since 1925

	Wheat		Maize		Cereals total	
	ha	prod	ha	prod	ha	prod
1925[a]	2 861	2 435				
1934–1938	2 537	2 600	3 879	4 032		
1948–1952	2 728	2 778	3 089	2 369	6 987	5 813
1961–1965	2 966	4 321	3 308	5 853	6 772	10 888
1969–1971	2 527	4 433	3 170	7 354	6 211	12 662
1919–1981	2 154	5 371	3 309	11 823	6 340	18 1O9
1989–1991	2 242	6 866	2 592	8 023	5 927	18 286
1995–1997	2 224	5 999	3 141	10 737	6 202	18 726

[a] 1925 data refer to the former Kingdom of Romania, including Bessarabia.
Sources: IIA and FAO statistics.

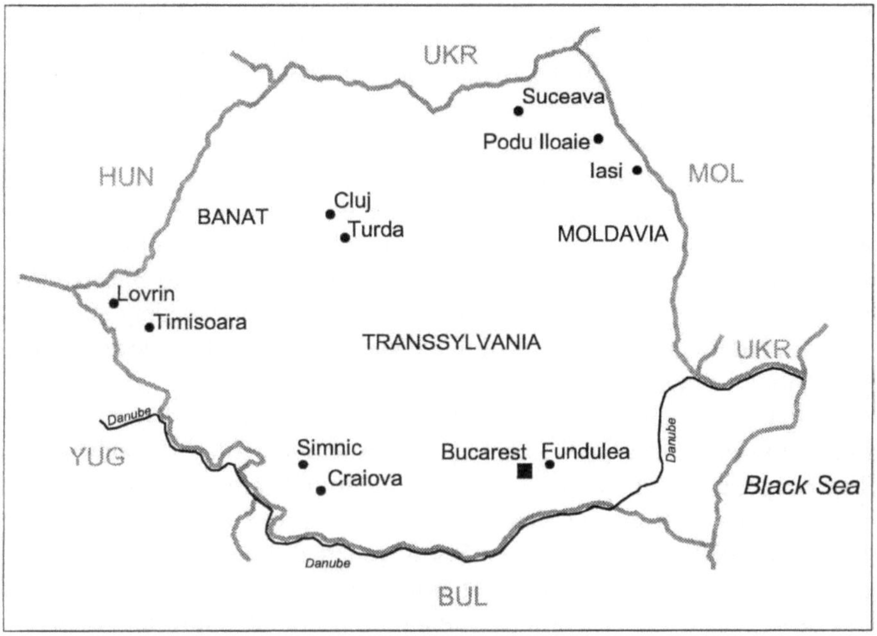

Figure RO.1. Wheat growing areas in Romania and the location of the most important breeding centres for bread wheat. (Design: Lucas Janssen)

Romania has a continental climate that is characterized by large differences in temperature between winter and summer, as well as between night and day during spring. Heavy snowfalls during the winter usually provide enough soil moisture to compensate for a lack of rain during the summer. The vast majority of the wheat grown in Romania is autumn-sown. The main production area is the Danube Plain in the south of the country, where wheat is partly grown under irrigation. Other important wheat growing areas are Transsylvania, the northern part of Moldavia and the Banat region (Figure RO.1).

2. Wheat breeding in Romania

2.1. The period before 1960

Because Romania was an exporter of wheat, the quality aspect was always an important issue and descriptions of it may be found in the early literature. An extensive survey, that included chemical analyses of wheat harvests from 1900 to 1908 was reported by Zaharia (1910).

For export purposes, wheats were divided into three groups (van Hissenhoven, 1938), ranging from the highest quality wheats originating from

Moldavia and Bessarabia (nowadays the Moldavian Republic), good quality wheats originating from the Danube Plain, to third class, but still sufficiently good quality wheats from the Banat area and west Transsylvania. Most of the wheat harvested in these regions consisted of mixtures of different types, such as the White Rumanian Wheat, Banat wheat, Red Wheat, Ghirca Wheat, Sandomir Wheat and White Wheat Ulka (Sakoff, 1960).

Organized scientific breeding in Romania goes back to 1900, when Munteanu started to improve wheat by individual plant selection from local varieties, which consisted mainly of awned winter types similar to Banatka (Vavilov, 1935). The work of two Austrian breeders at the breeding station of Cenad in the Banat region resulted in the variety Cenad-117, selected from a population of Banatka wheat.

In 1911, Ionescu-Sisesti started with the hybridization of local populations with squarehead types (e.g. Balan Laza × Squarehead Hohenheim). In close cooperation with professor Nilson-Ehle from Sweden, wheats collected from all parts of Romania were screened as possible sources for breeding (Saulescu, 1930). In 1928 Ionescu-Sisesti founded the Institute for Agricultural Research (ICAR) in Bucharest, with experimental stations in Cluj and Iasi. Several Cluj and Iasi varieties were released in the following years.

The most important achievement of Ionescu-Sisesti as a wheat breeder was the cultivar A-15, a pure line selection from the American cultivar Tenmarq. One of the parents of Tenmarq was Red Fife, a selection made in material originating from the Pannonic region, which is geographically linked to the Banat region in Romania. Thus, after a trip around the world, this wheat type returned to its 'roots'. Soon after its introduction in 1933, A-15, which combined productivity, quality and a wide adaptability, was grown on two-thirds of the area under wheat in Romania and continued to be grown for more than thirty years (Ceapoiu, 1985).

The most representative Romanian wheat varieties released until the 1960s are listed in Table RO.2 with their pedigree and year of release.

2.2. Wheat breeding after 1960

Beginning in the 1960s, Romania, along with other Eastern European countries, made concerted efforts to select new cultivars with higher yields. These were based mainly on Soviet (Bezostaya-1, Skorospelka-3B), Italian (San Pastore, Fiorello) and Bulgarian (No. 301) varieties. As a result, wheat production increased, but the newer cultivars were generally less cold resistant and had widely varying baking quality (Lasztity, 1995).

The main centre for wheat-breeding research during these years was and still is the Research Institute for Cereals and Industrial Crops Fundulea (ICCPT), with several experimental stations (SCA) throughout the country.

Table RO.2. Romanian bread wheat varieties released until 1965

Variety	Pedigree	Year
Odvos-241	selection from Champlain	1924
Cenad-117	selection from Banatka	1930
A-15	selection from Tenmarq	1933
Baragan-77	selection from A-15	1952
Cluj-650	Ridit/Odvos-241//Bankuti 1201	1957
Iasi-1	Iasi-116/Petkus Rye	1963
Bucuresti-1	Kanred/Tiganesti	1965

The most important stations working on bread wheat are located in Lovrin, Turda, Podu Iloaie, and Simnic (Figure RO.1). Several Romanian wheat varieties have been named after these stations. The first varieties from the Fundulea wheat-breeding programme were introduced in 1970. The area sown with varieties from Fundulea increased from 1% in 1971 to more than 50% since 1977, with a maximum of 65% in 1987 (Saulescu & Ittu, 1994). The rest of the area under wheat was mainly sown with cultivars released by the Agricultural Research Stations in Turda (18%) and Lovrin (15%).

Wheat-breeding research is also carried out by the Agronomic Institute 'Nicolae Balcescu' in Bucharest and the Agronomic Institute in Timisoara, close to the border with Hungary and Yugoslavia.

A list of the most important winter-wheat varieties released after 1970 is presented in Table RO.3.

2.3. Quality breeding at Fundulea after 1970

An extensive quality breeding programme was carried out during the 1970s by Ceapoiu and his co-workers. The most important aim with respect to bread-making quality was to obtain new wheat varieties with high protein content. Although some old Romanian wheat varieties such as Odvos-241, Cenad-117 and Cluj-650, and newer ones like Dacia and Favorit were known for their high protein content and good bread-making qualities (Ceapoiu et al., 1975), a further increase was pursued.

The protein content of some varieties, commonly grown in the 1970s, and of the line F26–70, is presented in Table RO.4. Bezostaya-1 and Kavkaz have been used as a reference. The line F 26–70 (obtained from a cross between the Austrian line Au 57–59, and Bezostaya-1) was used by Ceapoiu and his colleagues (as well as by Italian breeders) to further increase the protein content of their wheat material. They also used the soft red winter-wheat variety

Table RO.3. Romanian winter-wheat varieties released after 1970

Variety	Pedigree	Year
Dacia	Bucuresti-1/Skorospelka-3B	1971
Exelsior	Bucuresti-1/Skorospelka-3B	1971
Moldova	Bucuresti-1/Skorospelka-3B	1971
Favorit	Odvos-241/Bezostaya-1	1971
Lovrin-10	Abbondanza/Triumph//Bezostaya-2	1973
Lovrin-231	Bezostaya-1/Fiorello	1974
Iulia	Bezostaya-1/Belaya Tserkov-198	1974
Ceres	Michurinka/Bezostaya-1	1974
Lovrin-13	Heine VII/Skorospelka-3B	1974
Doina	Etoile de Choisy/Monon	1977
Lovrin-24	Lovrin-10/Lovrin-62	1977
Lovrin-32	Rannyaya-12/Nadadores-63//Lovrin-12	1979
Fundulea-29	Aurora/Riley-67	1979
Transilvania	US-(60)43/Aurora//Fiorello/Bezostaya-1	1981
Lovrin-34	Nadadores-63/Rannyaya-12/Heine VII/Bezostaya-1	1981
Moldova-83	Bezostaya-1/Bruckner-6111//Fiorello/Bezostaya-1	1983
Turda-81	Bezostaya-2/Novi Sad-611	1984
Fundulea-133	Mexican dwarf/Colorado line//Bezostaya-1/Odesskaya-16	1984
Flamura-80	Lovrin-32/3/Rannyaya-12/Nadadores-63//Lovrin-12	1984
Arieşan	Rubin//Bezostaya-1/Fiorello	1985
Aniversar	Lovrin-11/F 53–67	1986
Fundulea-4	Lovrin-32/Fundulea-29//Lovrin-32	1987
Flamura-85	Multiline from Flamura-80	1989
Dropia	Colotana/3/Fundulea-29/Lovrin-32//Flamura-80	1993

Source: Gh. Ittu (1997).

Atlas 66, as well as several Argentine varieties, such as Klein Atlas, Magnif-41, Agrolit Vigliano and Fontezuela, all known to have very high protein contents (Ceapoiu et al., 1975).

It is likely that the extremely high protein levels in 1971 were influenced by the weather of that year: wheat yields were far below average, probably due to extreme heat, resulting in high protein levels. Ceapoiu et al., however, do not mention these special circumstances.

The data in Table RO.5 may serve as an illustration to the characteristics of the most important varieties grown in Romania in the 1970s.

The release of Fundulea-29 in 1979 was a marked step forward in yielding capacity. In trials in all parts of the country it outyielded Bezostaya-1, Iulia

Table RO.4. Protein content (%) in the grain of winter-wheat varieties and lines tested at Fundulea during 1971–1973 (average of 6 locations)

Variety	Protein content in year:			Average
	1971	1972	1973	
Bezostaya-1	16.53	13.47	14.08	14.69
Kavkaz	17.06	13.66	13.64	14.78
Lovrin-10	17.04	13.94	14.19	15.09
Dacia	16.45	15.51	14.65	15.53
Favorit	17.69	15.27	14.60	15.86
Iulia	15.85	14.98	14.93	15.25
F 26–70	19.16	19.20	17.05	18.47

Source: Ceapoiu et al. (1974).

Table RO.5. Production and quality characteristics of seven Romanian varieties compared to Bezostaya-1 and Partizanka (harvest 1980)

	Year	Yield (t/ha)	Protein content	Sediment. index	Pelshenke index
Bezostaya-1	1961	4.06	14.9	54.70	65
Partizanka	1977	4.28	13.5	63.80	155
Dacia	1970	4.19	13.8	54.70	63
Ceres	1973	4.42	14.9	63.80	68
Iulia	1973	4.22	13.8	54.70	35
Lovrin-231	1974	4.12	13.1	38.70	35
Lovrin-32	1979	4.47	13.8	50.00	120
Diana	1976	4.45	13.8	54.70	55
Doina	1976	5.05	13.5	43.30	60

Source: Marinescu et al. (1982).

and Partizanka by 10–20% (Ceapoiu et al., 1982). The quality of Fundulea was reported to be equal to that of Bezostaya-1, and especially the crude protein content with an average of 14.8% was said to be high (Table RO.6).

Soon after the release of Fundulea-29, the variety Transilvania was released, which was equal in quality but had a higher yielding capacity than Fundulea-29. A number of bread-making characteristics of Fundulea-29 and other varieties in trial between 1981 and 1984, are presented in Table RO.7. The data are the average of three-year trials at some 25 locations throughout the country. For some varieties the *Glu-1* score has been added (after

Table RO.6. Quality characteristics of the variety Fundulea 29 compared to Bezostaya-1, Iulia and Partizanka

Variety	Protein content	Sediment. index	Pelshenke index	Loaf volume
Bezostaya-1	14.61	46.1	90.5	446
Partizanka	14.58	46.0	98.5	446
Iulia	15.42	39.1	55.5	451
Fundulea-29	14.78	30.1	88.8	435

Source: Ceapoiu et al. (1982).

Table RO.7. Bread-making characteristics (average 1981–1984) and relative yields (average 1980–1982) of 10 bread wheat varieties released between 1973 and 1984

Variety	TKW[a]	Protein content	Pelshenke index	Proportion wet gluten	Farin. value	Loaf volume	*Glu-1* score	Relative yield
Iulia	40.5	14.58	51	32.22	39	441		97
Lovrin-32	35.2	13.54	82	25.86	41	470		107
Turda-195	41.8	13.76	84	27.25	40	440	7	97
Fundulea-29	36.9	13.64	73	27.80	37	420	9	107
Transilvania	47.9	13.46	68	26.52	36	426	5	112
Turda-81	43.8	13.08	57	25.29	34	431	9	
Lovrin-34	45.3	13.64	70	27.63	39	467		
Moldova-83	39.5	14.30	76	31.15	40	454	10	
Fundulea-133	37.2	13.50	64	28.28	39	463	9	
Flamura-80	44.8	13.72	86	26.19	40	466	10	

[a] Thousand kernel weight (g).
Sources: Oproiu et al. (1986); Botezan et al. (1983); Hagima et al. (1989b).

Hagima et al., 1989b). From these data it does not seem obvious that there is a correlation between any of the quality parameters and the *Glu-1* score. This correlation will be discussed under 2.4.

The variety Aniversar, released in 1987, set new standards for yielding capacity: 10% higher yield than Fundulea-29 and better grain quality (especially gluten content) than Fundulea-29 and Partizanka (Zama et al., 1994). One of the most recent important Romanian-bred wheat varieties is Dropia, released in 1993. It has been reported to have superior bread-making quality and a high yielding capacity (Saulescu et al., 1995).

2.4. HMW glutenin and gliadin composition of Romanian wheats

Research on glutenin and gliadin composition of wheat has been carried out by Hagima and her co-workers at the ICCPT Fundulea. A positive correlation between *Glu-1* scores and farinograph values as well as loaf volumes was observed (Hagima et al., 1987). The correlation was even higher when the glutenin score was corrected for the gliadin composition. Especially the presence of gliadin block 1B3, derived from rye, was found to have a negative effect on quality. Highest correlations were found between sedimentation index and farinograph score ($r = 0.905$) and between sedimentation index and loaf volume ($r = 0.923$).

A survey of 16 cultivars commonly grown in southern Romania showed that 11 of these possessed the subunit *Glu-D1* (5+10), considered to be related to good bread-making quality. A similar result was found in a survey carried out by Hagima & Saulescu (1987), in which it also appeared that the less favourable allele *Glu-B1* (7+9) was present in many Romanian varieties.

From an analysis of the gliadin and glutenin composition of 24 varieties, Hagima et al. (1989) concluded that the gliadin blocks 1A2, 1A4, 1B1, 1D5, 6A3 and 6D2 (coded according to Poperelya & Sobko, 1987) could be associated with good bread-making quality. The classification of glutenin alleles corresponded with that of Payne et al. (1987).

The change in HMW glutenin composition of Romanian-grown winter-wheat varieties between 1980 and 1995 was analysed by Hagima et al. (1998). The results are presented in Table RO.8. They calculated that the average *Glu-1* score increased from 5.7 in 1980 to 8.6 in 1996. The latter score is very high and comparable to that of modern Finnish spring-wheat varieties, calculated by Sonntag-Strohm (1997).

Comparing the glutenin composition of the older Romanian varieties to the modern ones, it is apparent that the increased *Glu-1* score is mainly due to the disappearance of the nullisome on the *Glu-A1* locus and the increased occurence of the 7+8 allele on the *Glu-B1* locus. The 2+12 allele was only found in Aniversar and Lovrin 41, two varieties that consisted of a mixture of genotypes. Mixtures offer the possibility to chose the genotype with the highest score for seed multiplication. The subunit-pair 2+10, found in Transilvania-1, had not been described before and had never been given a score.

Hagima et al. (1998) do not comment on the effect of selection on the improved quality score. Reference is made to a publication by Graybosch (1992), who studied the glutenin composition as a function of year of release of 291 American cultivars. He suggested that the change in composition is more dependant on random factors than on selection for enhanced quality characteristics.

Table RO.8. HMW glutenin composition and *Glu-1* score of the most important winter-wheat varieties grown in Romania between 1980 and 1996

	Year	HMW glutenin subunit			*Glu-1*
		1A	1B	1D	score
Bezostaya-1	1961	2*	7+9	5+10	9
Dacia	1971	N	7	5+10	6
Libellula	1971	N	20	2+12	4
Iulia	1974	N	7+9	2+12	5
Ceres	1974	N	7+9	2+12	5
Doina	1977	N	7+8	5+10	8
Lovrin-24	1977	N	7+9	5+10	7
Partizanka	1977	2*	7+9	5+10	9
Fundulea-29	1979	2*	7+9	5+10	9
Lovrin-32	1979	N	7+9	5+10	7
Transilvania-1	1981	N	7+9	2+10	⩾4
Lovrin-34	1981	N	7+9	5+10	7
Moldova-83	1983	1	7+9	5+10	9
Turda-81	1984	2*	7+9	5+10	9
Fundulea-133	1984	2*	7+9	5+10	9
Ariesan	1985	2*/N	7+9	5+10	7/9
Aniversar	1986	1	7+8/7+9	5+10/2+12	8/10
Albota	1986	N	7+9	5+10	7
Lovrin-41	1987	2*	7+9	5+10	9
Simnic-30	1987	1	7+8	5+10	10
Fundulea-4	1987	N	7+9	5+10	7
Flamura-85	1989	2*/N	7+8	5+10	8/10
Apullum	1992	1	7+8	5+10	10
Rapid	1992	N	7+8	5+10	8
Dropia	1993	2*	7+8	5+10	10
Delia	1993	N	7+9	5+10	7

Source: Hagima et al. (1998).

Hagima et al. (1998) reported that 30% of the Romanian wheat crop in 1997 consisted of varieties with a *Glu-1* score of 10. In Scandinavian countries, varieties with similar glutenin composition, have been reported to cause problems due to overstrong gluten (see Chapter Finland, section 6.1). The Romanian bread-making industry, however, has not reported this type of problem. The difference in appreciation of similar varieties in Finland and Romania may be due to differences in processing techniques (e.g. intensity of kneeding) but may also be due to differences in growing conditions. It has been reported earlier that the same genotype grown under different conditions, resulted in different baking quality.

D.A.D.

Acknowledgment

The author wishes to thank Dr Gheorghe Ittu (Research Institute for Cereals and Industrial Crops, ICCPT, Fundulea) for reviewing this chapter and providing additional literature on wheat breeding in Romania.

References

Åkerman A (1929) La qualité boulangère des blés d'hiver de Suède et les possibilités de l'améliorer par croisement. Bulletin de l'Association Internationale des Sélectionneurs de Plantes de Grande Culture 2.

Botezan V, Moldovan V, Moldovan M (1983) Tehnologii, ameliorare, producere de seminte; soiul de griu de toamna Transilvania. Productia vegetala, cereale si plante tehnice 8: 3–11.

Ceapoiu N, Ittu G, Oproiu E (1974) Ereditatea continutului si calitatii proteinelor de griu (*T. Aestivum* spp. *vulgare*). Probleme de Genetica Teoretica si Aplicata 6(5): 321–343.

Ceapoiu N, Eustatiu N, Ittu G, Saulescu NN (1975) Wheat breeding for improved nutritional quality in Romania. Proc Intern Wheat Conf, Zagreb, pp 362–366.

Ceapoiu N, Saulescu NN, Ittu G, Tapu C, Eustatiu N, Oproiu E, Cernescu L (1978) Cercetari privind ameliorarea continutolui de proteine la griul comun. Probleme de Genetica Teoretica si Aplicata 10(5): 479–491.

Ceapoiu N, Saulescu NN, Eustatiu N, Tapu C, Ittu G, Negulescu F, Ionescu-Cojocaru M (1982) Soiul de griu de toamna Fundulea 29. Analele ICCPT 50: 100–105.

Ceapoiu N (1985) Achievements of Gheorghe Ionescu-Sisesti in the field of wheat breeding. Probleme de Genetica Teoretica si Aplicata 17(3): 183–203.

Craciun T, Pirvu T, Oproiu E, Cernescu L (1983) Linii noi de griu de toamna cu bobul bogat in proteina de calitate superioara. Lucrari Stiintifice, Institul Agronomic 'N. Balescu' 26: 5–13.

Graybosch RA (1992) High molecular weight glutenin subunit composition of cultivars, germplasm and parents of US Red Winter Wheat. Crop Sc 32: 1151–1155.

Hagima I, Oproiu E, Saulescu NN (1987) Corelatia intre compozitia subunitatilor gluteninice cu greutate moleculara mare si unii indici ai calitatii de panificatie la griu. Probleme de Genetica Teoretica si Aplicata 19(2): 73–86.

Hagima I, Saulescu NN (1987) Unelele aspecte ale utilizarii electroforezei gluteninelor in ameliorarea calitatii de panificatie la griu. Probleme de Genetica Teoretica si Aplicata 19(1): 1–12.

Hagima I, Saulescu NN, Mihailescu F (1989a) Cercetari privind determinismul genetie al proteinelor de rezerva din boabele de griu. Cercetari de Genetica Vegetala si Animala 1: 159–169.

Hagima I, Mihailescu FL, Saulescu NN (1989b) Caracterizarea unor soiuri de griu de toamna prin electroforeza gliadinelor si a gluteninelor. Analele ICCPT 57: 385–392.

Hagima I, Saulescu NN, Ittu G (1998) Changes in High Molecular Weight glutenin subunit composition of wheat (*Triticum aestivum* L.) grown in Romania from 1980 to 1995. Romanian Agricultural research (in press).

Hissenhoven P van (1938) Le mouvement des grains dans le monde. Brussels: Editions Ceres, 858 pp.

Ittu G (1997) Personal communication, Fundulea, Wageningen.

Lasztity R (1995) Wheat usage in eastern Europe. In Faridi H, Faubion JM (eds), Wheat end uses around the world. St Paul (MN) AACC.

Loewé ME (1923) Les pays exportateurs de blé dans le monde. In Le Blé; compte-rendu des travaux de la semaine nationale du blé, Paris, 1923.

Marinescu M, Cremenescu G, Enescu S, Grigore M, Ceausu C, Mihailescu D (1982) Influentia genotipului si a ingrasamintelor asupra calitatii griului. Analele ICCPT 49: 91–98.

Oproiu E, Cernescu L, Cretu M (1980) Calitate noilor soiuri Romanesti de griu comun de toamna extinse in productie. Productia vegetala, cereale si plante tehnice 32(11): 23–28.

Oproiu E, Cernescu L, Saulescu NN, Ittu G, Popescu S (1986) Cercetari privind calitatea de panificatie a noilor soiuri de griu de toamna. Productia vegetala, cereale si plante technice 38(6): 23–31.

Payne PI, Nightingale MA, Krattiger AF, Holt LM (1987) The relationship between HMW glutenin subunit composition and the bread-making quality of British grown wheat varieties. J Sc Food Agr 40: 51–65.

Poperelya FA, Sobko TA (1987) Ghenetika gliadina ozimoi miagkoi pşeniti. In Voprosi selektii i ghenetiki zernovih kultur, Praga, pp 231–242.

Sakoff AN (1960) Romania. In Anonymus (1960) Progressive wheat production. Geneva: Centre d'Etude de l'Azote, 338 pp.

Saulescu N (1930) Pflanzenzüchtung in Rumänien. Der Züchter 2: 22–30.

Saulescu NN, Ittu G (1994) Costs and returns of wheat-breeding programmes. Romanian Agricultural Research 1: 73–76.

Saulescu NN, Ittu G, Mustatea P, Tianu M (1995) Dropia, un nou soi de grau de toamna cu calitate superiora de panificatie. Analele ICCPT 62: 17–26.

Sontag-Strohm T (1997) Gliadin and glutenin subunit alleles on Group 1 chromosomes in Finnish Spring Wheats. Acta Agric Scand (Sect B, Soil and Plant Sci) 47: 98–105.

Vavilov NI (1935) The origin, variation, immunity and breeding of cultivated plants (transl. Starr Chester K) Chronica Botanica 13 (1949): 364 pp.

Zaharia AL (1910) Le blé Roumain. Bucarest: Ministry of Agriculture, 581 pp.

Zama E, Gaspar I, Bodea A (1994) Experimental results with winter wheat cultivars obtained at the Suceava Research Station. Cercetari Agronomice in Moldova 27(3–4): 55–60.

Poland

1. Introduction

With 8 500 000 ha of cereals, covering more then half of its arable land, Poland (PL) is one of the main cereal producing countries of Europe. The most important cereal crops are wheat and rye, occupying both 2 400 000 ha in 1995, and producing 8.7 and 6.6 Mt, respectively.

For centuries, and until very recently, rye was the main cereal crop in Poland. With a relatively short vegetation period, a low risk of winter killing and tolerance of soil acidity, rye is well adapted to the specific climatic and soil conditions in Poland. The introduction of triticale has been very successful in Poland and has replaced rye to some extent. Wheat production has gradually increased to take a leading position due to higher yields of wheat and a continuing increase in the consumption of wheat and rye-wheat bread, as opposed to rye bread.

Cereal cultivation in Poland has a long history and Poland has been an exporter of cereals since the 13th century. Large quantities of cereals were exported from the southeastern part of Poland to Constantinople during the 14th and 15th centuries. Whereas initially mainly rye was exported, exports of wheat increased from the 16th century onwards. The export was at its height during the late 17th and early 18th centuries, when large quantities of wheat were shipped down the river Vistula to Gdańsk and from there on to the market of Amsterdam (Kostecki & Wolski, 1963; Heyder, 1979). Much of this wheat originated from the Sandomierz region in southeastern Poland (Figure PO.1), a region that has given its name to Sandomirka wheat and Sandomirka bread, a sour type of rye-wheat bread.

Poland also has a long-standing tradition in plant breeding and seed production. The first plant breeding enterprise (Dańków) was established in the second half of the 19th century and seed export began soon after. The first four original Polish varieties were those of sugarbeet and cereals.

Through different periods of occupation by Germany, Austria and Russia, the territory of Poland has undergone substantial changes. Before World War II, Poland included a vast area in what is now Lithuania, Belarus and Ukraine. These changes have greatly influenced Polish agriculture and breeding work.

Figure PO. 1. Wheat growing areas and breeding centres in Poland. (Design: Lucas Janssen)

2. General remarks on wheat growing

2.1. Geographical situation and climate

Poland is situated in an area of transition between a maritime zone, influenced by the Atlantic Ocean and Baltic Sea, and a continental zone, affected by the vast area of Eurasia. The prevailing climate is continental but it is more temperate in the west than in the east of Poland. Rainfall is more abundant in the north and the south than in the centre, where in some areas annual rainfall is no more than 450 mm. About 60% of the arable land in Poland has light sandy soils of the 'rye-potato type'. These two crops covered 50% of the arable land in Poland up to the 1960s but dropped to approximately 30% in the 1990s.

2.2. Wheat growing areas and production

The main Polish wheat producing areas are (Kostecki & Wolski, 1963):

- the southeast, having the highest average soil fertility but also a continental climate with severe winters and hot summers; due to a short vegetation period, the possibilities of obtaining record yields in this region are limited;

- the southwest, a region in Silesia with a relatively mild climate; the highest yields are obtained in this area, partly due to a long vegetation period;
- two smaller regions in the north: the area around Gdańsk and the area around Szczecin, close to the German border.

Both winter and spring wheat are grown in Poland. Winter wheat, however, is more important, due to its better utilization of moisture. The areas and yields of wheat and rye in Poland since 1934 are presented in Table PO.1.

Table PO. 1. Area (1000 ha) and production (1000 t) of wheat and rye in Poland since 1934

Year	Wheat		Rye	
	area	prod	area	prod
1924–1925	1 083	1 229		
1934–1938	1 343	1 965	5 353	6 854
1948–1952	1 464	1 833	5 063	6 374
1961–1965	1 516	2 988	4 563	7 466
1969–1971	2 004	4 924	3 766	7 143
1979–1981	1 525	4 189	2 970	6 166
1989–1991	2 304	8 919	2 293	6 053
1995–1997	2 396	8 481	2 396	5 847

Source: IIA and FAO statistics.

Of the 25.9 Mt grain produced in Poland in 1995, by far the largest part was used as cattle feed. Only 5.9 Mt was used for human consumption, including 4.3 Mt wheat and 1.3 Mt rye (Jensen, 1992). Polish cereal consumption per head was 151 kg annually. Traditional Polish bread is made from a mixture of rye and wheat. Several types of mixed-flour breads are also produced, differing in rye-wheat flour ratio from 70:30 (Sandomierka type) and 50:50 (Zakopański type) to 20:80 (Mazowiecki type). 'French bread', made from wheat only, has become more popular recently (Lasztity, 1995).

3. Wheat breeding in Poland

The approach of Polish breeders has always been determined by the natural conditions of the country and the structure of its agriculture: light and sandy soils; a vegetation period that is shorter than in Western Europe; and a low amount of precipitation. Generally speaking, the value of Polish wheats for European breeders lies in the varieties originating from central and eastern

Figure PO. 2. Simplified pedigrees of winter-wheat varieties Grana, Luna and Malwa (names of foreign varieties are printed in italics).
Source: Wolski (1975).

Poland, with their adaption to severe growing conditions and their high level of winter-hardiness (Kostecki & Wolski, 1963).

3.1. Traditional varieties and wheat breeding before 1945

Three important traditional varieties in Poland are Sandomierka, Ostka Galicyjska and Banatka types, the latter originating from the Hungarian land variety Banat. These land varieties have been more or less valuable as a basis of subsequent varieties. Sandomierka proved to be of poor combining value and in spite of many crosses has not contributed to the pedigree of modern Polish varieties. However, selections from Ostka Galicyjska and Banatka, both representing a source of drought resistance, have resulted in several varieties, as may be seen in Figure PO.2. It is likely that the Canadian variety Red Fife has been selected from Ostka Galicyjska.

Plant breeding activities between 1850 and World War I were different for each of the three annexed sectors of Poland (Kostecki, 1962). In the German sector, local varieties were used and there were hardly any Polish breeders. In the Austrian sector, the most important breeders dealing with cereals were Zeliński, Turnau and Miczyński. The greatest breeding effect however, was made in the Russian sector by breeders such as Sempolowski, Peplowski and Bielawski. Peplowski produced Plocka and Sarnowska, the first Polish wheat varieties resulting from methodical selection. Bielawski, working in Wysokie Litewskie, bred the first wheat (and rye) varieties of the Wysokolitewka type.

The main efforts of Polish breeders towards the end of the 19th century were directed towards combining winter-hardiness and other positive qualities of the Polish winter wheat varieties with the high yielding capacity and lodging resistance of the squarehead type varieties. In the 1880s, Janasz at Dańków, the oldest breeding station in Poland, situated 40 km southwest of Warsaw, started to make crosses on a large scale. This resulted in several varieties bearing the name Dańkowska, such as Dańkowska Selekcyjna and Dańkowska Graniatka. Especially the latter gained a leading position in Poland in the period between the two world wars; it has also been cultivated in northwestern parts of the former Soviet Union (Wolski, 1975). Dańkowska 40, a mutant from Dańkowska Zachodnia (transl. = 'western Dańkowska') was grown for over 20 years in northern and central Poland, until the 1960s.

At the end of the 19th century, Miczyński started breeding in Austrian-occupied Poland. He produced several cultivars, some with improved baking quality. Through a selection from Ostka Galicyjska he obtained the variety Ostka Mikulicka, an early, bearded variety with good baking quality. A cross between this variety and Extra Squarehead resulted in Ostka Złotokłosa (transl. = 'with golden spikes') with improved lodging resistance. The variety Ostka Kazimierska was later obtained (by Ruebenbauer), from a cross between Kostromka, a very winter-hardy variety of Russian origin and the Hungarian variety Esterhazy, with very good baking quality.

In western Poland, mostly German varieties were grown. Some of the Polish spring wheat varieties grown before World War II were selections from German varieties: Pomorzanka (from Heines Kolben), Rokicka (from Heines Koga) and Opolska (from Jabo Janetzki).

3.2. Wheat breeding after 1945

3.2.1. Organisation

World War II was disastrous for Polish breeding work and seed production. The breeding companies that had developed since the beginning of the century had to start from the beginning after the country was liberated in 1945. The private breeding firms were put under state management and were incorporated in the Central Board of Plant Selection. The seed centres were transformed into the Association of Plant Breeders and Seed Producers. One of the oldest institutions of agricultural research is the State Research Institute of Agriculture (PINGW) in Puławy. With a cereal department in Kraków and laboratories in several regions, it has undertaken most of the technical work on grain quality. In 1951, the government initiated the Plant Breeding and Acclimatisation Institute (IHAR) in Warsaw, as well as a number of breeding stations: SHR Choryn (Poznań District), SHR Polanowice (Kraków District), SHR Borów and SHR Wiecławice. A Research Centre for Cultivar Testing

(COBORU) in Słupia Wielka (Poznań District) was founded in 1966. It is this institute that issues the annual List of Agricultural Cultivars.

3.2.2. General remarks

One of the major aims of the government was to become self-sufficient for cereals without extending the total area. This involved an expansion of wheat and barley at the cost of rye and oats. The main objectives of the breeders after the war were obviously related to government plans, which concentrated on a further improvement of winter-hardiness, yielding capacity, lodging resistance and resistance to diseases and pests. Although the screening on technical characteristics was integrated in the breeding programmes of both winter and spring wheat, baking quality was not a major issue. The first Polish spring-wheat varieties with good baking quality did not appear on the variety list until the late 1970s. The use of foreign material was integrated into Polish breeding programmes from an early stage and this has increased during the second half of the 20th century. From the pedigree of the Polish varieties, that have been grown since 1958 (Lupton, 1992), it is clear that material from Germany, Sweden, France and Italy has gradually found its way into breeding programmes.

3.2.3. Wheat breeding at Hodowla Roşlin (HR) Dańko

HR Dańko is the actual name of a group of breeding stations, including SHR Dańków where Janasz started developing Dańkowska wheats at the end of the 19th century. Apart from the Dańków station, HR Dańko includes SHR Choryn, SHR Laski and SHR Debina. All these stations are located in the Poznań District. The chief breeder is Wolski, who, together with his co-workers, is the author of an impressive number of Polish wheat, rye and triticale varieties. The development of a new generation of Polish wheat varieties, bred during the 1960s and released in the 1970s, has been described by Wolski (1975). With the intensification of nitrogen fertilization in the 1960s, foreign varieties with higher lodging resistance than the Polish varieties were introduced. They proved to be ill-adapted to the changeable Polish climate. However, the Swedish varieties Odin and Starke in the north, German varieties in the west and Etoile de Choisy in the southwest did fairly well. The Russian Bezostaya-1 did not do as well as in southeastern Europe. Mironovskaya-808 and Kavkaz, however, were very successful. Most of the wheats bred during the 1960s resulted from three-way crosses including French, German, Italian and Polish wheats. The most important were Jana, Luna, Aria, Grana, Dana and Malwa. (The pedigrees of Grana, Luna and Malwa are presented in Figure PO.2). These varieties covered 70% of the winter-wheat area in Poland in the 1970s. With Grana and Luna a new type of wheat was introduced in Poland: with a height of 1 m or even less they were resistant to lodging (and more drought tolerant). The breeding of Grana, Luna and Dana was an important step forward in terms of agronomic value. At the same time, however, it was a step backwards in terms of baking quality. Aria and Malwa were distinctly better in this respect, but there still was a need for quality milling and baking wheats, i.e. comparable to Mironovskaya-808. The Russian variety Kavkaz, which also possesses high grain quality, was under Polish conditions, rather susceptible to sprouting. In order to meet demands for improved quality, a special programme was started in the 1960s, based on a technological laboratory set up at the Dańków-Laski station. Some 10 years later, in 1978, several varieties with good bread-making characteristics were released, such as Begra, Panda and Modra, of which only the latter was grown for some years on a relatively large scale. Begra has very good baking characteristics and is still used as a reference variety, although it was never grown on a large scale.

Table PO. 2. Varieties released by SHR Kobierzyce between 1979 and 1991

Variety		Pedigree	Year
Winter wheat			
Beta		NS-116/Eros//NS-116/Eros	1979
Gama	Q[a]	Mironovskaya-808/Luna	1982
Delta		Joss C/Kavkaz//4013	1989
Rosa	Q	Nadieznaya-45//Grana/Luna	1990
Kobra	Q	M.Huntsman/Krasnodarskaya-39//Gama	1992
Spring wheat			
Henika	Q	CB 412/Selpek//Selpek	1986
Sigma	Q	Kalyan Sona/Sappo	1987
Alkora	Q	Kolibri/Alfa//Jara	1990
Jota		Hadmerslebener 38662.69/Jara	1991
Omega	Q	Kadett/Jara	1991

[a] Varieties with a Q are classified as quality wheats on the Polish list of cultivars.
Source: List of Agricultural Cultivars, Słupia Wielka.

3.2.4. Quality breeding at SHR Kobierzyce

An impressive number of Polish bread-wheat varieties (released during the past 20 years) is the result of the work of Bielawski, Bielawska and their co-workers at Kobierzyce Plant Breeding Station (Wrocław District). These varieties are presented in Table PO.2. The varieties with a 'Q' are classified as varieties with bread-making quality. The name of the most recent variety could suggest that it might be the last one of this group of breeders.

Table PO. 3. Varieties with good bread-making quality on the Polish variety list since 1978

Variety	Pedigree	Breeder	Year
Winter wheat			
Modra	Starke//Cappelle/Heine VII	SHR Modzurow	1978
Liwilla[a]	Etoile de Choisy/Mir-808//Perdix	SHR Kosieczyn	1978
Zeta[a]	Mir-808//Herold/Glutenowa	SHR Kosieczyn	1978
Begra	Grana/Bezostaya-1	SHR Choryn	1978
Alcedo	Rekord/Poros//Carstens 8	GDR	1978
Mir-808	selection from Artemovka	USSR	1979
Gama	Mir-808/Luna	SHR Kobierzyce	1982
Panda	Dana/Flevina	SHR Danków	1983
Rosa[a]	Nadieznaya-45//Grana/Luna	SHR Kobierzyce	1990
Spring wheat			
Sappo	W 177.62/W 176.62	Sweden	1976
William	WW 13.69/WW 41.69	Sweden	1982
Kolibri	Heine 2174/Peko//Koga II	Germany	1972
Kadett	V 439.66/Kolibri	Sweden	1983
Henika	CB 412/Selpek//Selpek	SHR Henryków	1987
Sigma	Kalyan Sona/Sappo	SHR Kobierzyce	1987
Alkora	Kolibri/Alfa//Jara	SHR Kobierzyce	1990
Omega	Kadett/Jara	SHR Kobierzyce	1991
Igna	Kadett/Jara	SHR Kobierzyce	1993
Banti	Timmo/jara//Kadett/WZ 6154	SHR Kobierzyce	1994
Jasna	Eta/Kokart	HRR Kobierzyce	1996
Santa	Henika///L.Mex.9/Selpek//Kolibri	HRR Kobierzyce	1996
Torka	William/RR 277–78//WW 19018	ZDHAR Konczewice	1996

[a] Liwilla, Zeta and Rosa were classified as quality wheats upon their release but were not classified as such according to the classification standards in 1998 (Kaczynski, 1998).
Source: Polish List of Cultivars, 1978–1996.

3.2.5. Varieties with bread-making quality

Varieties of winter and spring wheat that since 1978 have been classified as varieties with bread-making quality are listed in Table PO.3. As Etoile de Choisy is a parent of Dana, Grana and Luna, it appears that this French variety is a parent of several Polish winter-wheat varieties with bread-making

Table PO. 4. Loaf volume and protein content of Polish-grown quality winter and spring-wheat varieties compared to standard varieties

Winter wheat		Spring wheat	
variety	loaf volume	variety	loaf volume
Grana	559		
Begra	601		
after Subda & Biskupski (1982)			
Grana	578	Alfa	545
M.Huntsman (ctrl)	595	Jara	587
Modra	638	Kolibri	609
Begra	631	Sappo	596
after Biskupski et al. (1982)			
Begra	587	Kadett	715
Delta	614		
after Warchalewski et al. (1988)			
Begra	11.0	Alfa	11.1
Grana	10.1	Kadett	11.7
Panda	10.8	Sappo	13.2
Alcedo	11.4		
Gama	11.4		
Beta	11.2		
after Subda (1989)			

quality. Many German and Russian sources can also be found in the pedigrees of these varieties. It is apparent that most of the Polish quality wheats released since 1978 have been developed at HR Dańko (SHR Choryn and SHR Danków) and SHR Kobierzyce. Data on loaf volume and protein content of some important quality varieties are listed in Table PO.4.

3.2.6. Quality characteristics used in bread-wheat breeding

The laboratory baking test is the key factor in determining the baking quality of wheat flour in Poland. Determination of wet gluten content and its quality, as well as α-amylase activity (falling number), are also common measures (Lasztity, 1995). Quality research is carried out at the technological laboratory of the IHAR in Wrocław by Subda, Bogdanowicz, Biskupski et al. The parameters used at IHAR for the assessment of the quality of breadwheat (Subda, 1995) are:

grain: 1000 kernel weight, hectolitre weight, vitreousness of grain, size
milling traits: offal, flour extraction, bran extraction
gluten characteristics: wet gluten content, solution rate of gluten, sedimentation index
amylograph: viscosity, temperature, time
farinograph: water absorption, development of dough, stability of dough, tolerance index, valorimetric value
baking characteristics: loaf volume, porosity of the crumb

Baking tests (macro and micro) at IHAR are carried out according to the method of the Institute for Breeding and Genetics in Odessa (Bogdanowicz & Biskupski, 1984). Up until 1995, the Research Centre for Cultivar testing (COBORU) in Słupia Wielka used five quality classes for bread wheat (Table PO.5). Varieties were classified according to their 'theoretical baking volume' (Warchalewski et al., 1990), which was calculated from a regression analysis of three parameters: protein content of grains, sedimentation index, and loaf volume. The calculated value was related to the theoretical baking value of a standard variety, which was set at 100%. A new classification system was introduced in 1995, which is better adapted to the requirements of the milling and baking industry. It is similar to the German classification (see Chapter Germany, Section 3.2.1) and comprises quality Classes E (elite), A (quality wheat), B (normal bread wheat) and C (feed wheat). The classification is based on eight parameters:

technological properties: falling number, grain protein content, sedimentation index (SDS/Zeleny)
milling test: flour yield
farinogram: water absorption, softening of dough
extensogram: energy of dough
baking test: loaf volume

Table PO. 5. Quality classes for bread wheat used by COBORU until 1995

Class	Relative theoretical baking value
I very good	⩾ 100%
II good	99–94%
III average	93–90%
IV average/poor	89–87%
V poor	⩽ 86%

Source: Warchalewski et al. (1990).

The evaluation of varieties is based on comparison with a standard variety (in 1994, Begra for winter wheat and Sigma for spring wheat) for each of these parameters.

The new classication scheme is more rigorous than the old one, leaving Begra as the only winter-wheat variety with A quality in 1998. Several spring-wheat varieties were classified in Class A. Apparently no variety was assigned to Class E. Ten varieties of both winter and spring wheat are listed in Table PO.6.

Table PO. 6. Quality classification and grain yield (t/ha) of Polish winter and spring-wheat varieties in yield trials

Variety	year	class	yield	variety	year	class	yield
Winter wheat				*Spring wheat*			
Begra	1978	A	6,27	Eta	1986	B	5,89
Gama	1981	B	6,37	Henika	1986	B	5,75
Panda	1982	B	6,27	Sigma	1987	A	5,41
Juma	1992	B	6,60	Omega	1991	A	5,58
Roma	1992	B	6,81	Igna	1993	A	5,64
Kobra	1992	B	7,04	Ismena	1995	A	5,80
Mikon	1994	B	6,98	Santa	1996	A	5,85
Korweta	1997	B	6,81	Jasna	1996	A	5,91
Mobela	1998	B	6,61	Torka	1996	A	5,60
Rysa	1998	B	6,87	Hezja	1997	A	5,92

Source: Kaczyński (1998).

3.2.7. HMW glutenin composition

Research on HMW glutenin subunits of Polish wheats has not been done systematically. The composition of the winter-wheat varieties Begra, Gama and Koda was defined by Ng et al. (1991) and is presented in Table PO.7. Apparently the sample of Gama in the research consisted of a mixture of different biotypes. The *Glu-1* scores of these three varieties is rather high, especially that of Koda. Koda was released by Dańko in 1986 but was never grown on a large scale.

D.A.D.

Table PO. 7. HMW glutenin composition of three Polish winter-wheat varieties

Variety	Chromosome 1A	1B	1D	*Glu-1* score
Begra	N	7+9	5+10	7
Koda	2*	7+9	5+10	9
Gama	N	7+8	5+10	8
	N	6+8	2+12	4

Source: Ng et al. (1991).

Acknowledgments

The author wishes to thank Dr Lucjan Kaczynski (COBORU, Słupia Wielka) for his critical remarks on the manuscript of this chapter. Dr Ela Mietkiewska was very helpful with the translation of Polish literature and Dr Alicia Petcho assisted in providing information on Polish wheat varieties.

References

Anonymus (1962) 100th Anniversary of plant breeding in Poland. Biuletyn Iństytutu Hodowli i Akklimatyzacji Roşlin 2–3, 154 pp.

Barbacki S (1977) Institute of Plant Breeding and Acclimatisation; its organisation and activities for the 25 years 1951–1976.

Bogdanowicz M, Biskupski A (1984) Bewertung von Methoden des Laborbackversuchs für die Ziele der Weizenzüchtung. Int. Zeitschrift der Landwirtschaft 2: 149–151.

Bogdanowicz M (1984) The evaluation of baking quality of winter and spring wheat strains. Biuletyn Iństytutu Hodowli i Akklimatyzacji Roşlin 154: 67–74.

Heyder M (1979) Amsterdam: korenschuur van Europa. Amsterdam: Stadsdrukkerij van Amsterdam, 104 pp.

Jensen JD (1992) Input supplies in Polish agriculture. Statens Jordbrugsökonomiske Institut, Rapport No. 65.

Kaczyński L (1998) Odmiany pszenicy odpowiednie dla przetwórstwa. Przeglad Zbożowo Młynarski 1998(7): 2–7

Kostecki (1962) Historic outline of agricultural plantbreeding and seed production in Poland in the last century. Hodowla Roşlin, Akklimatyzacja i Nassienictwo 6(4/5): 411–412.

Kostecki J, Wolski T (1963) Winter wheat breeding in Poland. Euphytica 12(1): 81–89.

Lasztity R (1995) Wheat usage in eastern Europe. In Faridi H, Faubion JM (ed) Wheat end uses around the world. St Paul (MN): AACC.

Ng PKW, Słomiński E, Johnson WJ, Bushuk W (1991) Homogeneity and quality assessments of some Polish wheats by electrophoretic analyses. Hodowla Roşlin, Akklimatyzacja i Nassienictwo 35(3/4): 61–67.

Subda H (1989) Dependence of baking quality of wheat flour on its chemical composition. Hodowla Roşlin, Akklimatyzacja i Nassienictwo 33(1/2): 27–48.

Subda H, Tarkowski C, Karolini-Skaradzińska Z, Gil Z (1995) Effect of chromosomes 1BL/1RS on the chemical composition and technological value of grain in wheat. Hodowla Roşlin, Akklimatyzacja i Nassienictwo 39(1/2): 63–68.

Warchalewski JR, Klockiewicz-Kamińska E (1990) Baking quality versus some biological activities of proteins extracted from the grain of wheat varieties harvested in 1986. Die Nährung 34(5): 421–430.

Wolski T (1975) The actual state and main problems of winter wheat breeding in Dańków and Rogaczewo, Poland. Proc Intern Winter Wheat Conf, Zagreb, pp 114–124.

Wolski T (1994) Rye-the main cereal in Poland. Hodowla Roşlin, Akklimatyzacja i Nassienictwo 38(5): 5–10.

Russia and Ukraine

1. Introduction

In 1998, Russia and Ukraine (RU) together produced more than three-quarters of the wheat harvested on the territory of the former USSR, whilst another 10% was produced in the Republic of Kazakhstan. Until 1991, Russia (the Russian Federation) and Ukraine were part of the USSR, which now is divided in 15 independent states, including the Baltic countries (Lithuania, Estonia and Latvia), Belarus, Moldavia and Ukraine. The greater part of Russia is situated in Asia, with vast areas of spring wheat in Siberia. The term Russia in this chapter is also used to refer to the country as it existed before 1917. For practical reasons, the first part of the chapter (Sections 1.1–1.3) relates to the whole territory of the former USSR, unless mentioned otherwise. The rest of the chapter (Sections 2.1–6.3) focuses on wheat breeding in the European part of Russia and in Ukraine. Wheat breeders in these countries over the past century developed a large number of valuable winter-wheat varieties, ranking among the best in the world in terms of winter and drought resistance, in addition to milling and baking quality (Prutzkov, 1970).

1.1. The country and its climate

A schematic division of the former USSR into three zones according to Skoroparov (1977) comprises:

- a cold zone in the north (limited in the south by the isotherm with a thermal sum of 1600 degrees), where arable farming is very restricted;
- a temperate zone between 60 °N and 45 °N (see Figure RU.2), including 90% of the arable land of the USSR; the thermal sum increases and the precipitation decreases from north to south; grass podsoils alternate with forest soils and variations of black soils; the temperate zone includes a forest zone, a forest-steppe zone, a steppe zone and a dry steppe zone;
- a hot (subtropical) zone in the south, consisting mainly of the trans Caucasian republics of Armenia, Azerbaijan and Georgia.

Agriculture is concentrated in a triangular area extending from the western border between the Baltic and the Black sea, stretching eastward for more than 5 000 km to the Yenisey River, near the town of Krasnoyarsk. This region, commonly called 'the fertile triangle', has climatic conditions that allow the cultivation of a wide range of crops, including wheat, rye and barley. Within this agricultural heartland lies the Central Black Earth region, with extremely rich 'chernozem' soils. A major drawback of this area is the occurence of extensive droughts, on average once every three years. These droughts, spread over a vast territory, have lead to national calamities, necessitating large-scale importation of foodstuffs from abroad.

The climate in the greater part of the country is continental, under the predominant influence of dry polar airmasses; tropical or maritime influences are slight. Due to the northern location, summers are short and winters long and severe. The normally prevailing flow of dry arctic air over the greater part of the country, the distance from open, warm oceans, and the screening effect of high mountains in the east and south result in a general pattern of precipitation that is far from abundant.

1.2. Main areas of wheat growing and figures on wheat production

By far the most important wheat growing areas in the European part of the former USSR are Ukraine, the northern Caucasus and the Volga Basin. These regions, together with vast areas in Kazakhstan and West Siberia, traditionally supplied the USSR with remarkably strong and hard wheats that were the best of the world with regard to their vitreousness, protein content, and bread- (or macaroni-) making qualities. In the northern Caucasus, the Kuban region was well known for its wheat production and it was from there that wheat was exported at the beginning of the 20th century. Within the Volga Basin, the area around Saratov was traditionally renowned for producing wheats with excellent bread-making quality.

Whereas large quantities of wheat were exported at the beginning of the century (4.5 Mt annually between 1903 and 1912, according to Loewé, 1923), a drastic change occurred after the revolution in 1917. The area sown with wheat increased considerably from 33 to 48 million hectares between 1913 and 1953. In the European regions this was partly due to the replacement of other cereals by wheat and the extension of the wheat crop beyond the black soils (chernozem) where it had hardly been grown before. In spite of this increase in area, the USSR was often not in the position to satisfy its home requirements. Moreover, attention paid to the quality of the grain diminished and problems with its drying and storage, especially in the eastern parts of the USSR, appeared (Kretovich, 1960).

According to FAO statistics, in the years 1992–1996 the territory of the former USSR produced 13% of the world's wheat production on 20% of the world's wheat area, compared to almost 25% of the world's wheat production on 25% of the world's wheat area in 1934–1938. In spite of this fall in the contribution of Soviet wheat production over the past century, it is expected that it will return to the world's export market by 2010.

Large differences in productivity exist within the former USSR. Highest yields are obtained in Ukraine with 3043 kg/ha, compared to 1556 kg/ha in

Table RU. 1. Wheat production in the former USSR between 1934 and 1997

Year(s)	Area (1000 ha)	Yield (kg/ha)	Production (1000 t)
1934–1938	40 922	930	38 090
1948–1952	42 633	839	35 759
1961–1965	66 627	964	64 207
1969–1971	65 230	1 423	92 804
1979–1981	59 403	1 513	89 859
1989–1991	47 242	1 842	87 014
1995–1997[a]	46 193	1 475	68 143

[a] Total of Estonia, Latvia, Lithuania, Moldavia, Belarus, Ukraine and Russia in Europe, and Armenia, Georgia, Azerbaijan, Uzbekistan, Turkmenistan, Kyrgystan, Tajikistan, Kazakhstan in Asia.
Source: FAO statistics.

Russia and 829 kg/ha in Kazakhstan (averages 1992–1996). With its fertile chernozem soils and a temperate climate, Ukraine has a high potential for wheat production. At the end of the 1980s Ukraine produced 27 Mt of wheat, with an average yield of 5 t/ha (Anonymus, 1998). Structural reforms and investments are necessary to open up what has been called the wheat loft of the 21st century. Table RU.2 compares the five years' average wheat production for 1992–1996 in the countries that belonged to the former USSR with production in the same regions in 1965. Unfortunately it was not possible to find more recent data on the distribution of wheat in the different parts of Russia and Ukraine.

Within Europe, a vast area of more than 20 million hectares is grown in Russia, with the northern Caucasus as the most important area for winter wheat and the Volga district as the most important area for spring wheat. The second most important wheat growing country is Ukraine, with almost 10 million hectares of wheat, producing one-third of the total production within the European part ot the former USSR. The highest yields are obtained in Ukraine and Moldavia.

Wheat is the main cereal crop in each country of the former USSR. In Russia, in 1993, the bread-making cereals wheat and rye covered 42% and 10%, respectively, of the total area under cereals. In Ukraine, in the same year wheat and rye covered 44% and 4% of the total area, respectively.

Table RU. 2. Wheat area (and production) in the Baltic states, Moldavia, Belarus, Ukraine and the European part of Russia in the years 1992–1996 compared to the area in the same regions in 1965

Country	1965 area (1000 ha)			1992–1996	
	winter	spring	total	area (1000 ha)	production (1000 t)
Belarus	16	171	187	161	396
Estonia	18	11	29	45	90
Latvia	80	26	106	117	261
Lithuania	166	2	168	289	767
Moldavia	416	3	419	327	984
Russia	9 035	14 306	23 341	24 001	37 356
northwest	*64*	*106*	*170*		
central	*1 092*	*436*	*1 528*		
Volga-Vyatka	*186*	*751*	*937*		
central chernozem	*1 341*	*511*	*1 852*		
Volga	*1 336*	*8 153*	*9 489*		
northern Caucasus	*4 913*	*549*	*5 462*		
remaining districts	*103*	*3 801*	*3 904*		
Ukraine	7 346	493	7 839	5 587	17 003
Total	17 947	14 907	32 854	30 527	56 857

Sources: Broekhuizen (1969); FAO statistics (1998).

1.3. Winter versus spring wheat

Most of the wheat grown in former USSR was spring wheat. The proportion of winter wheat has fluctuated between 25% and 35% since 1913. The major winter-wheat growing areas are located in Europe, on the steppe and in forest-steppe regions of Ukraine and northern Caucasus, whereas spring wheat is mainly grown in the areas with a sharp arid continental climate, in the Volga region, Kazakhstan and Siberia (Lukyanenko, 1972).

Data on the distribution of spring and winter wheat in the European part of the former USSR have been collected by Broekhuizen (1969), who surveyed the distribution of various cereal crops in the regions of the European part of the former USSR for the year 1965. According to his data, the area in the European part of the USSR covered with winter and spring wheat was 18 and 15 million hectares, respectively (Table RU.2). The importance of winter wheat compared to spring wheat varied from one region to another, e.g. in Ukraine the winter-wheat area was about fifteen times that of the spring-wheat area, whereas in the Volga region the winter-wheat area was about one-sixth of that of spring wheat. In some districts, a spectacular change from spring wheat in favour of winter wheat took place during the first half of the century. Thus, the proportion of winter wheat in Ukraine increased from 35% to 95% in the Ukraine between 1913 and 1956 (Sakoff, 1960), and from 29% to 91% in White Russia (Belarus) between 1940 and 1965 (Broekhuizen, 1969). With the release of winter-hardy varieties such as Mironovskaya-808, the proportion of winter wheat has continued to increase.

2. Wheat breeding before 1945

2.1. Traditional and local varieties

Typical examples of local varieties grown on Russian territory at the beginning of the 20th century are Sandomirka wheat, originally introduced from Poland, and Krymka wheat, grown for hundreds of years in the Crimea. Many of the winter wheats grown in southern Russia at the beginning of the 20th century were derived from the Hungarian cultivar Banatka (Jakubziner, 1962). Banatka wheats were already being grown in the middle of the 19th century in the Ukraine, from where they penetrated Bessarabia. Various forms of Banatka, such as Banatka Podolskaya, Banatka Hersonskaya, Banatka Parhomovskaya and Banatka Kahovskaya, were grown in different locations in Ukraine. Banatka was one of the most important sources of winter wheat in Russia and later in the USSR. Important varieties that have been selected from Banatka were Triumph Podolii, Ukrainka, Durable, Zemka, Stavropolka, Stepnyachka and Moskovskaya-2411. Banatka was grown in Ukraine and northern Caucasus until the early 1960s. Another group of Hungarian wheats, widely grown in southern Russia until 1920 were Tiszavka wheats. In the trans Caucasian republics, several wheats of Turkish origin were commonly grown.

2.2. Breeding at experimental stations

In 1894 the Bureau for Applied Botany was created in St Petersburg (Leningrad), where a world collection of cultivated plants was assembled over many years. This became the initial stock for breeding wheat and other crops. The Bureau was based in the All-Union Scientific Research Institute for Plant Industry, later called the Vavilov Institute of Plant Industry (VIR), named after its most outstanding contributor. Vavilov started a genetic resources collection there, which has developed into one of the world's most important gene banks.

In his book on the origin of cultivated plants and the scientific basis of wheat breeding, Vavilov (1935) surveyed the early years of Russian wheat

Table RU. 3. Characteristics of winter- and spring-wheat varieties recommended for different regions in the USSR during the period 1925–1935

Variety	Breeding station[a]	Productivity	Grain-quality	Flour output	Loaf volume	Winter-hardiness[a]
Winter wheat						
Ferrugineum-2411	Tim	4-	4-	3-	3	–
Ferrugineum-2453	Tim	4-	4-	3+	4	–
Erythrospernum-917	Kha	3+	4-	4-	3	–
Ferrugineum-1239	Kha	4-	4-	4	4	–
Lutescens-329	Sar	2	2	4-	4-	4
Lutescens-1060/10	Sar	2+	3-	4-	3+	4
Hostianum-237	Sar	4-	3	4-	4-	–
Kooperatorka	Ode	4	4+	4	4	2
Stepnyachka	Ode	4	4-	4	3+	3
Ukrainka	Mir	4	4+	4-	4	3
Durable	Iva	3	4-	3+	2	–
Novokrymka-102	Sim	3+	4	4	4-	2
Erythrospernum-328	Sta	4	4-	4	3	2
Zarya	Nem	4	3+	4-	2	(2)
Erythrospernum-7201	Bez	3-	4-	4-	3+	–
Spring wheat						
Lutescens-62	Sar	4	3	3	3	
Erythrospernum-341	Sar	3	4	4	4	
Sarrubra	Sar	3	4	4	4	
Hordeiforme-432	Sar	4-	4	3	4	
Erythrospernum-841	Kra	3	4-	4	4	
Hordeiforme-189	Kra	3	4	3	3	
Melanopus-69	Kra	3	4-	3	3-	
Hordeiforme-10	Dne	4-	3+	3	3-	
Hordeiforme-27	Kub	4-	3+	3	3-	
Novinka	Det	3+	4	4	4-	
Standard cultivars						
Marquis (moist zone of N.Caucasus)	CND	3	3	3	4	
Strube (foothill regions Daghestan)	GER	3	3	3	3	
Garnet (northern region etc.)	CND	4-	4-	3	3-	

[a] Breeding station abbreviated as Tim for Timiryazev (Moscow region), Kha for Kharkov, Sar for Saratov, Ode for Odessa, Mir for Mironovka (Ukraine), Iva for Ivanovka (Kharkov region), Sim for Simferopol (Crimea), Sta for Stavropol (north Caucasus), Nem for Nemerchansk, Bez for Bezenchuk (Volga), Kra for Krasnokutzk (Volga), Dne for Dnepropetrovsk, Kub for Kubansk (north Caucasus) and Det for Detski (Institute of Plant Industry).
[b] All the scores are given on a 1 (poor) to 4 (best) scale.
Source: Vavilov (1935).

Persian wheat in regard to its milling and baking qualities

by **K. M. Chingo-Chingas.**

Summary.

Summing up all the baking qualities of «Persian Wheat» according to the formule of Saunders, the ultimate value of its baking strength may be expressed by 50 balls.

On the ground of his experiments with Tr. persicum the author comes to the following conclusions.

1) Persian wheat without any admixture of flour of other wheats is characterised by a low baking strength.

2) In some of its peculiarities, as for instance, softness of the crust and brown colour, as well as ready breaking into granulated flour when ground, Persian wheat approaches durum wheats.

3) Persian wheat possessing several high qualities, as for instance strong flintyness and hard texture of the grain, it may be successfully used for crossing purposes.

The present test was conducted with one sample obtained from Georgia and the grinding was carried out in a roller mill. Analogical results were obtained with a pure line of Tr. persicum var. fuliginosum, grown in Saratov and ground in a stone mill.

Figure RU. 1. Publication of K.M. Chingo-Chingas in the Bulletin of Applied Botany and Plant Breeding of 1925.

breeding, which included a list of recommended winter- and spring-wheat varieties with their characteristics and region of cultivation. In some cases these regions enclosed complete countries, such as Ukraine or Belarus, but mostly comprised regions like the Voronezh, Moskov or Saratov, or regions based on soil type, such as the non-chernozem soil strip of the European part of the USSR, the birch forest strip of Ukraine, or the steppes of the Lower Volga. His survey illustrated the complexity of the system of varietal recommendation in this very vast country, where latitude, soil type and ecological cirumstances are so important. A summary of the list of recommended varieties is presented in Table RU.3. It is beyond the context of this chapter to present recommendations for all the regions of cultivation.

According to Vavilov (1935), Russian wheat breeding started in 1902 with the work of Rudzinski at the experimental breeding station of Timiryazev under the Moscow Agricultural Academy. By individual selections from Hungarian and Polish wheats, Rudzinsky obtained excellent winter-wheat varieties, such as Ferrugineum-2411 and Ferrugineum-2453, the most important varieties for non-chernozem soils up to the mid 1930s. In view of the traditionally high demands on the quality of Russian wheat, in 1914 Rudzinsky founded a laboratory for the determination of technological charateristics for wheat breeding. In 1923, Vavilov and Chingo-Chingas set up a laboratory in

St Petersburg to evaluate the technological properties of agricultural crops. In 1937 the laboratory was moved from St Petersburg to Moscow. As an illustration of the work of Chingo-Chingas, a description of the quality of Persian wheat from the year 1925 is given in Figure RU.1.

Figure RU. 2. Location of wheat-breeding stations in the European part of the Russia and Ukraine. (Design: Lucas Janssen)

Between 1909 and 1912, wheat breeding was initiated at a number of experimental breeding stations (Figure RU.2) such as Kharkov, Saratov, Odessa, Mironovka, Ivanovka and Simferopol (Crimea). Sozinov (1979) wrote that special attention was given to the improvement of the technological quality of the wheat kernel during the first few decades of the 20th century. On the initiative of Vavilov, well-equiped laboratories were set up at the most important breeding centres in the country from 1919 onwards. Baking characteristics were amongst the most important selection criteria. In 1937, it was decided that one of the guidelines in agricultural policy would be the increase in grain production by every possible means. Together with the increase in grain production, it was decided that there should be an improvement in quality. Special attention was to be paid to the production of the valuable hard wheats.

A brief overview of the most important winter- and spring-wheat cultivars that were developed at the experimental breeding stations during this period is given in Sections 2.2.1 and 2.2.2, respectively.

2.2.1. Winter wheat

In 1909, the Kharkov Breeding Station was founded, where wheat breeding was carried out by Juryev. It became a major centre for winter- and spring-wheat breeding, initially by individual selection and later by hybridization. Important winter-wheat cultivars from Kharkov, released between 1929 and 1933 were Milturum-120, Albidum-0676 (also named Juryevka), Erythrospernum-917 and Ferrugineum-1239. Especially the latter two varieties could be grown under a wide range of ecological conditions.

In southern Ukraine wheat breeding was initiated in 1912 at the Odessa Breeding Station by the famous geneticist and breeder Sapegin. Individual selection in Krymka wheat, Banatka wheat and local wheats led to the first varieties: Kooperatorka, Stepnyachka and Zemka. These had a high level of drought resistance, good grain quality and productivity, but insufficient winter hardiness. The regions in which Kooperatorka was cultivated, were southern Crimea, the very dry steppes of northern Caucasus, Daghestan, Azerbaijan and Georgia. Stepnyachka was grown on the semi-arid steppes of northern Caucasus.

At the Mironovka Station, Mironovski (who had started his work at the Sugar Trust Station) selected the winter-wheat variety Ukrainka from Hungarian Banatka wheat. Ukrainka was widely distributed throughout southern Ukraine and northern Caucasus. Vavilov estimated that approximately half of all the Soviet winter wheat sown in the mid 1930s was Ukrainka. It was renowned for its high productivity, good milling and baking quality and universal adaptation.

At the Ivanovka Station (in the Kharkov region), where wheat breeding started in 1909, the variety Durable was obtained by individual selection from the local form Banatka Pahomovskaya. This winter-hardy variety with an exceptionally wide adaptation was grown up to the region of Leningrad. In the 1930s it occupied a significant amount of territory both in Ukraine and in the non-chernozem strip of the European part of the USSR. Other lines selected in local material were 20/430 (Golubaya Dama), 21/64 and 21/825. Crossing of the best lines resulted in Ivanovskaya (= Erytrospernum-2119), released in 1939 (Lastovich, 1991).

At the Simferopol station (Crimea), the variety Novokrymka was obtained by crossing Novokrymka-258 with Kooperatorka (a selection from the local winter wheat Krymka). Novokrymka was noted for its drought resistance, earliness and high quality grain. Apart from the Crimea region, it was also grown on the steppes of Ukraine.

The Saratov, Ulyanovskaya and Bezenchukskaya Breeding Stations, located in the Volga river basin, were founded in the years 1910–1912. Wheat breeding there began with the evaluation of and selection from local and foreign material. Thus, the variety Hostianum-237 was selected from Kharkov local wheats, the variety Lutescens-1060/10 from Volga local wheat, the variety Lutescens-329 from the Polish Sandomirka, the variety Bezenchukskaya from the local variety Grushevskaya and the variety Ulyanovka from Teyskay wheat. These and other wheats from this region were among the most winter-hardy Soviet wheats; they were widely grown and used in breeding programmes for many years. In the 1920s, the application of interspecific and intergeneric hybridization led to some very successful results, such as the awnless spring-wheat hybrid Sarrubra (see Section 2.2.2).

2.2.2. Spring wheat

The Volga region (Povolje) is one of the main spring-wheat producing regions of Russia. Spring-wheat breeding was started at the beginning of the century at the Saratov, Bezenchukskaya and Krasnokutskaya breeding stations. Most of the spring-wheat varieties recommended in the mid-1930s were derived from local populations of Poltavka, Beloturka, Rusaka and Khivinka. From 1925 onwards, breeders started to apply crosses, with particular attention being paid to disease and drought resistance, non-shattering of the grain, adaptation to mechanical harvesting, as well as milling and baking quality, and the analysis of the protein fraction of wheat grain (Vavilov, 1935).

At the Saratov Breeding Station the spring-wheat varieties Lutescens-62 and Erythrospernum-341 were developed by individual selection. Lutescens-62, selected from Poltavka, became a leading variety, accounting for more than 8 million hectares in 1938 and about 7 million hectares in 1956 (Ivanov, 1971). Another successful variety from Saratov Station, grown on hundreds of thousands of hectares in the 1930s, was Sarrubra, selected from a cross between Poltavka and the durum wheat Beloturka. The spring-wheat variety Saratovskaya-42, grown half a century later, resulted from a complex hybridization between four local varieties: Beloturka, Poltavka, Rusaka, and Khivinka, which contributed to the adaptation of this drought resistant variety to the severe continental climate of the region (Ilina, 1986).

At the Krasnokutsk Station (Volga), individual selection from local wheats gave some excellent drought resistant varieties, such as Erythrospernum-841, Hordeiforme-189 and Melanopus-69, the latter with a wide adaptability. Erythrospernum-841 was grown on 1.3 million hectares in 1966 (Ivanov, 1971).

At the All Union Institute of Plant Industry, Pissarev selected the variety Novinka in a population obtained from a cross between two Canadian cultivars. Novinka was described as having excellent bread-making characteristics and was renowned for its wide adaptation. It could be grown from 57 °N up to 65.5 °N. The adaptation to northern Russian conditions may be related to the fact that Ladoga, a variety originating from northern Russia (named after Lake Ladoga), figures in the pedigree of both parents of Novinka. The pedigree of Novinka (given in Figure RU.3) illustrates the international background of this variety.

A survey of the productivity and quality characteristics of the cultivars mentioned in this section and Section 2.2.1 is presented in Table RU.3. These figures indicate that many varieties with very good bread-making characteristics were available on the market. The best winter wheats in this respect were Ukrainka, Kooperatorka and Stavropolka, as well as local varieties from Crimea. The qualitatively best spring wheats were Cesium-111, Sarrubra, Erythrospernum-341 and -841 (Samsonov, 1960). The survey does not include the work initiated by Lukyanenko in Krasnodar in the 1930s. His work will be discussed in Section 3.2.

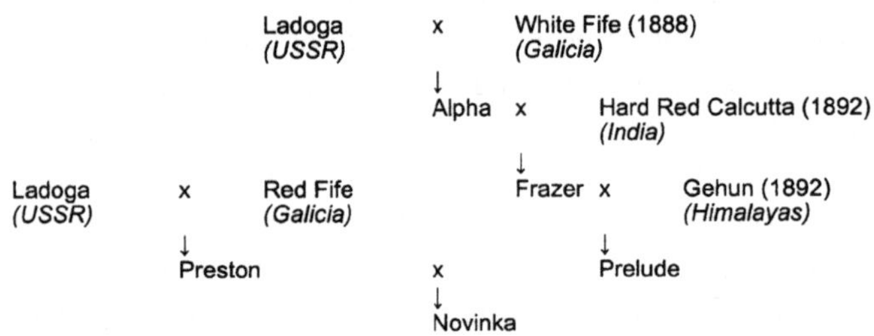

Figure RU. 3. Pedigree of the spring-wheat cultivar Novinka, bred in the 1920s by Pissarev in Leningrad (after Vavilov, 1935).

3. Wheat breeding after 1945

3.1. Ecological groups of wheat

The hybridization of ecologically and geographically remote groups of wheat has been applied by Russian wheat breeders since Michurin introduced it into fruit breeding at the beginning of the 20th century. Several of the high yielding winter-wheat varieties bred by Lukyanenko and others in Krasnodar during the 1960s could be grown in various ecological regions in the USSR, as well as in a number of foreign countries, due to their high ecological plasticity. Kirichenko, a wheat breeder in Odessa, objected strongly to the idea of 'universal wheat varieties'. He stated: 'There are no universal varieties which can be sown in all the climatic zones from the Baltic to the Black Sea and such an approach is abiological'. Nevertheless, the Odessa breeding programme was oriented towards combining the properties of wheats from different ecological sources.

For a better understanding of the wheat-breeding programmes discussed in the following paragraphs, the most important eco-geographical groups of wheat according to Prutskov (1970) are presented here, including a number of cultivars belonging to these groups that were grown in the USSR in 1970.

- *The northern Russian group*, generally maturing late and mid-season, tolerant to excessive snow cover, with a poor drought resistance. This group includes local varieties such as Wysokolitovskaya and Borovichskaya, as well as the selected varieties Lutescens-116 and Priyekulskaya-481.
- *The Volga steppe group*, characterized by high winter and frost resistance, drought resistance and good grain quality. Most of the varieties of this group form a valuable initial stock for breeding in the steppe zone of European USSR. This group includes Lutescens-230 (with excellent baking quality), Alabasskaya, Petrovskaya-7 and Saratovskaya jubileynaya.
- *The southern steppe group*, excelling in rapid maturity, length and density of the spike, and in size of the grain. Their winter-hardiness is low and the drought resistance is average. Included are the varieties: Bezostaya-1, Novukrainka-84, Rannyaya-12, Krasnodarskaya-16, Novokrymka-204 and Kooperatorka.
- *The southern forest-steppe group*, excelling in length of the ear and size of the grain. Winter resistance is average and the drought resistance is low. Included in this group are the varieties Ukrainka, Belotserkovskaya-198, Odesskaya-16, Mironovskaya-264 and Mironovskaya-808.

3.2. Ukraine

The breeding work carried out at three main breeding institutes (in Odessa, Mironovka and Kharkov) and two breeding stations (Belaya Tserkov and Ivanov) will be discussed hereafter.

3.2.1. The Plant Breeding and Genetics Institute in Odessa

Odessa is located in the steppe region of south Ukraine, where rich chernozem soils and a relatively dry climate offer favourable conditions for the production of high quality bread wheat. The port of Odessa provides Ukraine with a natural access to the Black Sea and traditionally this harbour has been one of the centres of Russian wheat export.

The Plant Breeding and Genetics Institute has played a leading role in Ukrainian wheat breeding since the 1930s. Kirichenko was one of its most outstanding breeders; his work was continued by Dolgushin, Lifenko and Litvinenko. For many years, the institute's technical laboratory was directed by Sozinov, who contributed significantly to the development of quality research there. Sozinov directed the institute between 1970 and 1979. Wheat varieties bred at the institute in Odessa have been grown in Ukraine and northern Caucasus.

The most important winter-wheat varieties grown in the steppe regions of Ukraine and northern Caucasus up to the 1970s were Odesskaya-3, -12 and -16, grown on a total area of 5–7 million hectares annually. In the late 1950s these cultivars began to be gradually replaced by new forest-steppe varieties, such as Belotserkovskaya-198, Mironovskaya-264 and later Mironovskaya-808. The massive acceptance of these varieties, which were characterized by lower winter-hardiness and insufficient drought resistance, brought a decline in the grain harvest. Especially during the winter 1968–1969, a high proportion of winter killing occured in Bezostaya-1, whereas a high proportion of Odesskaya-16 overwintered. On the other hand, Odesskaya varieties, especially Odesskaya-16, were subject to lodging in years with abundant rainfall.

Kirichenko intended to combine the properties of wheats from different ecological and geographical sources, including high yielding varieties from Bulgaria, Yugoslavia, Hungary, USA, Mexico, Canada (see also Section 3.1). Repeated back-crossing was used to improve a specific property. A first group of varieties resulting from this programme included Odesskaya-26, Odesskaya-28 and Stepova.

Odesskaya-26, selected from a cross between Odesskaya-3 and Lutescens-17, was a hard wheat, with grains that possessed good technological and baking qualities: a thousand kernel weight that was 6–10 g higher than that of Odesskaya-16; a higher flour strength; a high loaf volume; and excellent crumb porosity. With a frost resistance similar to Odesskaya-3, Odesskaya-26 was especially noted for its drought resistance, due to the development of a heavy root system that extended deep into the soil. Eleven years of testing at the Odessa Institute proved that the average grain yield was 4.2 t/ha, compared with 3.9 t/ha from Odesskaya-16 and 3.6 t/ha from Odesskaya-3. The variety Odesskaya-26 was designated to replace Odesskaya-3 and the varieties of the forest-steppe origin, Bezostaya-1 and Mironovskaya-808. Although Odesskaya-26 was selected in the steppe regions of Ukraine, it produced a high grain yield of superior quality in most years. It was recommended for the whole Odessa region in 1964. Thus it became possible 'to move steppe breeding to a qualitative new stage of work, making it possible to create intensive-type varieties, but of the steppe-ecological type' (Kirichenko, 1970).

The creation of the variety Stepova was described by Morgunov (1990) as an example of the selection procedure often used by Russian breeders. While 10–12 years were usually required to develop a new variety of winter wheat, Kirichenko set out to create a variety in 6 years on the basis of intervarietal hybrids. In the F_2 from a cross between Bezostaya-4 and Odesskaya-16, hundreds of the best uniform progenies of individual plants were selected. After threshing and laboratory culling of the grain, the offspring of 92 plants was selected. The seeds of the single plants were sown apart and at the same time a portion of the seed from each family was sown in boxes for winter freezing tests. In the third year, 16 families were selected, that had been given a good evaluation for frost resistance, productivity and disease resistance, as well as for morphological uniformity within the family. The selected families were grouped and directly included in the first competitive varietal testing under the breeding number 98/63, which was a very heterogeneous variety. Later on, the variety was named Stepova. Over 7 years of testing, Stepova yielded 0.7 t/ha more than Odesskaya-3 and 0.3 t/ha more than Odesskaya-16. Stepova was classified among the strong wheats and in the autumn of 1967 it was recommended for the whole Odessa region.

Stepova is an example of a variety consisting of different biotypes (see also Section 6.3), resulting from a selection in early generations without applying the pedigree method. This type of variety, which is very common in the former USSR, is believed to provide more stable yields in changeable environmental conditions, due to compensation for biotype-environment interaction (Morgunov et al., 1990).

Other varieties resulting from Kirichenko's hybridization programme were Chernomorskaya (from a cross between Bezostaya-4 and Odesskaya-16), Novostepnyachka (from a cross between Bezostaya-1 and Odesskaya-29) and Priboj (from a cross between Bezostaya-1 and Odesskaya-16), all three varieties of the intensive type, but having a steppe ecology, i.e. high yielding and a good technological quality.

A survey of the results of winter-wheat breeding for the Ukrainian steppes during the period 1929–1979 after Kirichenko et al. (1979) is presented in Table RU.4.

The increase in kernel yield from the land variety Krymka to Priboj is evident, as well as yield of crude grain protein. A trend of decline in grain protein content is also obvious, especially a decrease of 2 percentage points for Aurora compared with Krymka. Based on this and similar results, studies were made at several locations in the former USSR on the usefulness of certain wheat varieties as sources of quality in breeding programmes. In

Table RU. 4. Results of breeding winter-wheat varieties (listed according to age) for the Ukrainian steppe region during the period 1929–1979. Averages for 1972–79 harvest

Variety	Year	Grain yield		Crude protein	
		t/ha	rel	% d.m.	kg/ha
Krymka	1929	3.19	100	14.3	456
Kooperatorka	1930	3.02	95	12.9	390
Ukrainka	1930	3.55	111	14.4	509
Hostianum-237	1931	3.04	95	14.6	445
Odesskaya-3	1938	3.48	109	15.0	523
Odesskaya-12		3.21	101	14.7	472
Odesskaya-16	1952	3.45	108	14.1	486
Belotserkovskaya-198	195?	3.56	112	13.8	493
Bezostaya-l	1959	4.45	140	12.7	563
Mironovskaya-808	1960	4.01	126	13.9	557
Odesskaya-26	1964	3.94	124	14.5	570
Stepova	1964	4.13	130	15.5	638
Kavkaz	1964	4.47	140	12.6	561
Avrora	1964	4.42	139	12.3	541
Odesskaya-51	1969	4.58	144	13.0	595
Priboj	1972	4.63	146	13.0	598

Source: Kirichenko et al. (1979).

many cases, cultivars from the collection of the All-Union Institute of Plant Industry at Leningrad were studied in a specific ecological environment. For example, a study on the usefulness of 60 high protein winter-wheat cultivars as parent in crosses to improve protein content under conditions on the Ukrainian steppes was carried out by Kirichenko et al. (1979). Five varieties were identified that combined a high protein content with economically valuable traits such as winter hardiness and lodging resistance (see Table RU.5). From 1979 onwards, Ukrainian breeders attempted to combine these high protein donors with regionally approved varieties and with breeding lines of the highly intensive type.

From the mid-1970s onwards, wheat breeding in Odessa was characterized by the introduction of more dwarfing genes. The first group of semi-dwarf varieties included Odesskaya popukarlikovaya (translation: Odessa semi-dwarf) and Obriy (see also Section 3.3.5). Due to improved harvest indices, lodging resistance and tillering capacity, the yield level of these varieties increased considerably but they lacked sufficient winter-hardiness. This

Table RU. 5. High-protein winter-wheat varieties under conditions of the Ukrainian steppes: averages for 1976–78 harvest

Variety	Origin	Grain yield		Crude protein		Sedimentation
		t/ha	rel.	% d.m.	kg/ha	value
Bezostaya	SU	4.40	100	12.6	554	44
Priboj	SU	4.90	111	12.9	632	49
Forte	IT	5.79	132	13.5	782	43
Apache	US	4.58	104	13.6	623	45
Dunav (NS-60)	YU	5.50	125	13.7	754	36
Erythrosp.-127	SU	6.51	148	13.5	879	48
416/54[a]	SU	5.32	121	13.1	697	43

[a] 416/54 = Skorospelka/Odesskaya-16.
Source: Kirichenko et al. (1979).

deficiency was improved during a subsequent cycle of 'adaptive breeding', resulting in cultivars such as Albatros odesskiyi. This latter variety, released in 1990, was grown on 1.7 million hectares in 1995 (Litvinenko, 1998). Other important varieties, released since 1994 are Fedorovka, Odesskaya-132, Ukrainka Odesskaya, Fantaziya odesskaya.

Varieties with slow development in autumn, combined with stable frost resistance during winter and intensive root growth in early spring, have given the highest yields under Ukrainian steppe conditions, where drought resistance and winter-hardiness are essential features. With respect to selection for bread-making characteristics, Litvinenko (1998) mentioned that all breeding lines, starting from F_5, are evaluated by electrophoresis for storage proteins, gliadins and glutenins.

3.2.2. The Belaya Tserkov Breeding Station in Kiev

One of the varieties from the Belaya Tserkov Breeding Station that was widely grown in Ukraine during the 1950s and 1960s was Belotserkovskaya-198 (BTs-198). An improved variety derived from BTs-198, Belotserkovskaya-177 (pedigree: BTs-198/BTs-23//Bezostaya-1/BTs-198) was released in 1979. This high yielding variety had good milling and baking characteristics (Burdenyuk, 1991). The varieties BTs-47 and BTs-18, released thereafter, both have the same parentage: Bezostaya-1/BTs-198//BTs-21/ Uluchshennaya. The use of the parent Bezostaya-1/BTs-198 enabled reduction of the straw length of these varieties. Further reduction of the straw length was achieved with a single cycle of back-crossing to Donskaya Polukarlikovaya (see Section 3.3.2).

3.2.3. The V. N. Remeslo Institute of Wheat in Mironovka (Kiev region)
Starting in 1948, wheat breeding at this institute was carried out under the leadership of Remeslo, who was breeder of the second generation of Mironovka cultivars. By applying multiple group and bulk selection in autumn sown spring-wheat varieties, such as Artemovka, Ukrainka spring and Narodnaya, several winter-wheat varieties were developed, e.g. Lutescens-808, Kievskaya-893 and Mironovskaya-264. Lutescens-808 was released in 1963 under the name Mironovskaya-808. Due to its yielding capacity, winter hardiness, bread-making characteristics and wide-adaptability, Mironovskaya-808 spread rapidly throughout the Soviet Union as well as many European countries. The area upon which Mironovskaya-808 was grown, exceeded 9 million hectares annually in the 1960s and 1970s.

At the same time, Remeslo started hybridization of the best wheats from northern Caucasus, resulting in the varieties Mironovskaya jubileynaya, Ilichovka and Mironovskaya-25. These varieties of the 'third generation' of Mironovka varieties together covered about 2.5 million hectares annually in the 1970s and 1980s. Since the 1970s, hybridization between spring and winter forms, as well as intergeneric and interspecies crosses has been widely applied, as well as mutation breeding, resulting in a 'fourth' generation of cultivars. They cover about 25% of the area under wheat in Ukraine and are also grown in Russia and Belarus. The most widely spread variety of this group is Mironovskaya-61, which was developed in cooperation with breeders in Germany, where it is marketed under the name Mirleben. The average yield in trials of the fourth generation of Mironovka cultivars was 6.02 t/ha compared to 5.10 t/ha of the third, 4.42 t/ha of the second and 3.58 t/ha of the first generations. However, the quality of the latest generation is relatively poor due to the presence of the 1B/1R chromosome translocation from rye, which appears to be present in 60% of the varieties (Zhivotkov et al., 1996).

The development of semi-dwarf varieties (75–115 cm) was a major breeding objective at the Remeslo Institute in the 1980s. Wheat material from the UK, Japan and China was introduced to this end. In recent years the Remeslo Institute in Mironovka has worked in close cooperation with the breeding institutes in Odessa and Kharkov (Pospelowa, 1997).

3.2.4. The V. Ja. Juryev Institute (UNIIRSG) in Kharkov
Wheat breeding at this institute, where Juryev started his work in 1910 (see Section 2.2.1), has produced several important varieties, of which Kharkovskaya-63 and Kharkovskaya-81 have been the most successful. Kharkovskaya-63, selected from a cross between Bezostaya-1 and Mironovskaya-808, was released in 1963. It was outstanding for its winter hardiness and grain

quality. Most of the varieties released from Kharkov have been noted for their winter hardiness. Kharkov varieties covered 14 million hectares in the years 1987–1988 (Pospelowa, 1997).

3.2.5. The Ivanov Breeding Station (Kharkov region)

A large-scale programme of hybridization was initiated in 1951, resulting in one of the best breeding lines ever: Lutescens-317, selected in 1956. As in other breeding programmes, good results were obtained from crosses between fairly well-adapted but ecologically distant forms of high yielding wheats with various useful traits. Akhtyrchanka and Ivanovskaya-60 were bred from such material. Ivanovskaya-16, also named Vavilovka, was selected from a cross between Odesskaya-66 and Amfidiploid-206. This high yielding variety of the intensive type was submitted for state variety trials in 1986, followed some years later by Ivanovskaya-16a, which was selected from the same cross (Lastovich, 1991).

3.2.6. The quality of the Ukrainian wheat harvest

A survey of the most important wheat varieties grown in Ukraine in 1987, including some of their quality characteristics, is given in Table RU.6. Clearly, more than half the area under winter wheat was sown with the relatively old varieties Bezostaya-1, Odesskaya-51 and Mironovskaya-808. Pospelowa (1997) reported that these varieties are gradually being replaced by high yielding varieties such as Poleskaya-70 and Donskaya polukarlikovaya. Unfortunately, varieties with a higher quality level, such as Erythrospernum-127 and Obriy, are less popular, due to their lower yielding capactity. The average quality of the harvest is therefore declining. According to Pospelowa, two-thirds of the varieties registered in Ukraine belonged to the group of strong/improver wheats. The remaining varieties belonged to medium quality wheats or 'weak' wheats, which have to be mixed with improver wheat for bread-making.

3.3. Russia

3.3.1. The Lukyanenko Research Institute of Agriculture in Krasnodar

The town of Krasnodar is situated close to the river Kuban in northern Caucasus, one of the most important winter-wheat growing areas in Russia. The Kuban area was renowned for the export of wheat until World War I.

Lukyanenko started work on wheat breeding at the Krasnodar Research Institute of Agriculture in 1930. The most important improved winter-wheat varieties grown in this region around the 1930s were Ukrainka, Kooperatorka, Stavropolka and Stepnyachka, all varieties that were highly valued on the European grain market. They had large vitreous grains and excellent

Table RU. 6. Characteristics of the most important winter-wheat varieties grown in Ukraine in 1987

Variety	Year	Area sown in 1987 (ha)	Quality group	Protein content	Gluten content
Bezostaya-1	1959	620 300	strong	11.2–19.3	18.6–41.4
Mironovskaya-808	1963	529 400	strong	10.5–15.6	23.4–35.4
Odesskaya-51	1969	1 506 700	strong	12.6–18.9	21.9–38.9
Poleskaya-70	1974	765 600	–	12.5–15.2	26.3–34.3
Erythrospernum-127	1977	276 700	strong	13.1–18.0	27.6–39.2
Akhtyrchanka	1978	652 700	high quality	13.6–18.0	20.0–39.8
Dnjeprovskaya-846	1980	365 800	strong	11.6–18.5	23.4–41.6
Donskaya polukarlik.	1980	263 500	high quality	11.3–17.3	23.2–35.6
Kiyanaka	1981	296 500	high quality	10.9–16.0	23.8–36.3
Cayka	1982	250 600	high quality	12.5–16.6	20.0–37.0

Source: Pospelowa (1997).

Figure RU. 4. The pedigree of the winter-wheat variety Bezostaya-1, released in 1959 (after Prutskova & Ukhanova, 1972).

milling and baking qualities. However, they were tall (a strawlength of 140–160 cm) and susceptible to lodging. Moreover, Stavropolka suffered from grain shedding and was susceptible to loose smut; the other three cultivars were susceptible to rust and lacked sufficient winter hardiness.

The major aim of Lukyanenko and his colleagues was to develop varieties with improved harvest index and grain yield per ear, which were considered to be key components of the yield potential. The hybridization of distant eco-geographical forms was considered to be an effective method of achieving this; wheat varieties from Argentina, Italy, Canada, Sweden, England and Germany were widely used in the crossing programme. This resulted in the variety Bezostaya-1 (translation: Awnless-1) released in 1959. Bezostaya-1, the pedigree of which is presented in Figure RU.4., may be considered as one of the most outstanding results of Russian plant breeding.

Bezostaya-1 became a valuable gene source that was widely used in breeding programmes in the USSR and all over Europe. In 1972, 58 newly-developed Russian varieties of winter wheat had been obtained by crosses involving Bezostaya-1 or Bezostaya-4. Due to the combination of high yield potential with a quality level of the traditional varieties and a wide adaptation, the area under Bezostaya-1 increased rapidly after its release, covering 13 million hectares in 1971 (see also Section 4). In the Kuban region the release of Bezostaya-1, together with more intensive cultivation allowed yields to increase 1.5–2.5 fold. Yields of more than 5 t/ha were obtained on the best farms.

The variety Rannyaya-12, released shortly after Bezostaya-1 (pedigree in Table RU.8), was similar to Bezostaya, with very good baking quality and the flour strength of an improver variety.

To further increase the grain yield per ear, Western European ecological types were used as crossing parents. Thus the varieties Kavkaz and Aurora, released in 1967, were selected from the offspring of a cross with the German variety Neuzucht. The pedigree of both varieties is the same:

Neuzucht × Bezostaya-4
↓
Lutescens-314h147 × Bezostaya-1
↓
F_3 -population
↓
Kavkaz and Aurora

Rabinovich (1998) noted that the wheat-rye translocation was introduced into Russian material through Neuzucht, and through Kavkaz and Aurora was further transmitted to numerous varieties.

Kavkaz and Aurora were different from the available winter-wheat varieties because they had more productive ears and a high complex resistance to leaf, yellow and stem rust, as well as to powdery mildew. Both varieties greatly outyielded Bezostaya-1. In a five-year comparative variety trial (1967–1971), Kavkaz and Aurora yielded 11% and 13%, respectively, more than Bezostaya-1 (Lukyanenko, 1972). The stem length of Bezostaya-1, Aurora, Kavkaz and other newly-bred varieties ranged from 100 to 110 cm, i.e. 30–50 cm shorter than that of old varieties such as Ukrainka and Stavropolka-328.

With respect to their bread-making quality, Kavkaz and Aurora were inferior to Bezostaya-1. Three years of trials carried out between 1968 and

Table RU. 7. Baking quality of Kavkaz and Aurora compared to Bezostaya-1 and Mironovskaya-808 grown on three locations from 1968–1970

Variety	Years	No. of	Grain		Flour		Loaf	BMQ
location	of trial	analyses	vitreous-ness (%)	protein content (%)	gluten content (%)	strength (Joules)	volume (ml)	score
North Caucasian region								
Bez.-1	69–70	18	75	14.6	33.4	412	670	4.3
Kavkaz	69–70	18	81	15.8	34.7	225	566	3.3
Aurora	69–70	18	81	15.9	35.3	218	555	3.2
Steppe of Ukraine								
Bez.-1	68–70	13	57	13.4	28.5	322	660	4.3
Kavkaz	68–70	13	73	14.2	31.9	210	600	3.4
Aurora	68–70	13	72	14.2	31.9	211	589	3.4
Eastern Forest-Steppe of Ukraine								
Mir-808	68–69	22	95	12.6	28.7	309	650	4.3
Kavkaz	68–69	22	90	13.7	31.2	244	592	3.6
Aurora	68–69	22	90	13.7	31.2	278	589	3.6

Source: Prutskova & Ukhanova (1972).

1970 on three locations in Ukraine (Table RU.7) demonstrated that Kavkaz and Aurora had higher values for grain vitreousness, grain protein content and flour gluten content than Bezostaya-1. However, the quality of the gluten was such that the flour strength and baking quality of Kavkaz and Aurora were inferior to those of Bezostaya-1. It is very likely that this was related to the effect of the wheat-rye translocation. Several foreign publications made mention of this effect (see Chapter Yugoslavia).

Flour strength in these trials was measured by 'the micro method of raising in acetic acid' (probably a variation of the Zeleny sedimentation test), resulting in a 'raising index'. This parameter showed a wide range of variation, suggesting the possibility of improving the physical properties of Aurora and Kavkaz by selecting lines with high raising indices.

The variety Bezostaya-2, described as 'marked by good baking qualities and good flour strength', may have been the result of this type of selection.

One of the characters of Bezostaya-1 that needed to be improved was its winter-hardiness. For this purpose, Bezostaya-1 was crossed with varieties of the steppe ecological type from Ukraine and the Volga region, for example Odesskaya-16, Saratovskaya-3 and Lutescens-329. This resulted in the varieties Krasnodarskaya-39 (=Lutescens-39), Stepnaya-40, Severokubanskaya-43, Zagadka-44, Nadieznaya-45 and Krasnodarskaya-46. Several of these

Table RU. 8. Winter-wheat varieties bred at the Krasnodar Research Institute of Agriculture between 1934 and 1972

Variety	Pedigree	Year	BMQ
Ferrugineum-51	Ferrugineum-13/Kitchener	1939	insufficient
Ferrugineum-622	Ferrugineum-13/Marquis	1939	insufficient
Krasnodarka	sel.from Ferrugineum-622	1939	insufficient
Novukrainka-83	Ukrainka/Marquis		strong, improver
Novukrainka-84	sel. from Novukrainka-83		strong, improver
Skorospelka-3	Kanred/Fulcaster-266287//Klein-33	1955	
Kubanskaya-131	Kanred/Fulcaster-266287//Ferrugineum-622		
Kubanskaya-122	Stavropolka-328/3/Stavropolka-328//Kanred/Fulcaster-266287		
Bezostaya-4	Lutescens-17/Skorospelka-2	1955	strong, improver
Bezostaya-1	recurring individual selection from Bezostaya-4	1959	excellent
Krasnodarskaya-6	Odesskaya-3/Skorospelka-3		
Rannyaya-12	Bezostaya-4/Skorospelka-3	1963	excellent
Aurora	Lutescens-314h147/Bezostaya-1	1967	
Kavkaz	Lutescens-314h147/Bezostaya-1	1967	
Krasnodarskaya-39 (= Lutescens-39)	Saratovskaya-3/Bezostaya-1	1968	good,improver
Stepnaya-40	Odesskaya-16/Bezostaya-1		
Skorospelka-35	Erythrospernum-315h60/Bezostaya-1		
Bezostaya-2	Lutescens-314h147/Bezostaya-1	1969	good
Severokubanskaya-43	Lutescens-329/Bezostaya-1	1971	good/excellent
Zagadka-44	Bezostaya-1/Mironovoskaya 808	1971	good/excellent
Nadezhnaya-45	Mironovskaya-808/Bezostaya-1	1971	good/excellent
Krasnodarskaya-46	Bezostaya-1/Odesskaya-16//Bezostaya-1		

Source: Prutskova & Ukhanova (1972).

varieties had excellent bread-making characteristics (Table RU.8). Krasnodarskaya-39 was described as a cultivar with 'good qualities and satisfactory flour strength of an improver variety'. It has been widely grown in Russia since the 1970s. Severokubanskaya-43, Zagadka-44 and Nadezhnaya-45 were reported to have a high level of flour strength (374–440 Joules) and good to excellent baking qualities (BMQ score 4.1–5.0 points). The variety Krasnodarskii Karlik (transl. Krasnodar Dwarf) was frequently used as a parent in breeding winter wheat for the steppe zone of the Ukraine, resulting amongst others in the short straw varieties Salyut (KK/Odesskaya-26) and Progress (Krasnodarskii Karlik/Odesskaya-16) (Kirichenko et al., 1983). Numerous wheat varieties were developed over the past decades by breeders from Krasnodar. Varieties that are noted for their bread-making quality are Spartanka, Skifyanka, Yuna, Kolos, Rufa and Leda (see also Table RU.10).

3.3.2. Winter-wheat breeding in the Rostov region

The province of Rostov is situated in northern Caucasus, northeast of the Sea of Azov around 46 °N. Extensive-type wheat varieties were still commonly grown there after 1945; the average grain yield was 1.75 t/ha during the period 1938–1953. This changed with the release of the variety Odesskaya-3, which covered 50–90% of the wheat area between 1947 and 1962, and reached 6 million hectares in 1959. In the 1960s, Odesskaya-3 was replaced by Bezostaya-1, Odesskaya-16 and Mironovskaya-808. Mironovskaya-808 covered 43% of the winter-wheat area in 1970 and was still widely grown in the 1990s. Bezostaya-1 and Mironovskaya-808 were used for further breeding work, in combination with varieties from other regions as well as foreign cultivars. The most important varieties grown in the 1980s were described by Kalinenko (1990).

Rostovchanka was a variety with a high level of cold resistance and with grain yields varying between 5.6 and 5.75 t/ha. Donskaya Bezostaya (translation: Awnless from Don), selected from a cross between Bezostaya-1 and Mironovskaya-808, was resistant to some diseases and to lodging, and also had good bread-making quality. In ten years of trials (1977–1986) the average yield was 6.7 t/ha, with a maximum of 8.3 t/ha. It was sown on more than 2 million hectares in 1989, in Rostov as well as in other regions.

Donskaya Polukarlikovaya (translation: Semi-dwarf from Don), selected from a cross between the Bulgarian cultivar Rusalka and Severodonskaya, was resistant to lodging and various diseases, and it was drought tolerant. In ten years of yield trials (1977–1986) the average grain yield was 7.2 t/ha.

Tarasovskaya-29, selected from a cross between Mironovskaya Yubileinaya and Rostovchanka, had a straw length of 85–100 cm. In trials during 1975–1985 the grain yield was 5.9 t/ha. Don 85, with a complex parentage from local and other Russian cultivars, had good milling and baking qualities, and a high loaf volume (652 ml). The average yield in five years of trials (1984–1989) was 5.6 t/ha.

3.3.3. The Kuybyshev (Samara) Agricultural Institute in Bezenchuk

The institute is located some 350 km northeast of Saratov, at a longitude of 50 °East, where winter-hardiness is an important selection criterion. The wheat-breeding programme at this station was started in 1936. Among the first varieties released were Bezenchukskaya-51 and Bezenchukskaya-108, both selected in a cross between Lutescens-96 and Ferrugineum-474. A cross between Bezenchuk-108 and Bezostaya-1, back-crossed with Bezostaya-1, resulted in the variety Granit, a winter-hardy, drought-resistant, short-straw variety with a yield potential of more than 7 t/ha, and producing large grains with good technological quality.

Table RU. 9. Wheat varieties released from the Moscow Breeding Centre since 1980

Winter wheat	Pedigree
Moskovskaya-70	Mir-808/Krasnodarskii karlik-1//Mir-808/3/Zarya
Moskovskaya nizkostebelnaya	sib of Moskovskaya-70
Nemchinovskaya-2	Krasnodarskii karlik-1/Mir-808//Mir-808
Moskovskaya-642	Zarya/Akhtyrchanka
Inna	Nemchinovskaya-6/Zarya
Spring wheat	
Priokskaya	line 1833-6N427/Moskovskaya-35//Saratovskaya-54
Enita	line 1833-6N427/Mironovskaya yarovaya
Lyuba	line 30N1/Moskovskaya-35

Other varieties that have resulted from the hybridization of wheats from different ecological groups are Milturum-97 and Lutescens-114.

Bezenchuk-132, selected from a cross between Osctinskaya-3 and Bezostaya-1, is a winter-hardy variety, produducing large grains with good quality. It has a high yielding potential in spite of its slight susceptibility to brown rust and mildew. The spring wheat Zhigulevskaya, selected from a cross between a winter and a spring-wheat variety (Bezostaya-1/Bezenchukskaya-98), was classified as a strong wheat in terms of baking quality and gluten content.

3.3.4. The Research Institute of Agriculture in Central Regions of the Non-Chernozem at Nemchinovka (Moscow)

A combination of wheat research and wheat breeding at this institute, has resulted in a large number of winter and spring-wheat varieties (Table RU.9). Two important winter-wheat varieties, released in the late 1980s, were Moskovskaya-70 and Nemchinovskaya-52. Both varieties have similar pedigrees, including Krasnodarskii karlik (translation: Krasnodar dwarf) as a short-straw parent, and Mironovskaya-808. Moskovskaya-70 has, moreover, Zarya as a parent, which was described as a strong type quality wheat.

Moskovskaya-70 is a mid-season variety with a straw length of 85–100 cm that is resistant to the main fungal diseases. It shows profuse tillering and overwinters well in the Moscow region. Ears are 8–9 cm long and carry red grains with a thousand kernel weight of 36–45 g. and good baking quality (protein content 13.1%, crude gluten content of the flour 28.7% and loaf volume 950 ml per 100 g of flour). In trials in Moscow Province during 1984–1990 it yielded 6.14 t/ha. Moskovskaya nizkostebelnaya (translated: Moscow short-strawed) is a sib of Moskovskaya-70.

Nemchinovskaya-52, released in 1990, was described as a variety with resistance to lodging and shedding, and having good quality grain. The average yield varied between 5.87 (trials 1983–1989) and 5.36 t/ha (state trials 1986–88).

Two other winter-wheat varieties selected from crosses with Zarya are Moskovskaya-642 and Inna, both released in the early 1990s. Moskovskaya-642 is a variety producing large grains (thousand kernel weight 47 g) with good milling and baking characteristics (protein content 13.5%; crude gluten content 32.6% and loaf volume 1015 ml per 100 g flour). It overwintered well in the Moscow region and yielded 6.05 t/ha in trials during 1984–1990. Yields of Inna were even higher (7 t/ha in five years of trials), equaling Zarya in winter-hardiness and exceeding it in lodging resistance, due to 15–20 cm shorter straw. Inna produced large grains (thousand kernel weight 50–55 gram) with a baking quality and protein content that was better than that of Zarya.

Three spring-wheat varieties that have been released from the Moscow Breeding Centre since the 1980s are Priokskaya, Enita and Lyuba. All three are mid-season varieties with good lodging resistance. Priokskaya produces large uniform grains with good quality (gluten content 34–40%; loaf volume 900–1200 ml). The average yield (1987–1989) varied between 4.84 t/ha (Moscow region), 4.96 t/ha (Vladimir Province, 100 km east of Moscow) and 5.32 t/ha (Ryazan Province, 200–300 km southeast of Moscow); maximum yield was 8.29 t/ha.

Enita has a moderate baking quality (loaf volume of 800–1100 ml). It is an intensive type variety, recommended for various provinces of European Russia, with a high yield potential but moderate drought resistance. It has good yield stability and is responsive to fertilizers. The maximum yield obtained in state trials was 8.63 t/ha. Lyuba produces high-quality grains (thousand kernel weight 35–45 g; gluten content 30–40%; loaf volume 1000–1200 ml). It is responsive to fertilizer treatment and resistant to shedding and sprouting in the ear.

3.3.5. Recommended varieties in Russia

Half of the wheat varieties that were added to the Recommended List of Russia of 1994, 1995 and 1996 (Table RU.10) were bred at the Agricultural Research Institute in Krasnodar or one of its experimental stations. Those varieties were especially recommended for the northern Caucasus region. Nearly all the new varieties were recommended for the southern part of Russia: northern Caucasus (12x), the Volga Region (3x), the Central Region (1x). Three varieties were recommended for Ural and Siberia.

Table RU. 10. Pedigree, yield and quality characteristics of newly released Russian winter-wheat varieties 1994, 1995 and 1996

Variety	Br.[a]	Pedigree	Yield (t/ha)	Straw-length	BMQ
Rufa	Kra	Obriy/Mironovskaya-808	5.0	84–106	good
Yugtina	Kra	Zagore/Donskaya Polukarlikovaya// Krasnodarskaya-57-273/Erythrospernum-23OOg11313	5.1	80–106	satisfactory
Leda	Kra	Obriy/Lutescens-3161a29	4.2	82–98	very good
Gorlitza	Kra	Lutescens-3817h60/Lutescens-834h145-6	4.1	76–92	good/very good
Polovchanka	Kra	AD 206 (*Triticale*) /Rubin × 17t3 (*Triticale*)	4.2	83–98	satisfactory
Yeyka	SKu	Donskaya Polukarlikovaya/Erythrosp.-2300g11313	4.7	88–112	satisfactory/good
Krasnodarskaya-90	SKu	Obriy/Donskaya Bezostaya//Lutescens-2574h352	4.0	79–109	good
Nika kubani	SKu	Obriy/Krinitza	4.3	63–96	good
Ofeliya	SKu	Lutescens-1673h75/Pavlovka (ind.sel.F1,F5,F9)	4.0	72–97	good
Yershovskaya-10	Yer	Yershovskaya-8/Albidum-114	3.3	65–86	low/satisfactory
Donskaya jubileyn.	Don	Donskaya Bezostaya/Donskaya Polukarlikovaya	4.6	67–99	good/excellent
Komsomolskaya-75	Mir	selection from Albidum 2231-1		76–83	satisfactory/good
Bezenchukskaya-380	Sre	Lutescens-246//Mir.-808/Severokubanka/3/Mir.-808	4.6	84–108	good
Saratovskaya-90	Ely	Lutescens-36/3/Saratovskaya jubileynaya/ Bezostaya-1//Saratovskaya-5/4/F_2/Mironovskaya-10	5.7	69–98	satisf./very good
Imeni rapoporta	Pod	mutant of PPG-186/Mironovskaya-808	3.4	84–112	good
Bagrationovskaya	Sib	sel.from Mironovskaya jubileynaya	2.4	80–104	satisfactory/good
Kuludinka	Sib	selection from Albidum	2.5	74–78	good
Severodonskaya-12	Sev	Tarasovskaya-29/Zaporozhskaya ostistaya			

[a] Breeder abbreviated as Kra = Inst.Agr.Res.Krasnodar; SKu = Severokuban exp.st. of Krasnodar Inst.; Yer = Yershovska exp.st.; Don = Donskoi Selection Centre; Mir = Mironovskii Inst. of Wheat and Karabalikskaya exp.st.; Sre = NPO 'Srednevolzhskoye'; Ely = NPO 'Elyta Povolzyha'; Pod = Inst. of Biochem.Physics, NPO 'Podmoskovje'; Sib = Inst. Cytology and Genetics, Siberian Division of the Russian Ac.of Sc.; Sev = Severo-Donskaya experimental station.
Source: Ministry of Agriculture, Moscow.

A wheat variety that was frequently used as a parent in crosses was Obriy. This variety, released in 1983 from Odessa (see Section 3.2.1) was described as a strong winter wheat, highly resistant to lodging, with a height varying between 85 and 95 cm, and combining high yields (4.6–9.1 t/ha) with good baking quality, including high protein content. The pedigree of Obriy is presented here to demonstrate that only five steps of selection and/or crosses have been made since the original combination of two land varieties (Mestnaya is Russian for land variety).

LV from Kharkov		Mestnaya		
↓		↓		
Hostianum-237	×	Zemka		
	↓			
	Odesskaya-12			
	↓			
	Odesskaya-16		×	Bezostaya-1
			↓	
	Red River 68	×	Odesskaya 51	
			↓	
			F_1 × Odesskaya-51	
			↓	
			Obriy	

The average grain yield of the newly released winter-wheat varieties, based on the official trial reports, was 4.02 t/ha. According to FAO statistics the average wheat yield in the European part of Russia was 1.51 t/ha in the years 1993–1995. Comparison of these two figures suggests that the potential grain yield of the newly released varieties is on average 2.5 times higher than the actual harvest.

Considering the figures given in the trial reports for the maximum yield of the new varieties (average 7.54 t/ha), the potential yield of modern winter-wheat varieties would, under optimal conditions of husbandry, be five times that of the actual yield in 1993–1995. Thus, the results of recent breeding work in the European part of Russia have provided the basis for a major increase in the national wheat harvest, assuming the introduction of newly bred varieties and improved husbandry.

4. Fluctuations in the quality of the wheat harvest

The quality of the Russian wheat harvest between 1944 and 1968, calculated as the proportion of the USSR wheat area sown with strong cultivars, is

Table RU. 11. Proportion of the total area of winter and spring wheat sown with strong cultivars between 1944 and 1968

Year	Proportion of strong varieties	
	ww (%)	sw ($\times 10^6$ ha)
1940	47	3.4
1953	15	1.3
1958	15	2.4
1964	64	–
1965	67	–
1966	–	12.7
1967	–	19.0
1968	73	22.9

Source: Prutskov (1970) and Ivanov (1971).

presented in Table RU.11. It is obvious that the fluctuation of the quality of the wheat harvest is related to the release and success of certain cultivars. Two winter-wheat varieties that had a large impact in this respect were Bezostaya-1 and Mironovskaya-808, both released in the 1960s. Apart from their impact on the Soviet wheat production, they influenced wheat breeding all over Europe.

Whereas a considerable part of the Russian wheat area before 1940 was covered with varieties with excellent bread-making quality, in the post war period several of them were replaced by varieties with improved productivity but inferior bread-making quality. According to investigations by the Central Milling-Bread-Making Laboratory of the State Commission for Strain Testing of Agricultural Plants, the main varieties, acknowledged in 1960 (Samsonov, 1960) as improver wheats with good milling and bread-making qualities, were:

- *Winter wheat*: Bezostaya-1, Bezostaya-4, Novukrainka-83 (good milling and excellent bread-making quality), Priazovskaya (selection from Cheyenne). Other officially recommended varieties that produced flour of good bread-making strength included: Belotserkovskaya-198, Gibrid 343 and Kharkovskaya.
- *Spring wheat*: Bezenchukskaya-98 (Lutescens-DS-11–21–44 (USA)/Erythrospernum-b47), Lutescens-758 (Kitchener/Lutescens-62), Saratovskaya-29 (Lutescens-55–11/Albidum-24), Saratovskaya-210 (Lutescens-2074/Sarrosa). The presence of several North American parents in the pedigree of these varieties is remarkable.

Table RU. 12. Strong winter-wheat varieties harvested in 1965 and 1968 (area in 1000 ha and percentage of the total area grown with strong winter-wheat varieties)

Variety	1965		1968	
	abs	rel	abs	rel
Bezostaya-1	6 452	49.3	7 184	47.9
Belotserkovskaya-198	3 311	25.3	208	1.4
Mironovskaya-808	586	4.5	6 260	47.2
Mironovskaya-264	1 926	14.7	30	0.2
Ukrainka	156	1.2	43	0.3
Priazovskaya	507	3.9	305	2.0
Kooperatorka	73	0.6	59	0.4

Source: Prutskov (1970).

In 1968, one-third of the total spring-wheat area in the USSR was sown with strong wheat cultivars, of which Saratovskaya-29 and Bezenchukskaya-98 were the most important, covering 16.2 and 3.6 million hectares, respectively (Ivanov, 1971).

The rapid turnover of varieties that may take place is illustrated by the figures in Table RU.12, which show the success of Mironovskaya-808. The hectarage of Bezostaya-1 and Mironovskaya-808 continued to increase after the 1960s. In 1970 they covered about 40% of the harvested winter-wheat acreage in the USSR, whereas the total number of winter-wheat varieties available in the country was 115. Generally speaking, Bezostaya-1 was grown in the southern and Mironovskaya-808 in the northern and eastern parts of the USSR, including the areas of non-black soils and the Volga River region. Both varieties were grown up to the 1990s; in 1976–1987 they covered 14–16% of the Soviet winter-wheat area (Muzyka & Martynyuk (1991). Mironovskaya-808 has been grown on an area of 20 million hectares annually for more than 30 years; it was sown on 90% of the winter-wheat area in the mid-Volga region in 1995 (Boonman, 1996).

5. Methods and classification with respect to bread-making quality

5.1. Methods used in selecting for bread-making quality

A condition that is essential for selection work on improved baking characteristics is the availability of reliable methods for baking trials on laboratory scale. Sozinov (1979) stated that Russian breeders have met considerable difficulties in this respect, due to inaccuracy of methods, as well as to phenotypical variation. From many experiments it was known that optimal results in selection of genotypes with good bread-making quality could be obtained by Remix baking, under addition of sugar and various amounts of bromate. By doing so, a broad variation between varieties in loaf volume and in other characteristics may be detected. The wide range of this variation is shown in Table RU.13.

Table RU. 13. Varietal differences in baking characteristics after a baking trial according to the methods of the VSGI on samples from the 1976 harvest

Variety	Loaf volume	Porosity score	BMQ score
Odesskaja-51	1230	4.25	5.20
Erythrospernum line (Red River/Odesskaya-51)	1150	4.75	5.25
Odesskaya-66	1090	4.00	4.75
Priboj	1070	4.75	5.25
Erythrospernum-394/74 (Red River/Odesskaya-51)	1070	4.25	5.00
Bezostaya-1	1030	4.00	4.75
Erythrospernum-245/73 ('RPG'/Odesskaya-51)	970	3.75	4.00
Kavkaz	910	3.00	3.00
Lutescens-522/73 (Kavkaz/Odesskaya-51)	710	2.25	2.00
Erythrospernum-569/74 (V.S.1877/Kavkaz)	680	2.25	2.75
Erythrospernum-874/74 (Soti Lerma/Mirovskaya-808)	590	2.25	1.25

Source: Sozinov (1979).

Literature giving a clear description of the methods used for the selection on bread-making quality during the first stages of breeding is difficult to find. Apparently, different variants of the sedimentation test were applied during the late 1970s and 1980s. No indication was found of a test used in the first stages of selection during the 1930s or 1940s, when, for example, Pelshenke and Chopin tests were applied in Germany and France.

Bebyakin & Krupnova (1992) pointed out the importance of sedimentation analyses for assessing the grain quality in the primary stages of spring-wheat breeding, starting in the F_4 generation. They used a dodecylsulphate (DDS) sedimentation test, which is similar to the SDS sedimentation test. The DDS sedimentation value, referred to as SV_{DDS} was compared to three other variants of the sedimentation value: SV_{SE} (used at the Scientific Research Institute of Agriculture in the southeast); SV_{BG} (used at the All-Union Breeding and Genetics Institute); and SV_{PG} (used at the All-Union Scientific Research

Institute of Plant Growing). They were ranged according to their usefulness as a selection tool: $SV_{DDS} > SV_{SE} > SV_{BG} > SV_{PG}$.

Bebyakin & Krupnova studied the effectiveness of a selection index based on sedimentation and mixographic estimates. They recommended the use of a two-component index, consisting of the SDS sedimentation index and a parameter S, which is the area delimited by the mixograph curve (with maximum height h). It should be kept in mind that the use of such a selection index can hardly be called a simple method to select young breeding material with bread-making quality.

Pshenichnyi (1984) studied the efficiency of a sedimentation index to evaluate breeding material of winter wheat. He investigated hybrids of varieties with different flour strength (strong, weak and intermediate). The sedimentation value of the flour of the F_1, F_2, and F_3 and parents was compared. High estimates of heritability (up to 90%) for sedimentation values in crosses of contrasting parents indicated that selection would be effective. Selection for sedimentation value in F_2 led to simultaneous selection for flour strength and baking quality, but it was not a reliable method for obtaining a high gluten content.

5.2. Classification of technological quality

In a survey of the technological characteristics of wheat, related to different industrial purposes, Sozinov (1979) described five categories:

- *Hard wheats* are characterized by an improved protein content, and a vitreous kernel with good flour characteristics that produces an elastic dough, that absorbs water readily; hard wheats lead to a marked improvement of the baking characteristics in mixtures with weaker wheats. Typical examples of strong wheat varieties are Bezostaya-1, Mironovskaya-808, Odesskaya-51 and Saratovskaya-39.
- *Quality wheats* are characterized by a higher protein content, good milling characteristics, stable kernel development and high baking quality. Typical examples are the varieties: Odesskaya-16, Priboj, Ilyicevka and Dnjeprovskaya-775.
- *Wheats for cake making* should have a low protein content, a good flour and high technological characteristics for cake making. This type of wheat varieties is not bred in the USSR.
- *Fodder wheats* need a high content of biologicaly useful protein.
- *Macaroni wheats* are important for industry; the cultivation of durum wheat for pasta products is very important in the USSR.

Advances in the mechanization and automation of manufacturing processes in the milling and bread-making industries have resulted in a demand for higher standards for wheat, involving more complicated analyses and a large number of quality characteristics to be taken into account in the evaluation of varieties. Belousova (1991) mentioned a classification of bread wheat varieties into seven groups.

Vitreousness is becoming increasingly popular as an index in the selection of high quality wheats. This is based on the fact that the gluten content of the wheat grain depends more on geographical location and meteorological conditions than on the variety. Thus, the difference in gluten content of mealy and vitreous fractions may be as high as 10% within the same variety.

In addition to a direct baking test, alveograph and farinograph values are used as indirect measures for the physical properties of the dough. Characteristic alveogram indices such as dough stiffness (P), elasticity (L), P/L ratio and dough strength (W) are recorded. Farinogram indices recorded are time of formation, stability, degree of thinning and the mixing index according to the valorimeter. Farinograph tests are carried out by mixing 50 g flour samples.

6. Glutenin and gliadin composition of Russian wheats

6.1. High molecular weight glutenins

The HMW glutenin subunit composition of 128 Soviet varieties was studied by Morgunov et al. (1990). It appeared that the pattern of HMW glutenin subunits in Russian spring-wheat varieties was distinct from those observed in spring-wheat varieties from other countries. There was a similarity to Canadian and Yugoslav varieties in the low frequency of the null allele at *Glu-A1* and the relatively high frequency of the 7+9 allele at *Glu-B1*.

A striking conclusion from this study was that the HMW glutenin pattern in varieties with superior bread-making quality was hardly different from the pattern in varieties with lower quality levels (Table RU.14). The *Glu-1* score was even higher in the varieties classified as feeding wheats than in the quality wheats, due a lower frequency of subunits 5+10 in the latter.

From a comparison of the HMW glutenin composition of winter- and spring-wheat varieties from different regions (Table RU.15), it appeared that the frequency of *Glu-A1* subunits 1 and 2∗ in the winter-wheat varieties from Zernograd and those from Krasnodar was reverse. The most frequent *Glu-B1* subunits for all the locations were 7+9, with a clear difference between the winter-wheat varieties from Zernograd and Krasnodar and those from Odessa. The most frequent *Glu-D1* subunits were 5+10, except for the varieties from Saratov, where 2+12 prevailed. The unique combination of the *Glu-D1* subunits 5+12 was observed in some wheat samples from Saratov and Krasnodar. The varieties bred in Krasnodar and Odessa showed a high

Table RU. 14. Frequencies of HMW glutenin subunits in spring-wheat varieties with different levels of BMQ

Subunits	Superior quality	Good quality	Feed quality
Glu-A1			
N	6.4	12.5	0
1	6.4	12.5	30.0
2*	87.2	75.0	70.0
Glu-B1			
6+8	3.2	0	0
7+8	19.4	12.5	30.0
7+9	74.2	87.5	70.0
17+18	3.2	0	0
Glu-D1			
5+10	41.9	75.0	60.0
2+12	58.1	25.0	40.0
Number of cultivars	31	8	10
Average *Glu-1* score	7.97	8.37	8.50

Source: Morgunov et al. (1990).

frequency of subunits that are related to quality: 1 and 2∗ on *Glu-A1*, 7+8 and 7+9 on *Glu-B1* and 5+10 on *Glu-D1*, resulting in *Glu-1* scores 9 and 10.

Of the wheats from the breeding programme in Moscow, the varieties Moskovskaya-35 and Enita were mentioned as examples of varieties that have very high *Glu-1* scores (10 for Enita) but are not classified as strong wheats. The varieties from Saratov had relatively low *Glu-1* scores, mainly due to the low frequency of subunits 5+10. This is surprising as wheats from the Saratov region are traditionally renowned for their excellent bread-making quality (see Section 1.2).

Morgunov et al. (1990) found it difficult to explain the disagreement between *Glu-1* scores and bread-making quality in Russian wheats, as the *Glu-1* score is a direct measure of dough mixing strength, which is a parameter accepted as one of the most important in identifying bread-making quality in Russia. Morgunov et al. stated, moreover, that the disagreement cannot be explained by the presence of the 1BL/1RS translocation, because 'it is rare in Soviet varieties'. This is a remarkable statement bearing in mind the crosses with Kavkaz and Aurora, which have been reported to have the rye

Table RU. 15. Frequencies of HMW glutenin subunits in spring- and winter-wheat varieties from different regions of the (former) USSR

Subunits	Breeding programmes				
	Saratov	Moscow	Zernograd	Krasnodar	Odessa
	sw	sw	ww	ww	ww
Glu-A1					
N	6.1	11.7	0.0	3.6	10.0
1	8.2	20.0	69.9	31.8	50.0
2*	85.7	68.3	30.1	64.6	40.0
Glu-B1					
6+8	3.7	8.3	0.0	0.0	10.0
7+8	5.3	20.0	0.0	0.0	38.0
7+9	89.5	71.7	100.0	100.0	52.0
17+18	1.5	0.0	0.0	0.0	0.0
Glu-D1					
5+10	11.1	100.0	100.0	89.6	90.0
2+12	86.7	0.0	0.0	6.4	10.0
5+12	2.1	0.0	0.0	4.0	0.0
Number of varieties	27	5	14	11	10

Source: Morgunov et al. (1990).

translocation. Their conclusion was that the influence of certain groups of gliadins may play an important role in this respect.

6.2. Gliadins

In the 1970s, Sozinov and Poperelya, in Moscow and Odessa, respectively, were the first investigators to report on the inheritence of stable blocks of gliadin components in wheat. They identified more than 80 alleles coding for these gliadin blocks, located on chromosomes 1A, 1B, 1D, 6A, 6B and 6D. Metakovsky et al. (1984) designed a catalogue for the nomenclature of the different gliadin blocks. Sozinov & Poperelya (1982) evaluated the relation of allelic blocks of gliadin components to quality of the flour. They identified certain blocks that were associated with better grain quality than others and rated them as follows:

Gli 1A:	7>4>2>5>3>1>6	*Gli 6A*:	3>1
Gli 1B:	1>2≥7>5>4>3>6	*Gli 6B*:	2>1
Gli 1D:	4>5>1>2>3	*Gli 6D*:	2>1>3

The relation between the presence of certain gliadin blocks and technological and baking qualities of wheat has been investigated by several groups of researchers since the 1980s and research is still in progress. Sozinov & Poperelya (1984) noted that the allelic variants of gliadin blocks are used as genetic markers for characteristics such as frost or drought resistance, tillering and adaptation to certain growing conditions, as well as for quality characteristics. For instance, the *Gli 1B3* block is used as a genetic marker for the 1B/1R translocation. Sozinov (1993) reported on the possibility to use gliadins and glutenins as a genetic marker for adaptation of genotypes to certain growing conditions.

Pylnev (1994) suggested that in the course of breeding for increased grain yield in Ukraine and in the central regions of the non-chernozem zone of Russia, new gliadin blocks have evolved that are associated with improvements in yield, adaptation to environment and grain quality. Breeders have been advised to select forms with gliadin blocks characteristic of a given zone. Poperelya et al. (1998) noted that the identification of storage protein alleles with a positive effect on gluten strength has enabled breeders to improve the dough quality, for example in the goal-oriented making of super-strong wheat.

6.3. Heterogeneity within Russian wheat varieties

Morgunov et al. (1990) found that 28 of 128 investigated cultivars were heterogeneous for HMW glutenin composition. The heterogeneity was evident in old land races as well as in recent varieties. Three varieties had different alleles at all three *Glu-1* loci. Some varieties consisted of two biotypes in approximately equal proportions, while others had a predominant biotype with ratios of 4:1; 5:2; 7:3; 9:3; etc. Even more intra-varietal heterogeneity has been observed by using gliadin markers.

The heterogeneity encountered in Russian wheats was compared to other coutries. In countries with very strong requirements for complete uniformity of varieties, breeding programmes are adapted to this purpose. However, opinions on the desirability of such homogeneity differ. Several breeders in the former USSR believe that a variety consisting of different biotypes provides more stable yields in changeable environmental conditions, due to a compensation of biotype-environment interaction. Therefore they do not aim at creating completely homogeneous varieties. Selection in the early generations without applying the pedigree method leads to some variability within advanced lines for different traits. In many cases the selection procedure applied by Russian breeders is a very simple one. An example of such a procedure has been given in Section 3.2.1, for the breeding of Stepova in Odessa.

The intra-varietal heterogeneity offers opportunities for improving the variety. Two examples from literature, where such opportunities have been pursued, are the improvement of the leaf-rust resistance of Mironovskaya-808 (Nosenko & Rogozhinskiy, 1973) and the selection of a short-straw feature in Kharkovskaya-63-1 (Vasilenko, 1973). In the same way, differences in glutenin and/or gliadin composition within a variety may be used to improve the quality of that variety (Nettevich et al., 1991).

Finally, the presence of different biotypes within one variety may explain the varying performance that breeders have encountered in varieties with Russian parents (e.g. Yugoslav varieties selected from crosses with Kavkaz and Aurora).

J.M. & D.A.D.

Acknowledgments

The authors wish to thank the following persons for their contribution to this chapter: special thanks to Prof. A.A. Sozinov in Kiev for reviewing the manuscript and providing a large amount of additional information about wheat breeding in Ukraine; Prof. A.F. Merezhko and Prof. A. Konarev at the N.I. Vavilov All-Russian Institute of Plant Industry (VIR) in St. Petersburg, for reviewing the manuscript and for their additional remarks on wheat breeding in Russia; Dr E. Golovina at the Institute of Plant Fysiology (Russian Academy of Science) in Moscow for providing information on Russian wheat varieties; and ir. J. Boonman (†), former representative of the World Bank in Moscow, for critically reading the manuscript.

References

Anonymus (1998) Ukriane: the 21st century wheat loft? The sleeping beauty. Grain Magazine 2: 13.

Bebyakin VM, Krupnova OV (1992) Effectiveness of index selection of wheat baking qualities on the basis of sedimentation and mixographic estimates. Soviet Genetics 28 (6): 747–756.

Belousova EM (1991) Classification of bread wheat varieties for baking strength. Plant Breeding Abstracts 61: 174.

Boonman J (1996) Failing reforms keep farmers lethargic. Prophyta (the annual): 66–71.

Boonman J (1998) personal communication, Wageningen.

Broekhuizen S (ed) (1969) Atlas of the Cereal Growing Areas in Europe. Wageningen: Pudoc & Amsterdam, London, New York: Elsevier Publ Company.

Burdenyuk LA (1991) Methods and results of breeding intensive-type varieties of winter wheat at the Belaya Tserkov' Breeding Station. Plant Breeding Abstracts 61: p 1265.

Chingo-Chingas KM (1925) Persian wheat in regard to its milling and baking qualities. Bul Appl Bot Pl Br 15: 206

Ilina LG (1986) Breeding for drought resistance. Plant Breeding Abstracts 56: 277.

Ivanov PK (1971) Spring wheat. Joint Publ Res Serv, report nr JPRS 59968 Arlington (Virginia, USA).

Jakubziner MM (1962) Wheat species and varieties as resources in plant breeding. In Symposium on Genetics and Wheatbreeding, Martonvasar: p 405–422.

Kalinenko IG (1989) Breeding and grain production in winter wheat. Plant Breeding Abstracts 60: 752.

Kirichenko FG (1970) Methods and basic results of breeding winter soft wheat for the Ukrainian Steppe Regions. Odessa Voprosy Genetiki, Sel i Sem 9, 17–30 In: Joint Publ Res Serv, Arlington (Virginia, USA).

Kirichenko FG, Nefedov AV, Litvinenko NA (1979) Selekcija ozimoj mjagkoj pšenicy na juge Ukrainy. In Selekcija i sortovaja agrotechnika ozimoj pšenicy. Moscow 29–42.

Kirichenko FG, Litvinenko NA, Nefedov AV, Abakumenko AV, Gerzhov AF (1983) Achievements and problems of breeding winter bread wheat in the steppe zone of the Ukraine. Vestnik Sel'skokhozyaistvennoi Nauki 5: 102–105.

Kretovich VL (1960) Biochemistry of grain and bread making in relation to the 7-year plan for the development of the national economy of the USSR. In Kretovich VL (ed), Biochemistry of grain and of bread making. Moskva: Akademia Nauk SSSR (Transl: Israel Program for Scient Translations Ltd, 1965).

Lastovich AS (1991) Breeding winter wheat at the Ivanov Breeding and Experiment Station. Plant Breeding Abstracts 61: p 1265.

Litvinenko NA (1998) Breeding intensive winter bread wheat varieties for southern Ukraine. Euphytica 100: 7–14.

Loewé E (1923) Les pays exportateurs de blé dans le monde. In Le blé, production, industries de transformation et commerce; compte-rendu des travaux de la semaine nationale du blé, Paris: pp 351–362.

Lukyanenko PP, Timofeev VB (1972) Winter-wheat breeding for yield and the problem of utilizing heterosis. In The way ahead in plant breeding. Proc 6th Congr Eucarpia: 79–86, Cambridge.

Lukyanenko PP (1972) Some results of Winter-wheat breeding in the USSR and the trends of its progress. Proc Intern Winter-Wheat Conference. Ankara (June 1972): 13–21.

Matyukhina LA (1979) Production of winter-wheat varieties for irrigated conditions. Plant Breeding Abstracts 51 (1981): p 910.

Maslov IN, Shkvarkina TI, Kizina PM (1960) Breadmaking properties of individual wheat varieties cultivatied in the USSR. In Kretovich VL (ed), Biochemistry of grain and of bread making. Moskva: Akademia Nauk SSSR (transl: Israel Program for Scient Translations, Jerusalem, 1965) 22–28.

Metakovsky EV, Novoselskaya AY, Sozinov AA et al. (1984) Genetic analysis of gliadin components in winter wheat using two-dimensional polyacrylamide gel electrophoresis. Theor Appl Gen 69(1): 31–37.

Morgunov AI, Rogers WJ, Sayers EJ, Metakovsky EV (1990) The high-molecular-weight glutenin subunit composition of Soviet wheat varieties. Euphytica 51: 41–52.

Morgunov AI, Peña RJ, Crossa J, Rajaram S (1993) Worldwide distribution of *Glu-1* alleles in bread wheat. Journal of Genetics and Breeding 47: 53–60.
Muzyka VN, Martynyuk NS (1991) Economic effectiveness of winter-wheat breeding in the USSR. Plant breeding Abstracts 61 (11): 1266.
Nettevich ED, Morgunov AI, Rogers WJ, Bespalov AM, Metakovsky EV (1991) Polymorphism of native wheat varieties for loci of high molecular glutenins of grain. Soviet Agric. Sciences 9: 1–3.
Nosenko VV, Rogozhinskiy BI (1973) Increasing resistance to brown rust in variety Mironovka 808. Plant Breeding Abstracts 44: 45.
Perigaud M (1998) Russia comes back; slow recovery to the world market. Grain Magazine (Paris France) 2: 19–22.
Pissarev VE (1962) Breeding of high-protein and quality wheats in the wet non-chernozem belt in the USSR. In Symposium on Genetics and Wheat Breeding. Martonvasar (June 1962): 433–446.
Poperelya FA, Blagodarova OM (1998) Genetics of the bread-making quality for the first Ukrainian super-strong wheat genotypes. Cytology and Genetics 32(6): in press.
Pospelowa G (1997) Anbau und Züchtung von Weizen in der Ukraine. Osteuropastudien der Hochschule des Landes Hessen, Reihe I. Berlin: Duncker & Humbolt, 198 pp.
Prutskov FM (1970) Winter wheat. Joint Publ Res Serv (1973) report nr JPRS 60652 Arlington (Virginia, USA).
Prutskova MG, Ukhanova OI (1972) New varieties of winter wheat (translated from Russian) New Delhi (1976) Amerind Publ Co Pvt Ltd, 238pp.
Pshenichnyi AE (1984) Efficacy of using the sedimentation index in evaluating breeding material of winter wheat. Plant Breeding Abstracts 54: 877.
Puchkov YM et al. (1973) P.P. Lukyanenko Eulogized. Selektsiya i Semenovodstvo 5 In Translations on USSR Agriculture (1974) nr 777.
Pylnev VV (1994) Effect of long term wheat breeding on gliadin polymorphism. Izvestiya Timiryazevskoi Selskokhhozyaistvennoi Akademii 4: 57–68.
Rabinovich SV (1998) Importance of wheat-rye translocations for breeding modern cultivars of *Triticum aestivum* L. In Braun HJ et al. (eds), Wheat: Prospects for Global Improvement, 401–418. Dordrecht: Kluwer Academic Publishers.
Samsonov MM (1960) Technological Evaluation of wheat varieties in the USSR. In Kretovich VL (ed), Biochemistry of grain and of bread making. Moskva: Akademia Nauk SSSR (transl: Israel Program for Scient Translations Ltd, 1965): 10–21.
Skoropanov SG (1977) Aufgaben der Agrarforschung bei der weiteren sozialistischen Intensivierung der Pflanzenproduktion und bei der Durchsetzung industriemäszige Produktionsmethoden in Auswertung der Beschlüsse des XXV Parteitages der KPdSU. Tagungsbericht Akademie der Landwirtschafts Wissenschaften DDR, 154: 53–67.
Sozinov AA (1979) Die Züchtung von Weizen auf Kornqualität. Tagungsbericht Akademie der Landwirtschafts Wissenschaften DDR, 168: 37–48.
Sozinov AA, Poperelya FA (1980) Genetic classification of prolamins and its use for plant breeding. Ann Technol Agric 29: 229–245.
Sozinov AA, Poperelya FA (1982) Polymorphism of prolamins and variability of grain quality. Plant Foods Hum Nutr 31: 243–249.
Sozinov AA, Poperelya FA (1984) Blocks of gliadin components in wheat genetics, breeding and seed growing. Kulturpflanze 32: 165–170.
Sozinov AA (1993) Genetic markers in plants. Cytology and Genetics 27(5): 3–14.
Vasilenko II (1973) Breeding wheat for agricultural practices suitable for winter-wheat varieties. Plant Breeding Abstracts 48: p 349.

Various authors (information about new cultivars) Selektsiya-i-Semenovodstvo-Moskva (1991, 1992) Novye Sorta: -vyvedenie,-semenovodstvo,-vozdelyvanie (1992) Russian Federation State Register of new cultivars, Min Agr (1996).

Vavilov NI (1935) The origin, variation, immunity and breeding of cultivated plants (transl. Starr Chester K) Chronica Botanica 13 (1949), 364 pp.

Zhivotkov L, Vlassenko V, Kolyuchy V (1996) Achievement, initial material and methods of development of up-to-date winter wheat cultivars in Mironovka. Abstr 5th Int Wheat Conf Ankara (10–14 June 1996): 104–105.

List of varieties *

* *With special thanks to Andrea Haverkort*

MIX
Papier aus verantwortungsvollen Quellen
Paper from responsible sources
FSC® C105338

If you have any concerns about our products,
you can contact us on
ProductSafety@springernature.com

In case Publisher is established outside the EU,
the EU authorized representative is:
Springer Nature Customer Service Center GmbH
Europaplatz 3, 69115 Heidelberg, Germany

Printed by Libri Plureos GmbH
in Hamburg, Germany